T0360770

Fourier Methods in Science and Engineering

This innovative book discusses and applies the generalized Fourier Series to a variety of problems commonly encountered within science and engineering, equipping the readers with a clear pathway through which to use the Fourier methods as a solution technique for a wide range of differential equations and boundary value problems.

Beginning with an overview of the conventional Fourier series theory, this book introduces the generalized Fourier series (GFS), emphasizing its notable rate of convergence when compared to the conventional Fourier series expansions. After systematically presenting the GFS as a powerful and unified solution method for ordinary differential equations and partial differential equations, this book expands on some representative boundary value problems, diving into their multiscale characteristics.

This book will provide readers with the comprehensive foundation necessary for solving a wide spectrum of mathematical problems key to practical applications. It will also be of interest to researchers, engineers, and college students in various science, engineering, and mathematics fields.

Fourier Methods in Science and Engineering

Wen L. Li and Weiming Sun

CRC Press
Taylor & Francis Group
Boca Raton London New York

CRC Press is an imprint of the
Taylor & Francis Group, an **informa** business

First edition published 2023
by CRC Press
6000 Broken Sound Parkway NW, Suite 300, Boca Raton, FL 33487-2742

and by CRC Press
4 Park Square, Milton Park, Abingdon, Oxon, OX14 4RN

CRC Press is an imprint of Taylor & Francis Group, LLC

© 2023 Wen L. Li and Weiming Sun

Reasonable efforts have been made to publish reliable data and information, but the author and publisher cannot assume responsibility for the validity of all materials or the consequences of their use. The authors and publishers have attempted to trace the copyright holders of all material reproduced in this publication and apologize to copyright holders if permission to publish in this form has not been obtained. If any copyright material has not been acknowledged please write and let us know so we may rectify in any future reprint.

Except as permitted under U.S. Copyright Law, no part of this book may be reprinted, reproduced, transmitted, or utilized in any form by any electronic, mechanical, or other means, now known or hereafter invented, including photocopying, microfilming, and recording, or in any information storage or retrieval system, without written permission from the publishers.

For permission to photocopy or use material electronically from this work, access www.copyright. com or contact the Copyright Clearance Center, Inc. (CCC), 222 Rosewood Drive, Danvers, MA 01923, 978-750-8400. For works that are not available on CCC please contact mpkbookspermissions@ tandf.co.uk

Trademark notice: Product or corporate names may be trademarks or registered trademarks and are used only for identification and explanation without intent to infringe.

ISBN: 9781032048420 (hbk)
ISBN: 9781032048482 (pbk)
ISBN: 9781003194859 (ebk)

DOI: 10.1201/9781003194859

Typeset in Times LT Std
by KnowledgeWorks Global Ltd.

Contents

Preface

This book is intended for researchers, engineers, and college students who are interested in numerical and mathematical methods for solving differential equations and boundary value problems encountered in various scientific and engineering disciplines. Since the readers are expected to be familiar with Fourier series, advanced calculus, linear algebra, differential equations, and mathematical physics methods, the relevant knowledge, if needed, is only briefly reviewed for sake of totality. Instead, the attentions are mostly focused on the content which best represents the distinguishing features of this book, also the cultivation of our many years of combined research and experience with the subject matters. While we strive to write this book as rigorously and concisely as possible, certain parts may still be considered too advanced or exhaustive to some readers. In our humble opinion, the current theories, algorithms, and implementation procedures can be better comprehended through applications in the form of term, research or practical projects with varying degrees of challenges, which, in the meantime, provide ample opportunities for expanding and furthering some aspects of the described mathematical framework.

Acknowledgments

The authors are indebted to their respective families for their understanding and support during the long hours in preparing and writing this book. The second author would like to express his appreciations to his mentors, Professors Zimao Zhang, Guangsong Yang, Songlin Hao, and Dongxu Li, for their encouragement and guidance throughout his graduate studies.

Wen L. Li
Weiming Sun
Ningbo and Wuhan, China,
Aug. 22, 2022

Authors

Dr. Wen L. Li received his B.S. (1982) in Physics from Liaoning Teachers University in Dalian, China; M.Eng. (1984) in Vehicle Engineering from Beijing Institute of Technology in Beijing, China; and Ph.D. (1991) in Mechanical Engineering from University of Kentucky in Lexington, USA. From 1992 to 1995, he worked with the Case Corporation as Technical Specialist. From 1995 to 2004, he worked with the United Technologies Carrier Corporation as Sr. Staff Engineer and United Technologies Research Center as Principal Engineer. In 2004, he joined the Mississippi State University as an Associate Professor in Mechanical Engineering. In 2007, he moved to Wayne State University as an Associate Professor of Mechanical Engineering. In 2014, Dr. Li started the Advanced Engineering and Technologies company, and in 2016, the Advanced Information Services company in China as the founder and general manager. Dr. Li has author/co-authored about 60 journal papers, 1 book, and 3 book chapters. He is the inventor or co-inventors of more than 30 technical patents. He is the Editor-in-Chief of *Open Journal of Acoustics* (OJA), member of Editorial Board of several other journals, and co-chair of Prediction and Modeling Technical Committee of Institute of Noise Control Engineering (INCE) and member of Structural Vibration and Acoustics Committee of Acoustical Society of America (ASA). He has chaired/co-chaired dozens of technical sessions of international conferences. His research and experience are mostly related to numerical methods, computer modeling and simulations, dynamics systems, acoustics, and machinery designs.

Dr. Weiming Sun received his B.E. (1995) in Structural Strength of Spacecraft; M.E. (1998) in Solid Mechanics both from National University of Defense Technology in Changsha, China; and Ph.D. (2011) in Solid Mechanics from Beijing Jiaotong University in Beijing, China. Under the supervision of Prof. Zimao Zhang, his doctorate research was focused on proposing a set of general formulas for the Fourier series of higher order (partial) derivatives of one- and two- dimensional functions, developing the generalized Fourier series method for linear differential equations with constant coefficients, and applying it to boundary value problems commonly encountered in engineering applications. After receiving his Ph.D. degree, Dr. Sun became a full-time lecturer in the Department of Mathematics at Jianghan University in Wuhan, China. He has published six journal papers.

1 Introduction

The interactions of computer technology, numerical methods, natural science, and engineering have led to a new discipline – computational science. Computational science is devoted to the development of computer technology-based algorithms and software to solve a wide spectrum of scientific and engineering problems. The capacities potentially offered by computer technology and the demands from various industries and scientific sectors have been driving computational science into an unprecedented opportunity of growth. It can be said that the impressive achievements of computational science have already made computer simulations, together with physical experimenting and theoretical analysis, the three pillars of scientific and engineering investigations, and continue to play a critical role in promoting the advancements of science and technology.

The analysis methods in computational science can often be divided into analytical methods and numerical methods. The former is typically adopted in mathematical physics in which a specific solution algorithm is derived for each type of differential equations or boundary conditions. An analytical method is usually more suitable for parametric study of a physical phenomenon or process, and understanding its behaviors. In practice, the analytical methods can not only save a lot of computational time, but also facilitate assessing the influences on the results of the particular parameters of interest. However, the use of an analytical method is seriously limited by its stringent requirements on the shapes of the solution domains, the load conditions, the boundary conditions, or some other factors. In contrast, the numerical methods, such as, the finite difference methods, the finite element methods, the boundary element methods, and other discretization-based methods can better cope with these practical difficulties and complexities, and have been extensively used in almost all engineering and scientific fields for several decades (Zienkiewicz and Taylor 2000).

1.1 SCALES AND SCALE EFFECTS

Scale often refers to the spatial or temporal unit used in describing a physical event, or can be broadly understood as the scope and frequency of an object, phenomenon, or process in space and time.

The division of matter scales in the universe is usually made in an absolute sense, such as the use of the macroscopic (0.1 mm), microscopic (0.1 nm ~ 1 nm), and mesoscopic (1 nm ~ 100 nm) trichotomy. In comparison, the division of scientific scales is relatively more flexible: each scientific field may have its own material scales defined in accordance with the specific research scope, subject matters, and other concerns.

DOI: 10.1201/9781003194859-1

The limit scale for human activities has stretched out magnificently in the past century. In the direction of large-scale, with the inventions of, such as satellites, space shuttles, and space telescopes, mankind has laid their eyes onto the deeper space of the boundless universe; in the meantime, in the direction of micro scale, artificial superlattices, nano-materials, Micro-Electro-Mechanical Systems (MEMS), biochips and other technologies have taken our attentions to the mesoscopic world.

With expanding scientific limit scales, the scale effects have become increasingly evident and important. For example, on the mesoscopic scale, since the specific surface area of a material is very large, the surface tension, electrostatic force, bonding force, magnetic force, and the likes can all emerge as important factors which may be administered to modify the physical and chemical properties of the material.

The flow and heat transfer laws under micro scale can also deviate significantly from those under normal (macro) scale conditions. For example, in micro scale, heat conduction is often accompanied by the wave and radiation effects, and the thermal conductivity is greatly reduced. In micro scale convective heat transfer, the compressibility of fluid, especially gas, the surface effect of liquid and the rarefaction effect of the gas all need to be taken into account. On the micro scale, the thermal radiation is not only related to the phonon free path, but also dependent upon the photon wavelength and photon interference scale.

Similarly, the microminiaturization process of machinery also faces the issues of scale effects, such as the dependence of micro friction and micro lubrication mechanisms on the scale of micro machinery, the restriction of heat transfer and combustion on the micromechanical scale, etc.

The upsurges of scale effect studies and experimental verifications have spanned from the traditional fields of physics, chemistry, material science, engineering, machinery manufacturing into other scientific and social disciplines such as geography, ecology, environment, meteorology, economics, and information technologies.

1.2 MULTISCALE PHENOMENA AND MULTISCALE ANALYSIS METHODS

Recently, multiscale problems in science and engineering have attracted considerable attentions (Oden et al. 2003). The spatial, material, and/or physical parameters in a multiscale problem can vary easily by orders of magnitude, resulting in the appearance of boundary layers or other forms of local discontinuities with large gradients in the solution region (Onate 2003). In solving for multiscale problems, the traditional analysis methods tend to have the shortcomings of ineffectiveness and low accuracy. Therefore, renovating the traditional methods to make them better fit for solving multiscale problems has become a hot topic nowadays.

Solving multiscale problems often requires us to first have a good understanding of the scale dependences of the involved phenomena or processes, and define the appropriate scales accordingly. Of equal importance, we also need to pay attention to the scale coupling characteristics and skillfully deal with the problems concerning scale transformation and scale integration.

Here, our attention is specifically focused on a class of multiscale problems in which some interesting features such as local defects or abrupt changes of some physical variables are potentially embedded in an event primarily described on a larger scale. Such a situation will typically involve at least two scales: a smaller scale to resolve the local behavior, and a larger scale to capture the macroscopic characteristic of the event.

The multiscale coupling provides a new perspective to understand the complexity of a physical phenomenon or process. For example, the complex flow patterns formed by turbulence can be regarded as the result of vortex interactions on different scales. When the physical mechanisms are similar on different scales, the so-called similarity solution method is formulated for the multiscale phenomenon. When the coupling is considered weak at different scales, the small parameter perturbation method is more suitable. It is the hybrid use of the similarity solution method and small parameter perturbation method that has overcome the major hurdles such as boundary layer simulations in aerospace. By fully recognizing the diversity and the strong coupling of physical mechanisms on different scales, fluid turbulence research has entered a new era (Wang et al. 2004).

Multiscale phenomena are also prevalent in scientific and engineering simulations. In 1995, Hughes studied the external field problem of the Helmholtz equation from the perspective of near-field and far-field multiscale coupling (Hughes 1995). To this end, he first artificially selects a closed surface (such as a spherical surface), then the infinite solution domain can be decomposed into the near field (the area between the inner boundary and the selected closed surface) and the far field (region outside the selected closed surface). The near-field region can be regarded as a small scale and solved by using conventional numerical methods (such as finite element method). The far-field region is considered as a large scale, and the conventional numerical method is no longer suitable; instead, an analytical method (the Green's function method) is adopted. The near-field scale and far-field scale are then coupled together via the boundary conditions enforced on the interface (that is, the selected spherical surface), leading to a multiscale coupling phenomenon.

The solution space, which involves the scale coupling, via a differential equation, and/or its corresponding boundary/compatibility conditions, between the subspaces with different scales, also represents a familiar multiscale phenomenon in scientific and engineering simulations. The coupling between the subregions with different scales in the solution region, and the coupling between subspaces corresponding to different scales constitute two typical manifestations of multiscale phenomena in science and engineering. Multiscale phenomena are frequently associated with:

1. Boundary layer problems with convection-diffusion-reaction equation;
2. Singular perturbation problems;
3. Intermediate faults in the fracture processes of concrete, rock and soil materials;
4. Fluid turbulence;

5. Shear bands in solids;
6. Shock waves of compressible fluids;
7. Compressible, incompressible solid or fluid conversion problems.

Many techniques, such as the stabilized finite element methods (SFEMs), the bubble function methods, the multiscale finite element methods, the wavelet finite element methods, the meshless methods, the variational multiscale methods, have been proposed to better cope with one or more aspects of the multiscale characteristics (Donea and Huerta 2003), and have spurred a wave of research activities related to the multiscale analysis methods.

SFEM was originally proposed to solve multiscale problems in fluid dynamics (such as convection-diffusion equation and Stokes equation). Its basic idea is to introduce a stabilizing term into the Galerkin's weak formulation. The stabilizing term is the weighted residual of the differential equation to be solved on each element. The uses of different weighting functions will potentially have diverse effects, hence leading to the various versions of the SFEMs.

SFEMs have been successfully applied to a spectrum of scientific and engineering problems such as the convection-diffusion equation (Agarwal and Pinsky 1996), the convection-diffusion-reaction equation (Huerta and Donea 2002), the Stokes equation (Bochev et al. 2007), Darcy equation (Burman and Hansbo 2007), the generalized Stokes equation (Araya et al. 2008), Navier-Stokes equation (Whiting and Jansen 2001), incompressible flows (Codina 2001), viscoplastic flow problem (Maniatty and Liu 2003), plate and shell stability problems (Bischoff and Bletzinger 2004), and fluid-structure interactions (Thompson and Sankar 2001).

The bubble function methods represent another popular approach in solving multiscale problems in which special forms of bubble functions are employed to expand the finite element interpolation space. In the process, the proper selection of the bubble functions is critically important to ensuring the accuracy and stability of the resulting solution. The bubble function methods have demonstrated certain successes in solving the Helmholtz equation (Franca et al. 1997), the convection-diffusion equation (Sangalli 2004), the diffusion-reaction equation (Parvazinia and Nassehi 2007), the Navier-Stokes equation (Canuto et al. 1998), the Stokes equation (Franca and Oliveira 2003), the Brinkman equation (Parvazinia et al. 2006), etc.

Wavelet analysis is hailed as "digital microscope" with the multiscale and multi-resolution functionalities. Introducing wavelets as the interpolation functions into the traditional finite element solutions has led to another general type of multiscale solution methods, wavelet finite element methods, in science and engineering (Dahmen et al. 1997). The wavelet finite element methods can adaptively adjust to analysis scales, which tends to facilitate a better solution in terms of the numerical stability, efficiency and accuracy. In the last two decades, the wavelet finite element methods have emerged as a promising approach for solving various singularity and multiscale problems.

Most multiscale methods are built upon the existing numerical methods by considering a specific scale coupling characteristic of a multiscale phenomenon, or by integrating one or more suitable corrections into the corresponding solution scheme to strike a delicate balance between, such as accuracy and efficiency. More explicitly, since the existing multiscale methods still fall within the framework of the discretization-based solutions, they will inevitably inherit some, if not all, of their shortcomings such as high computational cost, comparatively low accuracy of high order derivatives of the field variables, and the difficulties with isolating out the influences of a specific model parameter on the results of concern. Thus, in order to utterly resolve these problems, we need to walk out the comfort zone outlined by the conventional approaches and explore for new paths for the groundbreaking advancements of the multiscale solution methods.

Computational science may still be at its primary stage of development. Insightfully revealing and simulating multiscale phenomena, although currently considered out of the reach, have already posted monumental challenges to computational science and computer technology for many years or even decades ahead.

A powerful multiscale solution method needs to have the following technical merits:

1. Feasibility: The traditional solution methods are typically only suitable for single-scale phenomena. However, when they are applied to multiscale problems, we will immediately face a dilemma: on the one hand, if a larger calculation scale is selected, the detailed information carried by the smaller scale tends to be lost in the solution process, which can adversely affect the accuracy and reliability of the results; on the other hand, if a smaller calculation scale is used, although the finer details can be recovered, the required computer memory and CPU time will become unreasonably high, or even technically impossible. Therefore, it becomes an inevitable trend to integrate the different scales into a single multiscale solution scheme to ease the struggle between computational accuracy and efficacy (Masud and Khurram 2004).

2. Efficacy: Accuracy, efficiency, and effectiveness of a given analysis method may be found sensitive to some model parameters. For a multiscale problem, there typically exist more such model parameters which vary in wider ranges and hence can affect the solution in more drastic ways. Therefore, a comprehensive use of the validation and verification methodologies has become an integral part of consistently improving the efficacy of a multiscale method.

3. Accuracy: For a multiscale phenomenon, the scale coupling effects can make the calculation errors be transferred from the larger scale to the smaller scale, which are often exhibited as violent oscillations, large gradients or near-singularities in a very small region. Therefore, there is a genuine need for the correct capture and comprehensive characterization of the multiscale structures in the solution domain in order to increase the accuracy and reliability of the corresponding multiscale technique.

1.3 FOURIER SERIES METHODS IN SCIENTIFIC AND ENGINEERING APPLICATIONS

Since the creation of calculus in the late 17th century, most scientific and engineering problems have been formulated in terms of differential equations (especially, partial differential equations) with corresponding boundary conditions. In 1807, in studying the heat conduction problem, Fourier expressed the temperature field in an object as a trigonometric series (often called as Fourier series nowadays), and separated the heat conduction equation into different parts according to the physical conditions (or formally, the method of separation of variables). The integration of the trigonometric series expansion and the separation of variables has led to a useful analysis method – Fourier series method. The Fourier series method has offered a simple approach for solving a variety of linear partial differential equations such as wave equations, diffusion equations, Laplace equation, etc.

The method of separation of variables has a long history, and its origin can be traced back to the study of string vibrations in the 18th century, which attracted the attentions of such mathematical giants as Euler, d'Alembert, Taylor, Bernoulli, Laplace, Lagrange, to name a few. In 1750, based on Taylor's results, d'Alembert first proposed the idea of separation of variables, and derived the solution as the linear combination of the eigenfunctions of the string equation. Later Bernoulli extended the eigenfunction solution to the initial value problem, which was questioned by Euler and Lagrange. Fourier is recognized as the pioneer of the Fourier series method for his prominent contribution of expressing the solution as an infinite series of the trigonometric functions to satisfy the initial conditions. However, the whole theory of Fourier series surprisingly took more than a century to be accomplished.

Although the method of separation of variables is widely discussed in mathematical physics as a primary means for solving various boundary value problems, its applications are typically limited to the cases when the partial differential equations are linear and homogeneous, and the boundary conditions are homogeneous as well. When a complete system of eigenfunctions is used to expand the solutions, the process of solving the boundary value problems is simplified to seeking a set of expansion coefficients to specially satisfy the differential equation and/or boundary conditions. It is important to point out that such a solution or solution process is valid only when the convergence and effectiveness of the series solution can be confirmed *a priori* or verified *a posteriori*.

In comparison with the method of separation of variables, the application scope of the Fourier series method is even smaller due to the concerns about the differentiability and convergence of the series solutions. Since its early successes in solving heat conductions and string vibrations, the most influential applications of the Fourier series method are probably associated with Navier solution for the elastic bending of thin plates with simply supported boundary conditions (Timoshenko and Woinowsky-Krieger 1959), and Levy solution for the plates with two opposite edges being simply supported and the others arbitrarily supported (Leissa 1993). It can be said that the applications of the Fourier series method in its original form are still severely hampered by its picky choice of only particular types of boundary conditions.

Nevertheless, because the trigonometric functions have certain desired characteristics such as, orthogonality, completeness, simplicity, and numerical stability, researchers have never ceased to pursue transforming the Fourier series method into a general means for solving a wide range of boundary value problems. Some of recent progresses include, such as: the superposition method by decomposing the boundary conditions into several suitable ones (Gorman 1982), direct Fourier series method with Stokes transformations (Wang and Lin 1996; Chaudhuri 2002; Kabir et al. 2007), analytical form of Fourier series method (Huang 1992; Sun and Yang 1998; Sun et al. 2004), the modified Fourier series method with supplementary terms (Li 2000; Li et al. 2009; Jin et al. 2015).

1.4 SCOPE OF THIS BOOK

Development of accurate and effective multiscale analysis methods is still a hot research topic in computational science. By aiming at developing a general and effective (multiscale) Fourier series method for solving a wide spectrum of boundary value problems in science and engineering, our attentions are here specifically focused on formulating systematic solution algorithms which are substantially different from the existing discretization-based numerical models. This goal will be achieved in three steps: (1) understanding the convergence characteristics of and other related issues with the traditional Fourier series solutions; (2) developing strategies, solution algorithms, and implementation procedures to effectively address these issues; and (3) applying the proposed solution method to various scientific and engineering problems to demonstrate and validate its reliability, effectiveness, and, in particular, multiscale capability. Accordingly, this book consists of three major topics.

The first one is focused on the fundamental theories about the Fourier series expansions and some of their modified versions. In Chapter 2, the basic knowledge is briefly reviewed, including: the continuity of a function, the concepts of periodic extensions, the relations between the smoothness of a periodic function and the convergence of its Fourier series expansions, etc. The rules about the integration and differentiations of the Fourier series are also discussed with sufficient details. In Chapter 3, a convergence acceleration scheme and a few corresponding forms of Fourier series expansions are described for a function defined on a compact interval. Specifically, a suitable supplementary function is incorporated into the original function to force the residual function to have the desired mathematical characteristics. The rapid convergence of the resulting Fourier series is demonstrated using examples. In Chapter 5, the Stokes transformation is employed to derive the recursive relations between the Fourier coefficients for the derivatives of different orders. The sufficient conditions are then established for the term-by-term differentiation of the Fourier series of a function to obtain its counterpart for the derivative of any order. Similar formulas and theorems are presented in Chapter 6 for two-dimensional cases, even though the corresponding results become much more complicated.

The second topic involves establishing a mathematical framework for expanding a sufficiently smooth function into the generalized Fourier series, and subsequently developing it into a general means for solving linear (partial) differential equations with arbitrary boundary conditions. In Chapter 4, the generalized Fourier series method is specifically applied to the Euler-Bernoulli beam problems with an emphasis on the procedures of determining the suitable supplementary polynomials to accommodate the arbitrary boundary conditions. In Chapters 7 and 8, the generalized Fourier series expansions are further elaborated in a more systematic and abstract manner; that is, they are described as a generic means for performing structural decompositions of a function in accordance with the sufficient conditions set forth for the involved Fourier series to be termwise differentiable. Such an approach makes it relatively easier to understand the specific forms of the generalized Fourier series for a two-dimensional function as discussed in Chapter 8. The excellent convergence characteristics and numerical accuracy of the generalized Fourier series are carefully validated using numerical examples. In Chapter 9, the generalized Fourier series are formulated into a general method for solving linear differential equations with constant coefficients. In order to enable the generalized Fourier series method to more effectively deal with multiscale phenomena, the structural decompositions are reformulated as the superposition of the general, particular and supplementary solutions to better capture the multiscale characteristics or information imbedded in the differential equation and/or loading condition.

Finally, the multiscale Fourier series method is applied to some of the widely studied problems in science and engineering: the convection-diffusion-reaction equations in Chapter 10, the elastic bending of a thick plate resting on an elastic foundation in Chapter 11, and wave propagation in elastic waveguides with rectangular cross-sections in Chapter 12. While the multiscale Fourier series method is repeatedly verified, through these examples, to be a reliable and effective means for solving general boundary value problems, we are equally fascinated by its competency of correctly capturing the multiscale phenomena which are deliberately created by substantially varying certain model parameters. These applications also serve to depict the various schemes often used to derive the final system of linear algebraic equations, and to examine the numerical accuracy of the corresponding solutions.

REFERENCES

Agarwal, A. N., and P. M. Pinsky. 1996. Stabilized element residual method (SERM): A posteriori error estimation for the advection-diffusion equation. *Journal of Computational and Applied Mathematics* 74: 3–17.

Araya, R., G. R. Barrenechea, and A. Poza. 2008. An adaptive stabilized finite element method for the generalized Stokes problem. *Journal of Computational and Applied Mathematics* 214: 457–479.

Bischoff, M., and K. U. Bletzinger. 2004. Improving stability and accuracy of Reissner-Mindlin plate finite elements via algebraic subgrid scale stabilization. *Computer Methods in Applied Mechanics and Engineering* 193: 1517–1528.

Bochev, P. B., M. D. Gunzburger, and R. B. Lehoucq. 2007. On stabilized finite element methods for the Stokes problem in the small time step limit. *International Journal for Numerical Methods in Fluids* 53: 573–597.

Burman, E., and P. Hansbo. 2007. A unified stabilized method for Stokes' and Darcy's equations. *Journal of Computational and Applied Mathematics* 198: 35–51.

Canuto, C., A. Russo, and V. Van Kemenade. 1998. Stabilized spectral methods for the Navier-Stokes equations: Residual-free bubbles and preconditioning. *Computer Methods in Applied Mechanics and Engineering* 166: 65–83.

Chaudhuri, R. A. 2002. On the roles of complementary and admissible boundary constraints in Fourier solutions to the boundary value problems of completely coupled *r*th order PDEs. *Journal of Sound and Vibration* 251(2): 261–313.

Codina, R. 2001. A stabilized finite element method for generalized stationary incompressible flows. *Computer Methods in Applied Mechanics and Engineering* 190: 2681–2706.

Dahmen, W., A. J. Kurdila, and P. Oswald. 1997. *Multiscale Wavelet Methods for Partial Differential Equations*. San Diego, CA: Academic Press.

Donea, J., and A. Huerta. 2003. *Finite Element Methods for Flow Problems*. Chichester: John Wiley & Sons Ltd.

Franca, L. P., C. Farhat, A. P. Macedo, and M. Lesoinne. 1997. Residual-free bubble for the Helmholtz equation. *International Journal for Numerical Methods in Engineering* 40: 4003–4009.

Franca, L. P., and S. P. Oliveira. 2003. Pressure bubbles stabilization features in the Stokes problem. *Computer Methods in Applied Mechanics and Engineering* 192: 1929–1937.

Gorman, D. J. 1982. *Free Vibration Analysis of Rectangular Plates*. New York: Elsevier-Science Publishing Company.

Huang, Y. 1992. *Theory of Elastic Thin Plates*. Changsha: Press of National University of Defense Technology (in Chinese).

Huerta, A., and J. Donea. 2002. Time-accurate solution of stabilized convection-diffusion-reaction equations: I-Time and space discretization. *Communications in Numerical Methods in Engineering* 18: 565–573.

Hughes, T. J. R. 1995. Multiscale phenomena: Green's functions, the Dirichlet-to-Neumann formulation, subgrid scale models, bubbles and the origins of stabilized methods. *Computer Methods in Applied Mechanics and Engineering* 127: 387–401.

Jin, G. Y., T. G. Ye, and Z. Su. 2015. *Structural Vibration: A Uniform Accurate Solution for Laminated Beams, Plates and Shells with General Boundary Conditions*. Beijing/ Heidelberg: Science Press/Springer.

Kabir, H. R. H., M. A. M. Hamad, J. Al-Duaij, and M. J. John. 2007. Thermal buckling response of all-edge clamped rectangular plates with symmetric angle-ply lamination. *Composite Structures* 79: 148–155.

Leissa, A. W. 1993. *Vibration of Plates*. Woodbury, NY: Acoustical Society of America.

Li, W. L. 2000. Free vibrations of beams with general boundary conditions. *Journal of Sound and Vibration* 237(4): 709–725.

Li, W. L., X. F. Zhang, J. T. Du, and Z. G. Liu. 2009. An exact series solution for the transverse vibration of rectangular plates with general elastic boundary supports. *Journal of Sound and Vibration* 321: 254–269.

Maniatty, A. M., and Y. Liu. 2003. Stabilized finite element method for viscoplastic flow: Formulation with state variable evolution. *International Journal for Numerical Methods in Engineering* 56: 185–209.

Masud, A., and R. A. Khurram. 2004. A multiscale/stabilized finite element method for the advection-diffusion equation. *Computer Methods in Applied Mechanics and Engineering* 193: 1997–2018.

Oden, J. T., T. Belytschko, I. Babuska, and T. J. R. Hughes. 2003. Research directions in computational mechanics. *Computer Methods in Applied Mechanics and Engineering* 192: 913–922.

Onate, E. 2003. Multiscale computational analysis in mechanics using finite calculus: An introduction. *Computer Methods in Applied Mechanics and Engineering* 192: 3043–3059.

Parvazinia, M., and V. Nassehi. 2007. Multiscale finite element modeling of diffusion-reaction equation using bubble functions with bilinear and triangular elements. *Computer Methods in Applied Mechanics and Engineering* 196: 1095–1107.

Parvazinia, M., V. Nassehi, and R. J. Wakeman. 2006. Multiscale finite element modelling of laminar steady flow through highly permeable porous media. *Chemical Engineering Science* 61: 586–596.

Sangalli, G. 2004. A discontinuous residual-free bubble method for advection-diffusion problems. *Journal of Engineering Mathematics* 49: 149–162.

Sun, W. M., and G. S. Yang. 1998. General analytic solution for elastic bending of Reissner plates. *Applied Mathematics and Mechanics* (English Ed.) 19(1): 85–94.

Sun, W. M., G. S. Yang, and D. X. Li. 2004. Exact analysis of wave propagation in an infinite rectangular beam. *Applied Mathematics and Mechanics* (English Ed.) 25(7): 768–778.

Thompson, L. L., and S. Sankar. 2001. Dispersion analysis of stabilized finite element methods for acoustic fluid interaction with Reissner/Mindlin plates. *International Journal for Numerical Methods in Engineering* 50: 2521–2545.

Timoshenko, S., and S. Woinowsky-Krieger. 1959. *Theory of Plates and Shells*. New York: McGraw-Hill Book Company.

Wang, H. Y., G. W. He, M. F. Xia, F. J. Ke, and Y. L. Bai. 2004. Multiscale coupling in complex mechanical systems. *Chemical Engineering Science* 59: 1677–1686.

Wang, J. T.-S., and C.-C. Lin. 1996. Dynamic analysis of generally supported beams using Fourier series. *Journal of Sound and Vibration* 196(3): 285–293.

Whiting, C. H., and K. E. Jansen. 2001. A stabilized finite element method for the incompressible Navier-Stokes equations using a hierarchical basis. *International Journal for Numerical Methods in Fluids* 35: 93–116.

Zienkiewicz, O. C., and R. L. Taylor. 2000. *The Finite Element Method* (5th ed.), vols. 1–3. Oxford: Butterworth-Heinemann.

2 Fourier Series Expansions of Functions

Fourier series is named after French mathematician Joseph Fourier for his break-through in 1807 by claiming that all functions could be expanded into trigono-metric series. As a matter of fact, the earliest development of Fourier theory can be traced back to the mid of 18th century, originating from the pioneering works of a number of prominent mathematicians, such as d'Alembert, Euler, Bernoulli, and Lagrange, who employed trigonometric series to solve heat transfer, string vibration, and wave propagation problems. However, the full formation of Fourier theory surprisingly took more than a century of endeavors as highlighted by the famous d'Alembert-Euler-Bernoulli controversy. Nevertheless, it did not take long before mathematicians and scientists came to appreciate the power and far-reaching implications of Fourier's claim regarding trigonometric series. Even today, it can be said that "by any yardstick, Fourier series is one of the greatest and most influential concepts of contemporary mathematics" (Iserles and Nørsett 2008).

2.1 PERIODIC FUNCTIONS AND THEIR FOURIER SERIES EXPANSIONS

Fourier series expansions have been extensively discussed in many books (e.g., Zygmund 1968; González-Velasco 1996; Hardy and Rogosinski 2013), and there is no need for a comprehensive presentation; instead, only some important results and conclusions will be briefly reviewed here for the sake of completeness.

Fourier series is also known as the trigonometric series expansion of a periodic function. A function $f(x)$ is referred to as periodic if it satisfies (see Figure 2.1)

$$f(x+T) = f(x), \tag{2.1}$$

where constant T is the period of the function $f(x)$.

A simple example of the periodic functions is the harmonic function

$$x(t) = A\sin(\omega t + \phi), \tag{2.2}$$

where A, ω, and ϕ are its amplitude, angular frequency, and phase angle, respectively.

In practice, the harmonic function is often used to represent an oscillatory variation of a physical quantity, such as the x coordinate of a point moving along a circle of radius A at a constant rotational speed ω; or if $A = m\omega^2 r$ is the con-centric force resulting from a rotating vehicle wheel with an unbalanced mass m (with an offset r from its center), then $x(t)$ can be interpreted as the horizontal or

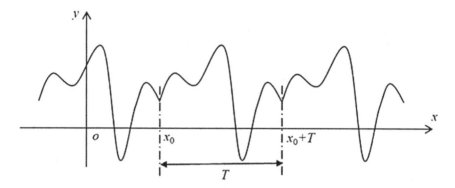

FIGURE 2.1 A T-periodic function.

vertical force component acting on the axial when the wheel rotates at speed ω. The period of the function $x(t)$ is $T = 2\pi/\omega$.

It is well known that the trigonometric functions, $\{1, \cos x, \sin x, \cos 2x, \sin 2x, \cos 3x, \sin 3x, \cdots\}$, constitute a complete system and can be used to expand any 2π-periodic function $f(x)$ as

$$f(x) \sim a_0/2 + \sum_{m=1}^{\infty}(a_m \cos mx + b_m \sin mx), \qquad (2.3)$$

where the expansion coefficients are calculated from

$$a_m = \frac{1}{\pi}\int_{-\pi}^{\pi} f(x)\cos mx \, dx, \qquad (2.4)$$

and

$$b_m = \frac{1}{\pi}\int_{-\pi}^{\pi} f(x)\sin mx \, dx. \qquad (2.5)$$

The trigonometric series (2.3) is commonly referred to as the Fourier series of function $f(x)$, and a_m and b_m as the Fourier coefficients.

If the function $f(x)$ has a period of T, (2.3)–(2.5) can be accordingly modified to

$$f(x) \sim a_0/2 + \sum_{m=1}^{\infty}\left(a_m \cos\frac{2\pi mx}{T} + b_m \sin\frac{2\pi mx}{T}\right), \qquad (2.6)$$

where

$$a_m = \frac{2}{T}\int_{0}^{T} f(x)\cos\frac{2\pi mx}{T} \, dx \qquad (2.7)$$

and

$$b_m = \frac{2}{T} \int_0^T f(x) \sin \frac{2\pi mx}{T} dx. \qquad (2.8)$$

In the process of determining the Fourier coefficients, it has been implicitly assumed that the function is absolutely integrable and the series can be integrated termwise.

EXAMPLE 2.1 Expand $f(x) = Ax^2 + Bx + C$ ($-\pi < x < \pi$), where A, B, and C are constants, into Fourier series.

The Fourier coefficients are readily calculated as:

$$a_0 = \frac{1}{\pi} \int_{-\pi}^{\pi} \left(Ax^2 + Bx + C \right) dx = \frac{1}{\pi} (Ax^3/3 + Bx^2/2 + Cx) \Big|_{-\pi}^{\pi}$$

$$= 2A\pi^2/3 + 2C,$$

$$a_m = \frac{1}{\pi} \int_{-\pi}^{\pi} \left(Ax^2 + Bx + C \right) \cos mx dx$$

$$= \frac{1}{m\pi} (Ax^2 + Bx + C) \sin mx \Big|_{-\pi}^{\pi} - \frac{1}{m\pi} \int_{-\pi}^{\pi} (2Ax + B) \sin mx dx$$

$$= \frac{1}{m^2\pi} (2Ax + B) \cos mx \Big|_{-\pi}^{\pi} - \frac{2A}{m^2\pi} \int_{-\pi}^{\pi} \cos mx dx$$

$$= \frac{4A(-1)^m}{m^2},$$

and

$$b_m = \frac{1}{\pi} \int_{-\pi}^{\pi} \left(Ax^2 + Bx + C \right) \sin mx dx$$

$$= \frac{-1}{m\pi} (Ax^2 + Bx + C) \cos mx \Big|_{-\pi}^{\pi} + \frac{1}{m\pi} \int_{-\pi}^{\pi} (2Ax + B) \cos mx dx$$

$$= \frac{-2B(-1)^m}{m} + \frac{1}{m^2\pi} (2Ax + B) \sin mx \Big|_{-\pi}^{\pi} - \frac{2A}{m^2\pi} \int_{-\pi}^{\pi} \sin mx dx$$

$$= \frac{-2B(-1)^m}{m}.$$

Thus, we have

$$Ax^2 + Bx + C \sim A\pi^2/3 + C + 4A\sum_{m=1}^{\infty}\frac{(-1)^m}{m^2}\cos mx - 2B\sum_{m=1}^{\infty}\frac{(-1)^m}{m}\sin mx. \quad (2.9)$$

If the function $f(x) = Ax^2 + Bx + C$ $(-\pi < x < \pi)$ is viewed as the linear combination of two functions $f(x) = f_1(x) + f_2(x)$, where $f_1(x) = Ax^2 + C$ is an even function and $f_2(x) = Bx$ is an odd function on the interval $(-\pi < x < \pi)$. The traces of these two functions in the Fourier series (2.9) are then clearly labeled by the constants A, B, and C, and reflected in the corresponding even and odd constituents of the series:

$$f_1(x) = Ax^2 + C \sim A\pi^2/3 + C + 4A\sum_{m=1}^{\infty}\frac{(-1)^m}{m^2}\cos mx \quad (2.10)$$

or

$$x^2 \sim \pi^2/3 + 4\sum_{m=1}^{\infty}\frac{(-1)^m}{m^2}\cos mx, \quad (2.11)$$

and

$$f_2(x) = Bx \sim -2B\sum_{m=1}^{\infty}\frac{(-1)^m}{m}\sin mx \quad (2.12)$$

or

$$x \sim -2\sum_{m=1}^{\infty}\frac{(-1)^m}{m}\sin mx. \quad (2.13)$$

This observation can be generalized into:

for an odd function $f(x)$ of period 2π, the series (2.3) becomes

$$f(x) \sim \sum_{m=1}^{\infty}b_m \sin mx, \quad (2.14)$$

where

$$b_m = \frac{2}{\pi}\int_0^{\pi}f(x)\sin mxdx; \quad (2.15)$$

for an even function $f(x)$ of period 2π, the series (2.3) reduces to

$$f(x) \sim a_0/2 + \sum_{m=1}^{\infty} a_m \cos mx, \qquad (2.16)$$

where

$$a_m = \frac{2}{\pi} \int_0^{\pi} f(x) \cos mx dx. \qquad (2.17)$$

These results can also be directly obtained from the original expressions, (2.3)–(2.5), by making use of $\int_{-\pi}^{\pi}(.)\,dx = \int_{-\pi}^{0}(.)\,dx + \int_0^{\pi}(.)\,dx$ and taking advantage of the parity of the corresponding integrands wherein.

For convenience, the series expansions given by (2.3), (2.14), and (2.16) are referred to as the full-range Fourier series (or simply, Fourier series), the half-range Fourier sine series (or simply, Fourier sine series or sine series), and the half-range Fourier cosine series (or simply, Fourier cosine series or cosine series), respectively.

Recall we started the discussion by claiming that a periodic function $f(x)$ could be generally expanded into a Fourier series. In the above example, however, the function $f(x) = Ax^2 + Bx + C$ is only explicitly defined on the interval $(-\pi,\pi)$; there is no additional information about its behavior elsewhere. As far as expanding a function into a Fourier series is concerned, it is no difference whether $f(x)$ is actually defined at the end points, $x = -\pi$ and $x = \pi$. As a matter of fact, the periodicity condition has already been implicitly enforced upon the function $f(x)$ by the Fourier series expansion itself. Thus, if $f(x)$ is indeed a 2π-periodic function, then the Fourier series will truthfully represent it almost everywhere on the whole x-axis. Otherwise, the Fourier series actually corresponds to its periodic extension $\overline{f}(x)$, which is defined as: $\overline{f}(x) = f(x)$ for $-\pi < x < \pi$ and $\overline{f}(x+2\pi) = \overline{f}(x)$ elsewhere. This process of reinstating the periodicity of a function, which is originally defined in an interval $(-\pi,\pi)$, is schematically illustrated in Figure 2.2.

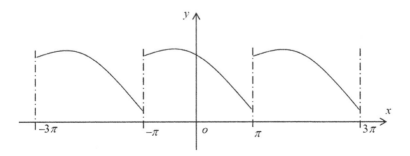

FIGURE 2.2 Periodic extension of a function defined in $(-\pi,\pi)$ onto the whole x-axis.

In addition to this periodic extension, there are other ways of extending a function (defined on an interval) into a periodic function. Let's describe them through a simple example.

EXAMPLE 2.2   Expand $f(x) = x$, $0 \leq x \leq \pi$, into Fourier series.

This function will be expanded into Fourier series in three different ways.

First, the function $f(x) = x$, $0 \leq x \leq \pi$ is viewed as the local display of the π-period function (refer to Figure 2.3a). Using (2.6)–(2.8) with $T = \pi$, we can easily obtain its Fourier series as

$$f(x) \sim \frac{\pi}{2} - \sum_{m=1}^{\infty} \frac{\sin 2mx}{m}, \quad 0 < x < \pi. \tag{2.18}$$

Next, as illustrated in Figure 2.3b, the function $f(x) = x$ is first extended onto $(-\pi, 0)$ to form an odd function on $(-\pi, \pi)$, which is then periodically extended onto the whole x-axis. The resulting Fourier series is the same as the previously obtained in (2.13), that is,

$$x \sim -2 \sum_{m=1}^{\infty} \frac{(-1)^m}{m} \sin mx, \quad 0 < x < \pi. \tag{2.19}$$

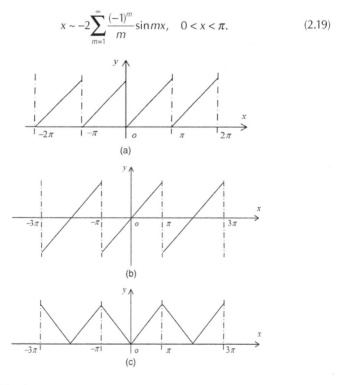

(a)

(b)

(c)

FIGURE 2.3   The function $f(x) = x$, $0 \leq x \leq \pi$ is periodically extended onto the whole x-axis via: (a) Periodic extension; (b) Odd-periodic extension; and (c) Even-periodic extension.

Finally, as illustrated in Figure 2.3c, the function $f(x) = x$ is first extended onto $(-\pi, 0)$ to form an even function on $(-\pi, \pi)$, which is then periodically extended onto the whole x-axis. The resulting Fourier series is then determined as

$$x \sim \frac{\pi}{2} - \sum_{m=1}^{\infty} \frac{4}{(2m-1)^2 \pi} \cos(2m-1)x, \quad 0 < x < \pi. \qquad (2.20)$$

While these three Fourier series expansions (2.18)–(2.20) all converge to the same function $f(x) = x$ on $(0, \pi)$, they actually correspond to three different functions on $(-\pi, \pi)$, namely:

$$f(x) = \begin{cases} \pi + x & -\pi < x < 0 \\ x & 0 < x < \pi \end{cases}, \qquad (2.21)$$

$$f(x) = x, -\pi < x < \pi, \qquad (2.22)$$

and

$$f(x) = |x|, \quad -\pi < x < \pi. \qquad (2.23)$$

Therefore, a function defined on a compact interval can always be made periodic by using any of these three schemes: periodic extension, even-periodic extension, and odd-periodic extension, and corresponding to them there exist three different forms of Fourier series expansions: the full-range Fourier series (2.3), the half-range Fourier sine series (2.14), and the half-range Fourier cosine series (2.16).

2.2 CONVERGENCE OF FOURIER SERIES EXPANSIONS

In (2.3), the sign ~ is used to indicate that the infinite series on the right-hand side is the Fourier expansion of the function $f(x)$ defined on $(-\pi < x < \pi)$. Strictly speaking, we cannot automatically replace the sign ~ with = without knowing, *a priori*, that the Fourier series indeed converges to the function correctly everywhere in the interval. To this end, let's first review a few basic concepts about functions.

A function $f(x)$ is said to be continuous at the point x_0 (or x_0 is a point of continuity), if $f(x_0)$ is defined and

$$\lim_{x \to x_0} f(x) = f(x_0). \qquad (2.24)$$

More explicitly, (2.24) implies that the left-hand limit

$$f(x_0 - 0) = \lim_{\varepsilon \to 0} f(x_0 - \varepsilon),$$

and the right-hand limit

$$f(x_0 + 0) = \lim_{\varepsilon \to 0} f(x_0 + \varepsilon),$$

both exist at the continuity point x_0, and

$$f(x_0 - 0) = f(x_0 + 0) = f(x_0). \tag{2.25}$$

If both of the left- and right-hand limits exist, but are not equal, that is,

$$f(x_0 - 0) \neq f(x_0 + 0), \tag{2.26}$$

we call x_0 a point of *discontinuity of the first kind*, or more specifically *jump discontinuity*.

Accordingly, function $f(x)$ is said to be *continuous* over $D = [a,b]$ if, for every point $x_0 \in D$, $f(x_0)$ is defined and

$$\lim_{x \to x_0} f(x) = f(x_0); \tag{2.27}$$

and *piecewise continuous* if $f(x)$ is continuous everywhere except at a limited number of discontinuity points x_i where

$$f(x_i - 0) \neq f(x_i + 0), \quad i = 1, 2, ..., N. \tag{2.28}$$

For example, the function

$$f(x) = \begin{cases} 1, & \text{if } 0 \leq x \leq 1 \\ x - 1, & \text{if } 1 < x \leq 2 \end{cases}$$

is piecewise continuous over $D = [0,2]$, having a discontinuity point at $x = 1$.

In geometrical terms, continuity is an indication whether a curve $y = f(x)$ jumps at some point(s). In studying the smoothness of a curve, we are also interested in other types of discontinuities, such as corners or sudden changes of directions of tangents. Mathematically, this is equivalent to determining if the derivative of a function is continuous; usually, a function is said to be smooth on an interval $D = [a,b]$ if its derivative is continuous at every point $x \in D$.

In order to measure the smoothness of a function in a broad sense, we use the notation $C^k(D)$ to represent a class of functions whose up to k-th order derivatives all exist and are continuous in the domain D; specifically, $C^0(D)$ denotes the space of continuous functions, and $C^1(D)$ the space of continuously differentiable functions. For convenience, the term "sufficiently smooth" or "differentiable" is sometimes loosely used to describe a class of functions $C^k(D)$ with k being as large as needed.

A function $f(x)$ is integrable over $[a,b]$, if

$$\int_a^b f(x)dx = C, \text{ where } C \text{ is a constant.}$$

If $\int_a^b |f(x)| dx$ exists, $f(x)$ is said to be absolutely integrable.

As in studying the convergence of an infinite series, we consider the truncated version of the Fourier series (2.3)

$$S_M(x) = a_0/2 + \sum_{m=1}^{M-1}(a_m \cos mx + b_m \sin mx). \tag{2.29}$$

The function $S_M(x)$ is sometimes referred to as the *partial sum* or the *trigonometric polynomial*.

If the partial sums have a finite limit, namely,

$$S(x) = \lim_{M \to \infty} S_M(x), \tag{2.30}$$

the Fourier series (2.3) is said to be convergent at point x. If (2.30) holds for every point x in (a,b), the Fourier series is said to be convergent in this interval.

A question is then naturally raised: when the Fourier series (2.3) converges everywhere in the interval, does this mean that $S(x)$ is equal to $f(x)$?

Theorem 2.1 *If $f(x)$ is an absolutely integrable piecewise smooth function of period 2π, then the Fourier series of $f(x)$ converges to $f(x)$ at points of continuity and to $\frac{1}{2}[f(x+0)+f(x-0)]$ at points of discontinuity.*

The proof of this theorem can be found in most books about Fourier series, and will not be given here for shortness.

In light of this theorem, we can now clarify some of the ambiguities left in the above discussions. First, in (2.3) the sign ~ (rather than =) has been used to connect the function $f(x)$ with its Fourier series expansion. As a matter of fact, the replacement of ~ with = can be done only when the Fourier series is known to converge correctly to the function for all x in the interval, except at a finite number of points of discontinuity. Second, this convergence theorem is established based on the assumption that the function $f(x)$ is a piecewise smooth function of period 2π. However, this will not prevent us from applying this theorem directly to a function defined on a compact interval $[-\pi,\pi]$ (or $[0,\pi]$). The reason is that the periodicity condition can be easily supplemented to a locally defined function by using one of the three periodic extension schemes described in Sec. 2.1. It must be noted, however, that after the periodic extension, an originally smooth function will typically lose some of its continuities at $x = \pi \pm 2m\pi$ ($m = 0,1,2,\cdots$), as illustrated in Figure 2.2. As a consequence, the Fourier series will only converge to the "means" at these discontinuity points.

To be more specific, let's consider the function $f(x) = x$, $0 \leq x \leq \pi$, which was previously discussed in Example 2.2. The function $f(x) = x$ is defined on the interval $[0, \pi]$. It can be verified that the Fourier series given in (2.18) will correctly converge to the function everywhere in $(0, \pi)$. This allows replacing the sign $\sim$ with $=$ in (2.18), that is,

$$x = \frac{\pi}{2} - \sum_{m=1}^{\infty} \frac{\sin 2mx}{m}, \quad 0 < x < \pi. \tag{2.31}$$

It is seen that the two end points, $x = 0$ and π, have been excluded from (2.31), although the function is fully defined there: $f(0) = 0$ and $f(\pi) = \pi$. Such exclusion, however, is certainly justified since the Fourier series actually converges to the mean value $[f(0) + f(\pi)]/2 = \pi/2$ at each end, rather than the correct values, 0 and π.

When the function $f(x)$ is even periodically extended (see Figure 2.3c), the resulting function is 2π-periodic and continuous everywhere. Thus, the Fourier series (2.20) can be safely rewritten as

$$x = \frac{\pi}{2} - \sum_{m=1}^{\infty} \frac{4}{(2m-1)^2 \pi} \cos(2m-1)x, \quad 0 \leq x \leq \pi. \tag{2.32}$$

In particular, at the two end points $x = 0$ and π, we have

$$\frac{\pi}{2} = \sum_{m=1}^{\infty} \frac{4}{(2m-1)^2 \pi}. \tag{2.33}$$

Generally, for a continuous function defined on an interval $[a,b]$, the continuity is retained at the end points after the even-periodic extension. In contrast, the continuity is typically lost in the periodic (or odd-periodic) extension, and the corresponding Fourier series invariably converges to the mean value, $[f(a) + f(b)]/2$ (regardless of the actual values of $f(a)$ and $f(b)$), at both ends $x = a$ and $x = b$.

As an observation from Example 2.2, when the continuity of the function $f(x) = x$, $0 \leq x \leq \pi$ is retained in the even-periodic extension, its Fourier series (2.20) converges at a faster rate of m^{-2} in comparison to m^{-1} for its counterparts, (2.18) and (2.19), respectively corresponding to the periodic and odd-periodic extensions.

One may then ask: (1) if this observation is generally applicable to all C^0 functions and (2) if there exists any relationship between the smoothness of a periodic function and the convergence characteristic of its Fourier series expansion? The answers to these questions will be described below.

2.3 FOURIER SERIES FOR THE DERIVATIVES OF FUNCTIONS

Once a function has been expanded into a Fourier series, we are interested in knowing if and how some of the basic mathematical operations (such as integration and differentiation), which are originally intended for the function, can be directly applied to the Fourier series?

Theorem 2.2 *If $f(x)$ is an absolutely integrable function over $(-\pi, \pi)$ and specified by giving its Fourier series*

$$f(x) \sim a_0/2 + \sum_{m=1}^{\infty} (a_m \cos mx + b_m \sin mx),$$

then the integral of $f(x)$ has the Fourier expansion as

$$\int_0^x f(x)dx$$

$$= \sum_{m=1}^{\infty} \frac{b_m}{m} + \sum_{m=1}^{\infty} \frac{-b_m \cos mx + [a_m + (-1)^{m+1} a_0] \sin mx}{m}, -\pi < x < \pi. \tag{2.34}$$

Proof. Define a function

$$F(x) = \int_0^x \left[f(x) - \frac{a_0}{2} \right] dx. \tag{2.35}$$

It is obvious that $F(x)$ is continuous and has an absolutely integrable derivative. In addition, it is 2π-periodic, as evident from

$$F(x + 2\pi) = \int_0^x \left[f(x) - \frac{a_0}{2} \right] dx + \int_x^{x+2\pi} \left[f(x) - \frac{a_0}{2} \right] dx$$

$$= F(x) + \int_{-\pi}^{\pi} \left[f(x) - \frac{a_0}{2} \right] dx$$

$$= F(x) + \int_{-\pi}^{\pi} f(x)dx - a_0\pi = F(x).$$

It should be mentioned that in the second step of the above expression, the function $f(x)$ is treated as a function of period 2π, which can actually be ensured by the periodic extension. Thus, $F(x)$ can be expanded into

$$F(x) = A_0/2 + \sum_{m=1}^{\infty} (A_m \cos mx + B_m \sin mx),$$

where
for $m \geq 1$,

$$A_m = \frac{1}{\pi} \int_{-\pi}^{\pi} F(x) \cos mx \, dx$$

$$= \frac{1}{\pi} F(x) \frac{\sin mx}{m} \Big|_{-\pi}^{\pi} - \frac{1}{\pi m} \int_{-\pi}^{\pi} \left[f(x) - \frac{a_0}{2} \right] \sin mx \, dx = -\frac{b_m}{m},$$

and similarly,

$$B_m = \frac{a_m}{m}.$$

Therefore, we have

$$\int_0^x f(x) \, dx = \frac{a_0 x}{2} + \frac{A_0}{2} + \sum_{m=1}^{\infty} \frac{-b_m \cos mx + a_m \sin mx}{m}. \tag{2.36}$$

Setting $x = 0$ in (2.36), we immediately have

$$\frac{A_0}{2} = \sum_{m=1}^{\infty} \frac{b_m}{m}. \tag{2.37}$$

Substitution of (2.37) and (2.19) into (2.36) will lead to (2.34).

Corollary. If $f(x)$ *is absolutely integrable over* $(-\pi, \pi)$ *and its Fourier coefficient* $a_0 = 0$, *then*

$$\int_0^x f(x) \, dx = \sum_{m=1}^{\infty} \frac{b_m}{m} + \sum_{m=1}^{\infty} \frac{-b_m \cos mx + a_m \sin mx}{m}; \tag{2.38}$$

that is, the Fourier coefficients of the integral can be obtained from the term-by-term integration of the series for $f(x)$.

Remark. If $f(x)$ is absolutely integrable over $(-\pi, \pi)$ and

$$f(x) \sim a_0/2 + \sum_{m=1}^{\infty} (a_m \cos mx + b_m \sin mx),$$

then the integration $\int_a^b f(x)dx$ can be obtained directly from the term-by-term integration of the Fourier series for $f(x)$, that is,

$$\int_a^b f(x)dx = \frac{a_0}{2}(b-a) + \sum_{m=1}^{\infty} \frac{a_m \sin mx - b_m \cos mx}{m}\Big|_a^b, \qquad (2.39)$$

where a and b represent any two points in the interval $(-\pi, \pi)$.

Equation (2.39) is readily available by setting $x = b$ and $x = a$, respectively, in (2.36) and subtracting the results.

As a matter of fact, even when a or b (or both) falls outside of the interval $(-\pi, \pi)$, (2.39) is still valid because $F(x)$ is a function of period 2π.

Now, let's turn to the differentiation of a function $f(x)$ and its Fourier series. First, assume that $f(x)$ is a continuous function of period 2π and its derivative $f'(x)$ is absolutely integrable over $(-\pi, \pi)$. The function $f'(x)$ then has a Fourier series as

$$f'(x) \sim a_0' / 2 + \sum_{m=1}^{\infty}(a_m' \cos mx + b_m' \sin mx), \qquad (2.40)$$

where

$$a_0' = \frac{1}{\pi} \int_{-\pi}^{\pi} f'(x)dx = \frac{1}{\pi}[f(\pi) - f(-\pi)] = 0, \qquad (2.41)$$

$$\begin{aligned} a_m' &= \frac{1}{\pi} \int_{-\pi}^{\pi} f'(x)\cos mx\, dx = \frac{1}{\pi} \int_{-\pi}^{\pi} \cos mx\, df(x) \\ &= \frac{1}{\pi} f(x)\cos mx \Big|_{-\pi}^{\pi} - \frac{1}{\pi}\int_{-\pi}^{\pi} f(x)d[\cos mx] \qquad (2.42) \\ &= \frac{1}{\pi} f(x)\cos mx \Big|_{-\pi}^{\pi} + \frac{m}{\pi}\int_{-\pi}^{\pi} f(x)\sin mx\, dx = mb_m, \end{aligned}$$

$$\begin{aligned} b_m' &= \frac{1}{\pi} \int_{-\pi}^{\pi} f'(x)\sin mx\, dx = \frac{1}{\pi} \int_{-\pi}^{\pi} \sin mx\, df(x) \\ &= \frac{1}{\pi} f(x)\sin mx \Big|_{-\pi}^{\pi} - \frac{1}{\pi}\int_{-\pi}^{\pi} f(x)d[\sin mx] \qquad (2.43) \\ &= \frac{1}{\pi} f(x)\sin mx \Big|_{-\pi}^{\pi} - \frac{m}{\pi}\int_{-\pi}^{\pi} f(x)\cos mx\, dx = -ma_m. \end{aligned}$$

Therefore, (2.40) can be rewritten as

$$f'(x) \sim \sum_{m=1}^{\infty} (mb_m \cos mx - ma_m \sin mx). \tag{2.44}$$

That is, the Fourier series of the derivative $f'(x)$ can be simply obtained from that of $f(x)$ via term-by-term differentiation.

At this point, we claim that the Fourier coefficients for an absolutely integrable function $f(x)$ approach to zero *as* $m \to \infty$. While this statement is often presented as a theorem in some Fourier series books, it suffices here to simply quote it as a fact to facilitate the current discussions.

Accordingly, from (2.44) we have

$$\lim_{m\to\infty} a_m m = \lim_{m\to\infty} b_m m = 0. \tag{2.45}$$

In other words, for a 2π-periodic function with an absolutely integrable derivative, its Fourier coefficients vanish faster than $1/m$ as $m \to \infty$.

This convergence assessment (2.45) can be generalized into an important theorem as given below.

Theorem 2.3 *Let $f(x)$ be a continuous function of period 2π, which has n derivatives, where $n-1$ derivatives are continuous and the n-th derivative is absolutely integrable (the n-th derivative may not exist at certain points). Then the Fourier series of all n derivatives can be obtained by term-by-term differentiations of the Fourier series of $f(x)$, where all the series, except possibly the last, converge uniformly to the corresponding derivatives. Moreover, the Fourier coefficients of the function $f(x)$ satisfy the relations*

$$\lim_{m\to\infty} a_m m^n = \lim_{m\to\infty} b_m m^n = 0. \tag{2.46}$$

The proof of this theorem is kind of straightforward; it simply involves applying (2.45) successively to all the Fourier series for the derivatives of up to the $(n-1)$-th order, and recognizing the fact that in the process all the involved derivatives are absolutely integrable and their corresponding series can be directly obtained from term-by-term differentiations.

More explicitly, (2.46) can be expressed as (Gottlieb and Orszag 1977)

$$a_m = b_m = \mathcal{O}(m^{-n-1}) \tag{2.47}$$

and

$$\max_{-\pi \le x \le \pi} |f(x) - S_M(x)| = \mathcal{O}(M^{-n}), \tag{2.48}$$

where $S_M(x)$ is the partial sum of the Fourier series of the function $f(x)$.

Theorem 2.3 is very useful in that it reveals the intrinsic connection between the smoothness of a periodic function and the convergence characteristic of its Fourier series expansion. In an ideal case, the Fourier series of a periodic analytic function can actually converge at an exponential rate (Tadmor 1986).

However, it must be emphasized that the periodicity is a prerequisite for this convergence theorem to hold. Once the periodicity condition is not met by a function, the convergence of its Fourier series can be seriously deteriorated, even when the function is defined sufficiently smooth on the interval.

To illustrate this point, let's assume that $f(x)$ is sufficiently smooth on $[-\pi, \pi]$. The periodic extension will transform it into a periodic piecewise smooth function potentially having a number of discontinuity points, $x = \pi \pm 2m\pi$ ($m = 0, 1, 2, \cdots$). Consequently, the resulting Fourier series converges to $f(x)$ for $x \in (-\pi, \pi)$, and to $[f(-\pi) + f(\pi)]/2$, rather than $f(-\pi)$ and $f(\pi)$, at $x = -\pi$ and $x = \pi$. For the special case when $f(-\pi) = f(\pi)$, the continuity is then retained during the periodic extension, and the corresponding Fourier series converges in accordance to $\lim_{m \to \infty} a_m m = \lim_{m \to \infty} b_m m = 0$ or, more explicitly, $a_m = b_m = \mathcal{O}(m^{-2})$. However, this convergence estimate does not hold anymore if $f(-\pi) \neq f(\pi)$; that is, when the continuity or the smoothness of a function is partially or completely lost in the process of a periodic extension, the convergence of the resulting Fourier series tends to deteriorate considerably.

An immediate consequence of a slow convergence appears to be that more terms will have to be included in the Fourier series for a desired degree of approximation accuracy. However, we are perhaps more concerned about the differentiability and convergence of the Fourier series for the function and its derivatives, as stated in the following theorem.

Theorem 2.4 *Let $f(x)$ be a continuous function defined on $[-\pi, \pi]$ with an absolutely integrable derivative (which may not exist at certain points). Then we have*

$$f'(x) \sim \frac{c}{2} + \sum_{m=1}^{\infty} \left[\left(mb_m + (-1)^m c \right) \cos mx - ma_m \sin mx \right], \qquad (2.49)$$

where a_m and b_m are the Fourier coefficients of $f(x)$ and $c = \frac{1}{\pi}\left[f(\pi) - f(-\pi) \right]$.

Proof. Let

$$f'(x) - a_0'/2 \sim \sum_{m=1}^{\infty} \left[a_m' \cos mx + b_m' \sin mx \right], \qquad (2.50)$$

where

$$a_0' = \frac{1}{\pi} \int_{-\pi}^{\pi} f'(x)dx = \frac{1}{\pi}[f(\pi) - f(-\pi)] = c. \qquad (2.51)$$

According to (2.38), the integral $F(x) = \int_0^x [f'(x) - a_0'/2]dx$ can be obtained from the term-by-term integration of (2.50), that is,

$$F(x) = \sum_{m=1}^{\infty} \frac{b_m'}{m} + \sum_{m=1}^{\infty} \frac{-b_m' \cos mx + a_m' \sin mx}{m}$$

$$= \sum_{m=1}^{\infty} \frac{-b_m'(\cos mx - 1) + a_m' \sin mx}{m}. \qquad (2.52)$$

On the other hand,

$$F(x) = \int_0^x \left[f'(x) - \frac{a_0'}{2} \right] dx = f(x) - \frac{a_0' x}{2} - f(0). \qquad (2.53)$$

Substituting $f(x) = a_0/2 + \sum_{m=1}^{\infty}(a_m \cos mx + b_m \sin mx)$, $f(0) = a_0/2 + \sum_{m=1}^{\infty} a_m$, and (2.19) into (2.53) results in

$$F(x) = \sum_{m=1}^{\infty} [a_m \cos mx + b_m \sin mx] - \sum_{m=1}^{\infty} a_m - \frac{a_0'}{2}(-2) \sum_{m=1}^{\infty} \frac{(-1)^m}{m} \sin mx$$

$$= \sum_{m=1}^{\infty} \left\{ a_m[\cos mx - 1] + [b_m + (-1)^m a_0'/m] \sin mx \right\}. \qquad (2.54)$$

By equating (2.52) with (2.54), we have

$$b_m' = -m a_m, \qquad (2.55)$$

and

$$a_m' = m b_m + (-1)^m a_0' = m b_m + (-1)^m c. \qquad (2.56)$$

Thus, (2.49) is directly obtained from (2.50).

Theorem 2.4 tells that given a function $f(x)$ on $[-\pi, \pi]$, the Fourier series of its derivative $f'(x)$ cannot be generally obtained from the term-by-term differentiation of the Fourier series for $f(x)$. However, if $f(\pi) = f(-\pi)$ or $c = 0$, then we have from (2.49):

$$f'(x) \sim \sum_{m=1}^{\infty} [m b_m \cos mx - m a_m \sin mx]. \qquad (2.57)$$

Since the Fourier coefficients for an absolutely integrable function approach to zero as $m \to \infty$, it is clear from (2.49):

$$\lim_{m \to \infty} m a_m = 0. \qquad (2.58)$$

and

$$\lim_{m \to \infty} [(-1)^{m+1} mb_m] = c. \tag{2.59}$$

If $f(\pi) = f(-\pi)$, (2.59) reduces to

$$\lim_{m \to \infty} mb_m = 0. \tag{2.60}$$

The above discussions are primarily focused on the integrations and differentiations of the Fourier series of a function $f(x)$ defined on $[-\pi, \pi]$, implying that the function is actually considered to be 2π-periodic. Giving a function $f(x)$ defined on $[0, \pi]$, there exist, at least, three ways to extend it periodically onto the whole x-axis: the periodic extension, the odd-periodic extension, and the even-periodic extension.

If the periodic extension is adopted, the above discussions regarding the integration and differentiation of the Fourier series are readily applicable since the function $f(x)$ on $[0, \pi]$ can be easily transformed into $\overline{f}(\overline{x}) = f[(\overline{x} + \pi)/2]$ for $\overline{x} \in [-\pi, \pi]$.

The even-periodic extension will create an even function over $[-\pi, \pi]$ whose Fourier series is in the form of (2.16). Thus, if $f(x)$ is continuous in the interval $[0, \pi]$, then the even-periodic extension will simply lead to a continuous function of period 2π.

In comparison, even if $f(x)$ is continuous over $[0, \pi]$, the odd-periodic extension will result in a 2π-periodic function, which typically has an infinite number of jump discontinuities at $x = \pm n\pi$ ($n = 0, 1, 2, \cdots$), and its Fourier series in the form of (2.14) will invariably converge to zero at these locations regardless of the actual values of $f(0)$ and $f(\pi)$.

Thus, if the function $f(x)$ is continuous and its derivative is absolutely integrable over $[0, \pi]$, then its Fourier cosine series is always termwise differentiable; and its Fourier sine series is termwise differentiable only if $f(\pi) = f(0) \equiv 0$.

If the condition $f(\pi) = f(0) \equiv 0$ cannot be met, the differentiation of the Fourier sine series will have to be conducted in accordance to the following theorem.

Theorem 2.5 *Let $f(x)$ be a continuous function on $[0, \pi]$ with an absolutely integrable derivative. If $f(x)$ is expanded into the Fourier sine series (2.14),*

$$f(x) = \sum_{m=1}^{\infty} b_m \sin mx \quad (0 < x < \pi),$$

then we have

$$f'(x) \sim \frac{[f(\pi) - f(0)]}{\pi} + \sum_{m=1}^{\infty} \left(mb_m + \frac{2}{\pi}[(-1)^m f(\pi) - f(0)] \right) \cos mx. \tag{2.61}$$

Proof. Since the derivative is absolute integrable, it can be expanded into the Fourier cosine series as

$$f'(x) \sim a_0'/2 + \sum_{m=1}^{\infty} a_m' \cos mx. \tag{2.62}$$

Define an auxiliary function

$$F(x) = \int_0^x \left[f'(x) - \frac{a_0'}{2} \right] dx = f(x) - \frac{a_0' x}{2} - f(0), \tag{2.63}$$

which is obviously continuous and has an absolutely integrable derivative.

Since $f(x)$ is already given in the form of Fourier sine series, the other two terms on the right-hand side of (2.63) will also be expanded into Fourier sine series.

The Fourier sine series for a constant can be directly obtained from the term-by-term differentiation of (2.20), that is:

$$1 = \sum_{m=1}^{\infty} \frac{4}{(2m-1)\pi} \sin(2m-1)x, 0 < x < \pi, \tag{2.64}$$

or

$$1 = \frac{2}{\pi} \sum_{m=1}^{\infty} \frac{[1-(-1)^m]}{m} \sin mx. \tag{2.65}$$

Thus, by making use of (2.19) and (2.65), (2.63) can be rewritten as

$$
\begin{aligned}
F(x) &= \sum_{m=1}^{\infty} b_m \sin mx - a_0' \sum_{m=1}^{\infty} (-1)^{m+1} \frac{\sin mx}{m} - \frac{2}{\pi} f(0) \sum_{m=1}^{\infty} [1-(-1)^m] \frac{\sin mx}{m} \\
&= \sum_{m=1}^{\infty} \left[mb_m - \frac{2}{\pi} f(0) + \left(a_0' + \frac{2}{\pi} f(0) \right)(-1)^m \right] \frac{\sin mx}{m}.
\end{aligned}
\tag{2.66}
$$

On the other hand, $F(x)$ can directly be obtained from (2.62) by term-by-term integration, namely,

$$F(x) = \int_0^x \left[f'(x) - \frac{a_0'}{2} \right] dx = \int_0^x \left[\sum_{m=1}^{\infty} a_m' \cos mx \right] dx = \sum_{m=1}^{\infty} \frac{a_m'}{m} \sin mx. \tag{2.67}$$

Equating (2.66) with (2.67) and comparing the coefficients for the like terms will lead to

$$a'_m = mb_m - \frac{2}{\pi} f(0) + \left(a'_0 + \frac{2}{\pi} f(0) \right)(-1)^m, \qquad (2.68)$$

which proves (2.61) by recognizing $a'_0 = \frac{2}{\pi} \int_0^x f'(x)dx = \frac{2}{\pi}[f(\pi) - f(0)]$.

Theorem 2.5 (also Theorem 2.3) can also be proven in a seemingly much simpler way as described below.

The Fourier coefficients in (2.62) are determined from

$$a'_0 = \frac{2}{\pi} \int_0^{\pi} f'(x)dx = \frac{2}{\pi}[f(\pi) - f(0)], \qquad (2.69)$$

and

$$a'_m = \frac{2}{\pi} \int_0^{\pi} f'(x) \cos mx\, dx$$

$$= \frac{2}{\pi} f(x) \cos mx \Big|_0^{\pi} + \frac{2m}{\pi} \int_0^{\pi} f(x) \sin mx\, dx \qquad (2.70)$$

$$= \frac{2}{\pi}[(-1)^m f(\pi) - f(0)] + mb_m.$$

However, this approach potentially has a minor problem; that is, when an absolutely integrable piecewise smooth function of period 2π is defined on the interval $[a,b]$, the Fourier series may fail to converge at $x = a$ and $x = b$ unless the length of the interval is equal to 2π (Tolstov 1965).

It is obvious from (2.61) that, if $f(\pi) = f(0) = 0$, the Fourier sine series of $f(x)$ can then be simply differentiated, term-by-term, to obtain the Fourier series of $f'(x)$.

Based on the fact that the Fourier coefficients of an absolutely integrable function approach to zero as $m \to \infty$, (2.61) gives rise to an interesting result: the end values, $f(\pi)$ and $f(0)$, of the function $f(x)$ can be determined from its Fourier coefficients:

$$f(\pi) - f(0) = -\frac{\pi}{2} \lim_{m \to \infty} mb_m, \quad \text{for even } m, \qquad (2.71)$$

and

$$f(\pi) + f(0) = \frac{\pi}{2} \lim_{m \to \infty} mb_m, \quad \text{for odd } m. \qquad (2.72)$$

The significance of the above relations is that even though the Fourier sine series does not generally converge to the function $f(x)$ correctly at the ends of the interval, its Fourier coefficients can be used to recover such information. Take $f(x) = x$ for example. From (2.19), we have

$$f(\pi) - f(0) = -\frac{\pi}{2} \lim_{m \to \infty} m b_m = \pi, \quad \text{for even } m,$$

and

$$f(\pi) + f(0) = \frac{\pi}{2} \lim_{m \to \infty} m b_m = \pi, \quad \text{for odd } m;$$

that is, $f(\pi) = \pi$ and $f(0) = 0$.

REFERENCES

González-Velasco, E. A. 1996. *Fourier Analysis and Boundary Value Problems*. San Diego, CA: Academic Press.

Gottlieb, D., and S. A. Orszag. 1977. *Numerical Analysis of Spectral Methods: Theory and Applications*. Philadelphia, PA: SIAM.

Hardy, G. H., and W. W. Rogosinski. 2013. *Fourier Series*. Mineola, NY: Dover Publications.

Iserles, A., and S. P. Nørsett. 2008. From high oscillation to rapid approximation I: Modified Fourier expansions. *IMA Journal of Numerical Analysis* 28: 862–887.

Tadmor, E. 1986. The exponential accuracy of Fourier and Chebyshev differencing methods. *SIAM Journal on Numerical Analysis* 23: 1–10.

Tolstov, G. P. 1965. *Fourier Series*. Englewood Cliffs, NJ: Prentice-Hall.

Zygmund, A. 1968. *Trigonometric Series*, vol. I. Cambridge: Cambridge University Press.

3 The Generalized Fourier Series with Accelerated Convergence

Fourier series will lose much of its luster in expanding a nonperiodic function, even when this function is defined sufficiently smooth on a compact interval. It is well known that a continuous function can always be expanded into a Fourier series *inside* the interval; the word "inside" is highlighted to emphasize the fact that the two end points shall be excluded, unless proven otherwise. While this may not seem to be a big deal at the first glance, it actually has profound implications to the validity and efficacy of the Fourier series when used to approximate a function or solve a boundary value problem. In what follows, a strategy will be discussed on how to improve the convergence of the Fourier series expansions of a locally-defined sufficiently smooth function.

3.1 IMPROVING THE CONVERGENCE OF FOURIER SERIES

The convergence of Fourier series is perhaps the most important concern in practical applications. Faster convergence implies that a relatively small number of terms in the series expansion may lead to a decent approximation to the original function regardless of whether the function is explicitly known or not. Perhaps more importantly, the convergence rate dictates if the term-by-term differentiations can be directly applied to the Fourier series of a function or solution to obtain the series representations for its derivatives, as often encountered in solving differential equations or boundary value problems.

By recognizing the slower convergence of sine series than its cosine counterpart, a modified Fourier series was proposed as (Iserles and Nørsett, 2008)

$$f(x) = a_0/2 + \sum_{m=1}^{\infty} \left[a_m \cos mx + b_m \sin\left(m - \frac{1}{2} \right) x \right]. \tag{3.1}$$

If $f(x)$ is differentiable and its derivative has bounded variation, the expansion coefficients, a_m and b_m, in (3.1) will both decay like $\mathcal{O}(m^{-2})$ (Adcock 2011), which is still considered slow in many cases.

To illustrate the possibility of improving the convergence of a Fourier series, let's start with an example. Suppose that the function $f(x)$ has the Fourier series

$$f(x) = \sum_{m=2}^{\infty} (-1)^m \frac{m^3}{m^4 - 1} \sin mx, \quad 0 < x < \pi. \tag{3.2}$$

DOI: 10.1201/9781003194859-3

Evidently, this Fourier series converges, if it does, at a rate of $1/m$ as $m \to \infty$. Since

$$\frac{m^3}{m^4 - 1} = \frac{m^4 - 1 + 1}{m(m^4 - 1)} = \frac{1}{m} + \frac{1}{m(m^4 - 1)},$$

(3.2) can be rewritten as

$$f(x) = \sum_{m=2}^{\infty} (-1)^m \frac{\sin mx}{m} + \sum_{m=2}^{\infty} (-1)^m \frac{1}{m(m^4 - 1)} \sin mx. \qquad (3.3)$$

In this way, the original Fourier series is split into two parts: one converges essentially at an equal rate of $1/m$ and the other at a much faster rate of $1/m^5$.

However, by recognizing the first term in (3.3) converges to $\sin x - x/2$ (refer to 2.19), we have

$$f(x) = -\frac{x}{2} + \sin x + \sum_{m=2}^{\infty} (-1)^m \frac{1}{m(m^4 - 1)} \sin mx, \quad 0 < x < \pi. \qquad (3.4)$$

This example has shed some lights on how to improve the convergence of a Fourier series expansion: by splitting the original Fourier series into two or more and replacing the slower converged with an appropriate (closed-form) function(s), the residual Fourier series is then better shaped to converge at a faster rate. Such an approach can be mathematically described as

$$f(x) = h(x) + \sum_{m=1}^{\infty} (\alpha_m \cos mx + \beta_m \sin mx). \qquad (3.5)$$

It needs to be pointed out that the convergence rates for the sine and cosine terms are typically different, and the slower one actually dictates the convergence of Fourier series as a whole. Therefore, if the Fourier series (3.5) is said to have an improved convergence, it means that the slower Fourier coefficients of (3.5) decay faster than their counterparts in (2.3).

More explicitly, suppose

$$\left(a_m, b_m\right) \propto \left(m^{-N_a}, m^{-N_b}\right) \text{ and } \left(\alpha_m, \beta_m\right) \propto \left(m^{-N_\alpha}, m^{-N_\beta}\right). \qquad (3.6)$$

Then what we say (3.5) converges faster than (2.3) actually implies

$$\min\left(N_\alpha, N_\beta\right) > \min\left(N_a, N_b\right). \qquad (3.7)$$

While this example demonstrates the feasibility of improving the convergence of a Fourier series expansion via manipulating the Fourier coefficients, such an approach is practically of limited use in view of the fact that the Fourier coefficients have to be known explicitly and be manipulatable.

From a different perspective, however, (3.5) has indeed pointed a direction for accelerating the convergence of the involved Fourier series expansion; it is possible to decompose a function into an appropriate function and a Fourier series with improved convergence. As a matter of fact, we can restate it in a more affirmative tone: given a sufficiently smooth function $f(x)$ defined on a compact interval, by properly selecting the subtraction (or supplementary) function $h(x)$, the Fourier series of the residual function $f(x) - h(x)$ can be obtained to have better convergence, as described below.

Theorem 2.3 in Sec. 2.3 tells that the smoothness of a periodic function actually determines the convergence rate of its Fourier series expansion, as formally established by (2.46). In practice, however, the function is often only explicitly defined on a compact interval. Thus, when the function is sought in a form of the Fourier series, the function has already been implicitly extended onto the whole x-axis via, such as, one of the periodic-extension schemes discussed in the previous chapter. This is evident from the series expansions (2.19) and (2.20): the same function $f(x) = x$ can have different series representations on the interval $(0, \pi)$, which respectively correspond to two different 2π-periodic functions, as illustrated in Figure 2.3(b) and (c). While the periodic function in Figure 2.3(c) is continuous everywhere, its counterpart in Figure 2.3(b) is not, and this difference has been reflected in the convergence rates for the resulting Fourier series.

It is now clear that for a given function $f(x)$ defined on the interval $[a,b]$, the convergence of its Fourier series can be accelerated by subtracting from it an appropriate supplementary function so that the residual function, when it is periodically extended, will have a desired smoothness and its Fourier series can accordingly converge at a faster rate.

This scheme was originally used to improve the convergence of the Fourier series expansion of a function which has a finite number of discontinuity points on the interval, and polynomials are often used as the subtraction functions (Lanczos 1966; Jones and Hardy 1970; Baszenski et al. 1995; Eckhoff 1998).

To be specific, we here assume that the function $f(x)$ is sufficiently smooth (e.g., having K continuous derivatives and the $(K + 1)$-th derivative being absolutely integrable) in the interval $[-\pi, \pi]$. When the function $f(x)$ is periodically extended, the resulting 2π-periodic function and its derivatives will potentially have discontinuities at $\pi \pm 2n\pi, n = 0,1,2,\ldots$.

In the polynomial subtraction techniques, the original function $f(x)$ is decomposed into the polynomial $h(x)$ and the residual function $F(x)$ as

$$f(x) = F(x) + h(x). \tag{3.8}$$

In particular, the subtraction polynomial $h(x)$ is such selected that

$$F^{(k)}(\pi) = F^{(k)}(-\pi) \quad (k = 0,1,2,...,K), \tag{3.9}$$

or

$$f^{(k)}(\pi) - f^{(k)}(-\pi) = h^{(k)}(\pi) - h^{(k)}(-\pi) \quad (k = 0,1,2,...,K). \tag{3.10}$$

In other words, the residual function $F(x)$, if periodically extended, will have K continuous derivatives and its $(K + 1)$-th derivative is absolutely integrable. Thus, according to Theorem 2.3, its Fourier coefficients, if denoted by α_m and β_m, are to satisfy

$$\lim_{m\to\infty} \alpha_m m^{K+1} = \lim_{m\to\infty} \beta_m m^{K+1} = 0. \tag{3.11}$$

In comparison, the convergence rate for the Fourier series of the original function $f(x)$ is only known as

$$\lim_{m\to\infty} a_m = \lim_{m\to\infty} b_m = 0. \tag{3.12}$$

Practically, the polynomial $h(x)$ can be constructed using a set of linearly independent or orthogonal polynomials as

$$h(x) = \sum_{l=0}^{K} c_{l+1} p_{l+1}(x). \tag{3.13}$$

As an example, consider the Lanczos polynomials:

$$p_1(x) = x/\pi, \tag{3.14}$$

$$p_l'(x) = p_{l-1}(x), \quad l = 2,3,... \tag{3.15}$$

and

$$p_{2l+1}(0) = p_{2l+1}(\pi) = 0, \quad l = 1,2,.... \tag{3.16}$$

Essentially, (3.16) is supplemented to uniquely determine the integration constants with the polynomials of higher orders introduced from using the recursive relations (3.15). The first few Lanczos polynomials can be explicitly given as

$$p_1(x) = x/\pi, \tag{3.17}$$

$$p_2(x) = (3x^2 - \pi^2)/6\pi, \tag{3.18}$$

$$p_3(x) = (x^3 - \pi^2 x)/6\pi, \tag{3.19}$$

$$p_4(x) = (15x^4 - 30\pi^2 x^2 + 7\pi^4)/360\pi. \tag{3.20}$$

The Lanczos polynomials with even (odd) subscripts are obviously even (odd) functions.

By making use of (3.13) and (3.10), we have

$$f^{(k)}(\pi) - f^{(k)}(-\pi) = \sum_{l=0}^{K} c_{l+1}[p_{l+1}^{(k)}(\pi) - p_{l+1}^{(k)}(-\pi)] \quad (k = 0,1,2,...,K). \tag{3.21}$$

Assuming K is an even integer, the first equation of (3.21), corresponding to $k = 0$, can be rewritten as

$$f(\pi) - f(-\pi) = c_1[p_1(\pi) - p_1(-\pi)] + \sum_{l=1}^{K/2} c_{2l+1}[p_{2l+1}(\pi) - p_{2l+1}(-\pi)]$$
$$+ \sum_{l=1}^{K/2} c_{2l}[p_{2l}(\pi) - p_{2l}(-\pi)]. \tag{3.22}$$

Since $p_{2l}(x)$ are even functions and $p_{2l+1}(\pi) = 0$ for $l = 1,2,...$, the 2nd and 3rd terms on the right-hand side of (3.22) will both vanish, leading to

$$c_1 = \frac{f(\pi) - f(-\pi)}{p_1(\pi) - p_1(-\pi)} = \frac{f(\pi) - f(-\pi)}{2}. \tag{3.23}$$

Similarly, for $k = 1$, (3.21) becomes

$$f'(\pi) - f'(-\pi) = c_1[p_1'(\pi) - p_1'(-\pi)] + \sum_{l=1}^{K/2} c_{2l+1}[p_{2l+1}'(\pi) - p_{2l+1}'(-\pi)]$$
$$+ \sum_{l=1}^{K/2} c_{2l}[p_{2l}'(\pi) - p_{2l}'(-\pi)]$$
$$= \sum_{l=1}^{K/2} c_{2l+1}[p_{2l}(\pi) - p_{2l}(-\pi)] + \sum_{l=1}^{K/2} c_{2l}[p_{2l-1}(\pi) - p_{2l-1}(-\pi)]$$
$$= c_2[p_1(\pi) - p_1(-\pi)] \tag{3.24}$$

or

$$c_2 = \frac{f'(\pi) - f'(-\pi)}{2}. \qquad (3.25)$$

By repeating this process for $k = 2, 3, \cdots, K$, the polynomial $h(x)$ can be finally determined as

$$h(x) = \sum_{k=0}^{K} c_{k+1} p_{k+1}(x), \qquad (3.26)$$

where

$$c_{k+1} = \frac{f^{(k)}(\pi) - f^{(k)}(-\pi)}{2}. \qquad (3.27)$$

To highlight the clear dependence of polynomial coefficients upon the orders of derivatives, the index l has been replaced with k for the summation in (3.13).

In the polynomial subtraction scheme, it is the residual function $F(x)$ that is expanded into Fourier series, that is:

$$f(x) = h(x) + a_0/2 + \sum_{m=1}^{\infty} (a_m \cos mx + b_m \sin mx), \quad -\pi \le x \le \pi, \qquad (3.28)$$

where the Fourier coefficients a_m and b_m are calculated from

$$a_m = \frac{1}{\pi} \int_{-\pi}^{\pi} [f(x) - h(x)] \cos mx \, dx \qquad (3.29)$$

and

$$b_m = \frac{1}{\pi} \int_{-\pi}^{\pi} [f(x) - h(x)] \sin mx \, dx. \qquad (3.30)$$

For distinction, (3.28) is here referred to as the generalized Fourier series of the function $f(x)$.

To help visualize the improved convergence of the generalized Fourier series, let's consider the calculations of its expansion coefficients using (3.29)

$$\pi a_m = \int_{-\pi}^{\pi} [f(x) - h(x)] \cos mx \, dx$$

$$= \int_{-\pi}^{\pi} \left[f(x) - \sum_{k=0}^{K} c_{k+1} p_{k+1}(x) \right] \cos mx \, dx$$

$$= \frac{\sin mx}{m} \left[f(x) - \sum_{k=0}^{K} c_{k+1} p_{k+1}(x) \right]\Big|_{-\pi}^{\pi} - \frac{1}{m} \int_{-\pi}^{\pi} \left[f'(x) - \sum_{k=0}^{K} c_{k+1} p'_{k+1}(x) \right] \sin mx \, dx$$

$$= \frac{\sin mx}{m} \left[f(x) - \sum_{k=0}^{K} c_{k+1} p_{k+1}(x) \right]\Big|_{-\pi}^{\pi} + \frac{\cos mx}{m^2} \left[f'(x) - \sum_{k=0}^{K} c_{k+1} p'_{k+1}(x) \right]\Big|_{-\pi}^{\pi}$$

$$- \frac{1}{m^2} \int_{-\pi}^{\pi} \left[f''(x) - \sum_{k=0}^{K} c_{k+1} p''_{k+1}(x) \right] \cos mx \, dx$$

$$= \frac{\sin mx}{m} \left[f(x) - \sum_{k=0}^{K} c_{k+1} p_{k+1}(x) \right]\Big|_{-\pi}^{\pi} + \frac{\cos mx}{m^2} \left[f'(x) - \sum_{k=0}^{K} c_{k+1} p'_{k+1}(x) \right]\Big|_{-\pi}^{\pi}$$

$$- \frac{\sin mx}{m^3} \left[f''(x) - \sum_{k=0}^{K} c_{k+1} p''_{k+1}(x) \right]\Big|_{-\pi}^{\pi}$$

$$+ \frac{1}{m^3} \int_{-\pi}^{\pi} \left[f^{(3)}(x) - \sum_{k=0}^{K} c_{k+1} p^{(3)}_{k+}(x) \right] \sin mx \, dx$$

$$= \frac{\sin mx}{m} \left[f(x) - \sum_{k=0}^{K} c_{k+1} p_{k+1}(x) \right]\Big|_{-\pi}^{\pi} + \frac{\cos mx}{m^2} \left[f'(x) - \sum_{k=0}^{K} c_{k+1} p'_{k+1}(x) \right]\Big|_{-\pi}^{\pi}$$

$$- \frac{\sin mx}{m^3} \left[f''(x) - \sum_{k=0}^{K} c_{k+1} p''_{k+1}(x) \right]\Big|_{-\pi}^{\pi} - \frac{\cos mx}{m^4} \left[f^{(3)}(x) - \sum_{k=0}^{K} c_{k+1} p^{(3)}_{k+1}(x) \right]\Big|_{-\pi}^{\pi}$$

$$+ \frac{1}{m^4} \int_{-\pi}^{\pi} \left[f^{(4)}(x) - \sum_{k=0}^{K} c_{k+1} p^{(4)}_{k+1}(x) \right] \cos mx \, dx.$$

(3.31)

While the first and third terms on the right-hand side of (3.31) vanish obviously, the second and fourth terms can be proven to be equal to zero; for instance,

$$-\frac{\cos mx}{m^4}\left[f^{(3)}(x)-\sum_{k=0}^{K}c_{k+1}p_{k+1}^{(3)}(x)\right]\Bigg|_{-\pi}^{\pi}$$

$$=-\frac{\cos mx}{m^4}\left[f^{(3)}(x)-c_3p_3^{(3)}(x)-c_4p_4^{(3)}(x)-\sum_{k=4}^{K}c_{k+1}p_{k+1}^{(3)}(x)\right]\Bigg|_{-\pi}^{\pi} \qquad (3.32)$$

$$=-\frac{\cos mx}{m^4}\left[f^{(3)}(x)-c_3p_1'(x)-c_4p_1(x)-\sum_{k=4}^{K}c_{k+1}p_{k-2}(x)\right]\Bigg|_{-\pi}^{\pi}=0.$$

In the second step of the above equation, (3.15) has been used, and the last step is based on (3.16) and (3.27) and the parity of Lanczos polynomials.

Thus, (3.31) reduces to

$$\pi a_m = \int_{-\pi}^{\pi}[f(x)-h(x)]\cos mx dx$$

$$=\frac{1}{m^4}\int_{-\pi}^{\pi}\left[f^{(4)}(x)-\sum_{k=0}^{K}c_{k+1}p_{k+1}^{(4)}(x)\right]\cos mx dx \qquad (3.33)$$

$$=\frac{1}{m^4}\int_{-\pi}^{\pi}\left[f^{(4)}(x)-\sum_{k=3}^{K}c_{k+1}p_{k+1}^{(4)}(x)\right]\cos mx dx.$$

As a matter of fact, (3.33) can be integrated, by part, continuously up to the $(K+1)$-th derivative, that is,

$$\pi a_m = \int_{-\pi}^{\pi}[f(x)-h(x)]\cos mx dx$$

$$=\begin{cases} \dfrac{(-1)^{(K+1)/2}}{m^{K+1}}\displaystyle\int_{-\pi}^{\pi}[f^{(K+1)}(x)-c_{K+1}p_{K+1}^{(K+1)}(x)]\cos mx dx & \text{if } K \text{ is odd} \\[20pt] \dfrac{(-1)^{K/2+1}}{m^{K+1}}\displaystyle\int_{-\pi}^{\pi}[f^{(K+1)}(x)-c_{K+1}p_{K+1}^{(K+1)}(x)]\sin mx dx & \text{if } K \text{ is even.} \end{cases}$$

$$(3.34)$$

Since the $(K+1)$-th derivative of $f(x)$, and hence $[f^{(K+1)}(x) - c_{K+1}p_{K+1}^{(K+1)}(x)]$, is absolutely integrable, its Fourier coefficients are guaranteed to approach to zero as $m \to \infty$; in other words, (3.34) implies

$$\lim_{m \to \infty} a_m m^{K+1} = 0. \tag{3.35}$$

Even though the above discussions are focused on the Fourier coefficients for the cosine terms, it is obvious that the conclusions are directly applicable to the Fourier coefficients b_m.

It should also be pointed out that the subtraction polynomial is constructed using $K+1$ Lanczos polynomials to maximize the convergence rate to $a_m m^{K+1} \to 0$ as $m \to \infty$. This maximum convergence is to take full advantage of the assumed C^K continuity of the function $f(x)$ on the interval. However, the Fourier series here can actually be tailored to converge at any desired slower rate by simply setting $h(x) = \sum_{l=0}^{\hat{K}} c_{l+1}p_{l+1}(x)$ for $\hat{K} < K$.

Now, by assuming that the function $f(x)$ has $(2K+1)$ continuous derivatives and the absolutely integrable $2(K+1)$-th derivative on $[0,\pi]$, let's expand it into the generalized Fourier cosine series as:

$$f(x) = h(x) + a_0/2 + \sum_{m=1}^{\infty} a_m \cos mx, \quad 0 \le x \le \pi, \tag{3.36}$$

where the Fourier coefficients a_m are calculated from

$$a_m = \frac{2}{\pi} \int_0^\pi [f(x) - h(x)] \cos mx\, dx. \tag{3.37}$$

The polynomial $h(x)$ can still be given in the form of (3.13). However, as evident from the previous discussions, regardless of the selection of $h(x)$, the continuities of $F(x)$ and its derivatives of even orders, at $x = \pm n\pi\,(n = 0,1,2,...)$, can always be ensured by the even-periodic extension. As a result, the polynomial $h(x)$ only needs to be selected to secure the continuities of the odd-order derivatives of $F(x)$ at these locations. In the meantime, notice that the odd-order derivatives are odd functions over the interval, then the only way for them to be continuous at $x = \pm n\pi\,(n = 0,1,2,...)$ is to set

$$F^{(2k+1)}(0) = F^{(2k+1)}(\pi) \equiv 0, \quad (k = 0,1,\cdots,K), \tag{3.38}$$

or

$$f^{(2k+1)}(\pi) - h^{(2k+1)}(\pi) = f^{(2k+1)}(0) - h^{(2k+1)}(0) = 0, (k = 0,1,\cdots,K). \tag{3.39}$$

Such a polynomial is easily constructed as

$$h(x) = \sum_{k=0}^{K} \left[f^{(2k+1)}(\pi) p_{2k+2}(x) - f^{(2k+1)}(0) p_{2k+2}(\pi - x) \right], \qquad (3.40)$$

and it can be readily verified that (3.39) has been satisfied according to the definitions of Lanczos polynomials, (3.14)–(3.16).

Similarly, the residual function $F(x)$ can be expanded into Fourier sine series over $[0, \pi]$:

$$f(x) = h(x) + \sum_{m=1}^{\infty} b_m \sin mx, \quad 0 \le x \le \pi, \qquad (3.41)$$

where the coefficients b_m are calculated from

$$b_m = \frac{2}{\pi} \int_0^{\pi} [f(x) - h(x)] \sin mx \, dx. \qquad (3.42)$$

In such a case, the polynomial $h(x)$ is sought to satisfy

$$F^{(2k)}(0) = F^{(2k)}(\pi) \equiv 0, \quad (k = 0, 1, \cdots, K), \qquad (3.43)$$

or

$$f^{(2k)}(\pi) - h^{(2k)}(\pi) = f^{(2k)}(\pi) - h^{(2k)}(0) = 0, \quad (k = 0, 1, \cdots, K). \qquad (3.44)$$

Accordingly, the polynomial $h(x)$ can be constructed as

$$h(x) = \sum_{k=0}^{K} \left[f^{(2k)}(0) p_{2k+1}(\pi - x) + f^{(2k)}(\pi) p_{2k+1}(x) \right]. \qquad (3.45)$$

In summary, assume that $f(x)$ is C^{n-1} continuous on $[-\pi, \pi]$ or $[0, \pi]$ and its n-th derivative is absolutely integrable. Then the residual function $F(x)$ defined by (3.8) can be extended into a 2π-periodic function of:

1. C^K continuity ($K \le n - 1$), if $h(x)$ is given by (3.26);
2. C^{2K+2} continuity ($2K + 2 < n$), if $h(x)$ is given by (3.40);
3. C^{2K+1} continuity for ($2K + 1 < n$), if $h(x)$ is given by (3.45).

Therefore, their corresponding Fourier series expansions will be able to converge at faster rates in accordance with Theorem 2.3.

3.2 THE GENERALIZED FOURIER COSINE SERIES EXPANSION WITH ACCELERATED CONVERGENCE

For a sufficiently smooth function $f(x)$ defined on a compact interval $[0, \pi]$, it can always be expanded into Fourier cosine series as

$$f(x) = a_0/2 + \sum_{m=1}^{\infty} a_m \cos mx, \quad 0 \le x \le \pi,$$

and the expansion coefficients a_m are known to decay like $\mathcal{O}(m^{-2})$.

In the previous section, a Fourier series convergence acceleration scheme has been described in which an appropriate polynomial $h(x)$ is purposely subtracted from the original function $f(x)$ so that the residual function $F(x)$ can be expanded into a Fourier series with a better convergence rate.

While the polynomial $h(x)$ can be easily constructed, as described in Sec. 3.1, using the Lanczos polynomials or others, the combined use of the trigonometric functions and polynomials may appear to be non-traditional, among some other undesired characteristics. To keep a close similarity to the conventional Fourier series, while maintaining the improved convergence characteristic of the generalized Fourier series, the polynomial $h(x)$ in (3.36) will be replaced with trigonometric terms, leading to a "pure" trigonometric series in the form of

$$f(x) = \mathcal{F}_{\infty,2P}[f](x) = a_0/2 + \sum_{m=1}^{\infty} a_m \cos mx + \sum_{p=1}^{2P} b_p \sin px, \quad 0 \le x \le \pi, \quad (3.46)$$

where

$$a_m = \frac{2}{\pi} \int_0^{\pi} \left[f(x) - \sum_{p=1}^{2P} b_p \sin px \right] \cos mx \, dx, \quad (3.47)$$

and the coefficients b_p are determined as follows.

Theorem 3.1 *Let $f(x)$ be of C^{n-1} continuity on the interval $[0,\pi]$ and its n-th derivative is absolutely integrable (the n-th derivative may not exist at certain points). If $n \ge 2$, then the Fourier coefficient a_m, as defined in (3.47), decays at a polynomial rate as*

$$\lim_{m \to \infty} a_m m^{2P} = 0 \quad (2P \le n), \quad (3.48)$$

provided that

$$\sum_{p=1}^{P} b_{2p}(2p)^{2q-1} = (-1)^{q-1}[f^{(2q-1)}(\pi) + f^{(2q-1)}(0)]/2 \quad (3.49)$$

and

$$\sum_{p=1}^{P} b_{2p-1}(2p-1)^{2q-1} = (-1)^q [f^{(2q-1)}(\pi) - f^{(2q-1)}(0)]/2 \quad (q=1,2,...,P). \quad (3.50)$$

Proof. By integrating by part, we have

$$(\pi/2)f_m = \int_0^\pi f(x)\cos mx\,dx$$

$$= \frac{\sin mx}{m} f(x)\Big|_0^\pi - \frac{1}{m}\int_0^\pi f'(x)\sin mx\,dx$$

$$= \frac{\cos mx}{m^2} f'(x)\Big|_0^\pi - \frac{1}{m^2}\int_0^\pi f''(x)\cos mx\,dx$$

$$\cdots\cdots$$

$$= \sum_{q=1}^{Q} \frac{(-1)^{q-1}\cos mx}{m^{2q}} f^{(2q-1)}(x)\Big|_0^\pi + \frac{(-1)^Q}{m^{2Q}}\int_0^\pi f^{(2Q)}(x)\cos mx\,dx$$

$$= \sum_{q=1}^{Q} \frac{(-1)^{q-1}}{m^{2q}}\left[(-1)^m f^{(2q-1)}(\pi) - f^{(2q-1)}(0)\right] + \frac{(-1)^Q}{m^{2Q}}\int_0^\pi f^{(2Q)}(x)\cos mx\,dx. \quad (3.51)$$

Denote $h(x) = \sum_{p=1}^{2P} b_p \sin px$, then

$$(\pi/2)h_m = \sum_{p=1}^{2P} b_p \int_0^\pi \sin px \cos mx\,dx$$

$$= \sum_{p=1}^{2P} b_p \int_0^\pi \frac{1}{2}[\sin(p+m)x + \sin(p-m)x]dx$$

$$= -\frac{1}{2}\sum_{p=1}^{2P} b_p \left(\frac{\cos(p+m)x}{p+m} + \frac{\cos(p-m)x}{p-m}\right)\Big|_0^\pi$$

$$= \sum_{p=1}^{2P} b_p \frac{p}{m^2 - p^2}[(-1)^{p+m} - 1] \qquad (3.52)$$

$$= \sum_{p=1}^{2P} [(-1)^{p+m} - 1]b_p \left[\frac{p}{m^2} + \frac{p^3}{m^4} + \frac{p^5}{m^6} + \cdots\right]$$

$$= \sum_{q=1}^{Q} \frac{1}{m^{2q}}\left(\sum_{p=1}^{2P} [(-1)^{p+m} - 1]p^{2q-1}b_p\right) + \mathcal{O}(m^{-2Q-2}).$$

Subtracting (3.52) from (3.51) leads to

$$
(\pi/2)a_m = (\pi/2)(f_m - h_m) = \sum_{q=1}^{Q} \frac{(-1)^{m+q-1}}{m^{2q}} \left[f^{(2q-1)}(\pi) - \sum_{p=1}^{2P} b_p p^{(2q-1)}(-1)^{q+p-1} \right]
$$

$$
- \sum_{q=1}^{Q} \frac{(-1)^{q-1}}{m^{2q}} \left[f^{(2q-1)}(0) - \sum_{p=1}^{2P} (-1)^{q-1} b_p p^{(2q-1)} \right]
$$

$$
+ \frac{(-1)^Q}{m^{2Q}} \int_{0}^{\pi} f^{(2Q)}(x) \cos mx \, dx + \mathcal{O}(m^{-2Q-2}). \tag{3.53}
$$

The first two terms in (3.53) vanish, if

$$
\sum_{p=1}^{2P} b_p p^{2q-1}(-1)^{q+p-1} = f^{(2q-1)}(\pi) \tag{3.54}
$$

and

$$
\sum_{p=1}^{2P} b_p p^{2q-1}(-1)^{q-1} = f^{(2q-1)}(0), \quad q = 1, 2, ..., Q. \tag{3.55}
$$

In order to have a unique and smallest set of coefficients, b_p, we set $Q = P$ in (3.54) and (3.55), or equivalently, in (3.49) and (3.50). The convergence estimate, (3.48), becomes evident from (3.53) and Theorem 2.3.

If $2P < n$, the relationship, (3.48), in Theorem 3.1 can be modified to

$$
\lim_{m \to \infty} a_m m^{2P+1} = 0 \tag{3.56}
$$

or, more explicitly,

$$
a_m \sim \mathcal{O}(m^{-2P-2}). \tag{3.57}
$$

Alternatively, (3.47) can be expressed as

$$
a_m = \frac{2}{\pi} \int_{0}^{\pi} f(x) \cos mx \, dx - \sum_{p=1}^{2P} \alpha_{mp} b_p, \tag{3.58}
$$

where

$$\alpha_{mp} = \frac{2}{\pi} \int_0^{\pi} \sin px \cos mx dx$$

$$= \begin{cases} \dfrac{2p[(-1)^{m+p}-1]}{\pi(m^2-p^2)} & \text{for } m \neq p. \\ 0 & \text{for } m = p \end{cases} \tag{3.59}$$

(3.49) and (3.50) can be rewritten in matrix form as

$$\begin{bmatrix} \mathbf{A}_1 & \mathbf{0} \\ \mathbf{0} & \mathbf{A}_2 \end{bmatrix} \begin{Bmatrix} \mathbf{B}_1 \\ \mathbf{B}_2 \end{Bmatrix} = \begin{Bmatrix} \mathbf{F}_1 \\ \mathbf{F}_2 \end{Bmatrix}, \tag{3.60}$$

where

$$[\mathbf{A}_1]_{p,q} = (2p-1)^{2q-1}, \tag{3.61}$$

$$[\mathbf{A}_2]_{p,q} = (2p)^{2q-1}, \tag{3.62}$$

$$\mathbf{B}_1 = \{ b_1 \quad b_3 \quad \cdots \quad b_{2p-1} \quad \cdots \quad b_{2P-1} \}^{\mathrm{T}}, \tag{3.63}$$

$$\mathbf{B}_2 = \{ b_2 \quad b_4 \quad \cdots \quad b_{2p} \quad \cdots \quad b_{2P} \}^{\mathrm{T}}, \tag{3.64}$$

$$\mathbf{F}_1 = \{ -f_-^{(1)} \quad f_-^{(3)} \quad \cdots \quad (-1)^q f_-^{(2q-1)} \quad \cdots \quad (-1)^P f_-^{(2P-1)} \}^{\mathrm{T}} \tag{3.65}$$

and

$$\mathbf{F}_2 = \{ f_+^{(1)} \quad -f_+^{(3)} \quad \cdots \quad (-1)^{q-1} f_+^{(2q-1)} \quad \cdots \quad (-1)^{P-1} f_+^{(2P-1)} \}^{\mathrm{T}}, \tag{3.66}$$

in which

$$f_+^{(2q-1)} = [f^{(2q-1)}(\pi) + f^{(2q-1)}(0)]/2 \tag{3.67}$$

and

$$f_-^{(2q-1)} = [f^{(2q-1)}(\pi) - f^{(2q-1)}(0)]/2. \tag{3.68}$$

Determining the coefficient matrices, $\mathbf{B}_1$ and $\mathbf{B}_2$, involves the inversion of the Vandermonde-like matrix

$$\mathbf{X} = \begin{bmatrix} x_1 & x_2 & \cdots & x_j & \cdots & x_P \\ x_1{}^3 & x_2{}^3 & \cdots & x_j{}^3 & \cdots & x_P{}^3 \\ \vdots & \vdots & \vdots & \vdots & \vdots & \vdots \\ x_1{}^{2i-1} & x_2{}^{2i-1} & \cdots & x_j{}^{2i-1} & \cdots & x_P{}^{2i-1} \\ \vdots & \vdots & \vdots & \vdots & \vdots & \vdots \\ x_1{}^{2P-1} & x_2{}^{2P-1} & \cdots & x_j{}^{2P-1} & \cdots & x_P{}^{2P-1} \end{bmatrix} \quad (i,j = 1,2,3,...,P),$$

(3.69)

which is known to be invertible if $x_k \neq x_j$ for $j \neq k$.

Consider a polynomial of degree $2P-1$

$$\phi_i(x) = \frac{x}{x_i} \prod_{\substack{k=1 \\ k \neq i}}^{P} \frac{x^2 - x_k^2}{x_i^2 - x_k^2} = \sum_{k=1}^{P} c_{ik} x^{2k-1}.$$

(3.70)

Then it is obvious that

$$\phi_i(x_j) = \sum_{k=1}^{P} c_{ik} x_j^{2k-1} = \delta_{ij},$$

(3.71)

where δ_{ij} is Kronecker's symbol.

According to (3.71), matrix $\mathbf{C} = [c_{ik}]$ is actually the inverse of matrix $\mathbf{X}$. To find an explicit expression for matrix $\mathbf{C}$, let

$$\prod_{\substack{q=1 \\ q \neq i}}^{P} (x^2 - x_q^2) = \sum_{q=0}^{P-1} \beta_q x^{2q} = \frac{1}{x^2 - x_i^2} \prod_{q=1}^{P} (x^2 - x_q^2) = \frac{1}{x^2 - x_i^2} \sum_{q=0}^{P} \alpha_q x^{2q}, \quad (3.72)$$

where

$$\alpha_q = \sum_{1 \leq j_1 < \cdots < j_{P-q} \leq P} (-1)^q x_{j_1}^2 x_{j_2}^2 \cdots x_{j_{P-q}}^2 \quad (\alpha_P = 1)$$

(3.73)

and

$$\beta_q = -\frac{1}{x_i^{2q+2}} \sum_{s=0}^{q} \alpha_s x_i^{2s}.$$

(3.74)

(3.74) can be easily derived from (3.72) (that is, $(x^2 - x_i^2) \sum_{q=0}^{P-1} \beta_q x^{2q} = \sum_{q=0}^{P} \alpha_q x^{2q}$) by comparing the coefficients for like terms, starting with the lowest orders.

Thus, we have

$$\phi_i(x) = -\frac{1}{\displaystyle\prod_{\substack{j=1 \\ j\neq i}}^{P}(x_i^2 - x_j^2)}\sum_{q=1}^{P}\left(\frac{1}{x_i^{2q+1}}\sum_{s=0}^{q-1}\alpha_s x_i^{2s}\right)x^{2q-1}. \tag{3.75}$$

Comparing (3.75) with (3.70) leads to

$$c_{ik} = -\frac{\displaystyle\sum_{s=0}^{k-1}\alpha_s x_i^{2s}}{x_i^{2k+1}\displaystyle\prod_{\substack{j=1 \\ j\neq i}}^{P}(x_i^2 - x_j^2)} \tag{3.76}$$

or

$$c_{ik} = \frac{\displaystyle\sum_{\substack{1\leq j_1<\cdots<j_{P-k}\leq P \\ j_1,\cdots,j_{P-k}\neq i}}(-1)^{k+1} x_{j_1}^2\cdots x_{j_{P-k}}^2}{x_i\displaystyle\prod_{\substack{j=1 \\ j\neq i}}^{P}(x_j^2 - x_i^2)}. \tag{3.77}$$

In light of (3.77), the coefficients b_p $(p = 1, 2, \cdots, 2P)$ can be obtained as

$$b_p = \sum_{k=1}^{P}(-1)^p[f^{(2k-1)}(\pi) + (-1)^p f^{(2k-1)}(0)]\frac{\displaystyle\sum_{\substack{1\leq j_1<\cdots<j_{P-k}\leq P \\ j_1,\cdots,j_{P-k}\neq i}}x_{j_1}^2\cdots x_{j_{P-k}}^2}{x_i\displaystyle\prod_{\substack{j=1 \\ j\neq i}}^{P}(x_j^2 - x_i^2)}, \tag{3.78}$$

where

$$p = x_i = \begin{cases} 2i-1 & \text{if } p \text{ is odd} \\ 2i & \text{if } p \text{ is even} \end{cases} \quad (i = 1, 2, ..., P). \tag{3.79}$$

By making use of (3.78), the first few coefficients, for example, are readily found as: for $P = 1$,

$$b_1 = -f_-^{(1)}, \tag{3.80}$$

$$b_2 = f_+^{(1)}/2; \tag{3.81}$$

for $P = 2$,

$$b_1 = -9f_-^{(1)}/8 - f_-^{(3)}/8, \tag{3.82}$$

$$b_2 = 2f_+^{(1)}/3 + f_+^{(3)}/24, \tag{3.83}$$

$$b_3 = f_-^{(1)}/24 + f_-^{(3)}/24, \tag{3.84}$$

$$b_4 = -f_+^{(1)}/12 - f_+^{(3)}/48; \tag{3.85}$$

and for $P = 3$,

$$b_1 = -75f_-^{(1)}/64 - 17f_-^{(3)}/96 - f_-^{(5)}/192, \tag{3.86}$$

$$b_2 = 3f_+^{(1)}/4 + 13f_+^{(3)}/192 + f_+^{(5)}/768, \tag{3.87}$$

$$b_3 = 25f_-^{(1)}/384 + 13f_-^{(3)}/192 + f_-^{(5)}/384, \tag{3.88}$$

$$b_4 = -3f_+^{(1)}/20 - f_+^{(3)}/24 - f_+^{(5)}/960, \tag{3.89}$$

$$b_5 = -3f_-^{(1)}/640 - f_-^{(3)}/192 - f_-^{(5)}/1920 \tag{3.90}$$

and

$$b_6 = f_+^{(1)}/60 + f_+^{(3)}/192 + f_+^{(5)}/3840. \tag{3.91}$$

Example 3.1   Expand the function $f(x) = Ax^2 + Bx + C$ $(0 \le x \le \pi)$ into the generalized Fourier cosine series.

For purpose of comparison, its conventional Fourier series expansions are first given below:

$$f(x) = \frac{\pi^2 A}{3} + \frac{\pi B}{2} + C + \sum_{m=1}^{\infty} \frac{A}{m^2} \cos 2mx - \sum_{m=1}^{\infty} \frac{(\pi A + B)}{m} \sin 2mx, \quad 0 < x < \pi, \tag{3.92}$$

and

$$f(x) = \frac{\pi^2 A}{3} + \frac{\pi B}{2} + C + \sum_{m=1}^{\infty} a_m \cos mx, \quad 0 \le x \le \pi, \tag{3.93}$$

where

$$a_m = \frac{-2B + (-1)^m(4\pi A + 2B)}{m^2\pi}. \tag{3.94}$$

Under the current framework, this function can be expanded into:
for $P = 1$,

$$f(x) = \mathcal{F}_{\infty,2}[f](x) = A + \frac{\pi B}{2} + C + \sum_{m=1}^{\infty} a_m \cos mx \tag{3.95}$$

$$- \pi A \sin x + \frac{1}{2}(\pi A + B)\sin 2x, \quad 0 \le x \le \pi,$$

where

$$a_m = \begin{cases} \dfrac{16(\pi A + B)}{\pi m^2(m^2 - 4)} & \text{if } m \text{ is odd} \\[3mm] \dfrac{-4A}{m^2(m^2 - 1)} & \text{if } m \text{ is even} \end{cases} ; \tag{3.96}$$

for $P = 2$,

$$f(x) = \mathcal{F}_{\infty,4}[f](x) = \frac{1}{9}(20 + 3\pi^2)A + \frac{\pi B}{2} + C + \sum_{m=1}^{\infty} a_m \cos mx$$

$$- \frac{9\pi A}{8}\sin x + \frac{2}{3}(\pi A + B)\sin 2x + \frac{\pi A}{24}\sin 3x - \frac{1}{12}(\pi A + B)\sin 4x, \quad 0 \le x \le \pi, \tag{3.97}$$

where

$$a_m = \begin{cases} \dfrac{-256(\pi A + B)}{\pi m^2(m^2 - 4)(m^2 - 16)} & \text{if } m \text{ is odd} \\[3mm] \dfrac{36A}{m^2(m^2 - 1)(m^2 - 9)} & \text{if } m \text{ is even} \end{cases} ; \tag{3.98}$$

for $P = 3$,

$$f(x) = \mathcal{F}_{\infty,6}[f](x) = \frac{(518 + 75\pi^2)A}{225} + \frac{\pi B}{2} + C + \sum_{m=1}^{\infty} a_m \cos mx$$

$$- \frac{75\pi A}{64}\sin x + \frac{3}{4}(\pi A + B)\sin 2x + \frac{25\pi A}{384}\sin 3x - \frac{3}{20}(\pi A + B)\sin 4x$$

$$- \frac{3\pi A}{640}\sin 5x + \frac{1}{60}(\pi A + B)\sin 6x, \quad 0 \le x \le \pi, \tag{3.99}$$

where

$$a_m = \begin{cases} \dfrac{9216(\pi A + B)}{\pi m^2 (m^2 - 4)(m^2 - 16)(m^2 - 36)} & \text{if } m \text{ is odd} \\[4mm] \dfrac{-900A}{m^2 (m^2 - 1)(m^2 - 9)(m^2 - 25)} & \text{if } m \text{ is even} \end{cases} \qquad (3.100)$$

The convergence rates for these different forms of series expansions are readily seen from the expressions, (3.92), (3.94), (3.96), (3.98), and (3.100), used to calculate the expansion coefficients, and the results are shown in Figure 3.1 for $A = 4$, $B = 2$, and $C = 1$. The corresponding truncation errors are plotted in Figures 3.2 and 3.3.

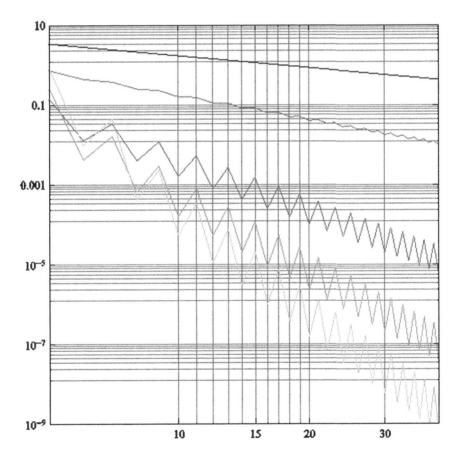

FIGURE 3.1 Decays of Fourier expansion coefficients: _____ b_m in (3.92), _____ a_m in (3.93), _____ a_m in (3.95), _____ a_m in (3.97), and _____ a_m in (3.99).

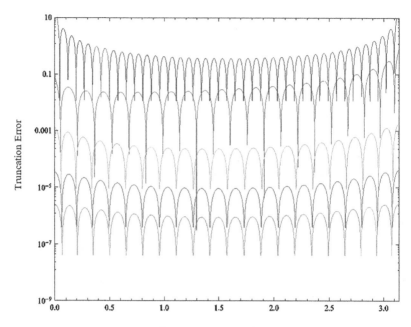

FIGURE 3.2   Truncation errors, $|f(x) - \mathcal{F}_{M,2P}[f](x)|$, for the series expansions: ____ (3.92), ____ (3.93), ____ (3.95), ____ (3.97), and ____ (3.99). M=20.

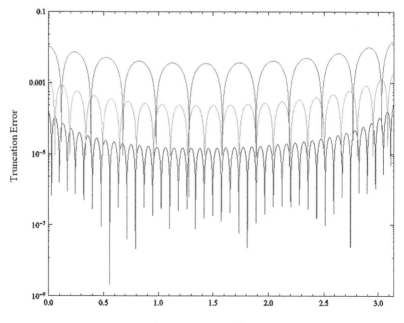

FIGURE 3.3   Truncation Errors, $|f(x) - \mathcal{F}_{M,2}[f](x)|$, for series expansion (3.95): ____, M=10, ____, M=20 and ____ M=40.

3.3 THE GENERALIZED FOURIER SINE SERIES WITH ACCELERATED CONVERGENCE

Similarly, the function $f(x)$ can also be expanded into the generalized Fourier sine series as

$$f(x) = \mathcal{F}_{\infty,2P}[f](x) = \sum_{m=1}^{\infty} a_m \sin mx + \sum_{p=1}^{2P} b_p \cos px, 0 \le x \le \pi, \qquad (3.101)$$

where b_p are the expansion coefficients to be determined, and

$$a_m = \frac{2}{\pi} \int_0^\pi \left[f(x) - \sum_{p=1}^{2P} b_p \cos px \right] \sin mx\, dx. \qquad (3.102)$$

Theorem 3.2 *Let $f(x)$ be of C^{n-1} continuity on the interval $[0,\pi]$ and the n-th derivative is absolutely integrable (the n-th derivative may not exist at certain points). Then for $2P \le n+1$ the Fourier coefficients of $f(x)$ defined in (3.101) decay at a polynomial rate as*

$$\lim_{m \to \infty} a_m m^{2P-1} = 0 \quad (2P \le n+1), \qquad (3.103)$$

provided that

$$\sum_{p=1}^{P} b_{2p}(2p)^{2q-2} = (-1)^{q-1}[f^{(2q-2)}(\pi) + f^{(2q-2)}(0)]/2 \qquad (3.104)$$

and

$$\sum_{p=1}^{P} b_{2p-1}(2p-1)^{2q-2} = (-1)^q [f^{(2q-2)}(\pi) - f^{(2q-2)}(0)]/2 \qquad (3.105)$$

$$(q = 1, 2, ..., P).$$

Proof. By integrating by part, we have

$$
(\pi/2)a_m = \int_0^\pi \left[f(x) - \sum_{p=1}^{2P} b_p \cos px \right] \sin mx \, dx
$$

$$
= - \left[f(x) - \sum_{p=1}^{2P} b_p \cos px \right] \frac{\cos mx}{m} \Bigg|_0^\pi + \frac{1}{m} \int_0^\pi \left[f'(x) - \sum_{p=1}^{2P} b_p \cos^{(1)} px \right] \cos mx \, dx
$$

$$
= \sum_{q=1}^Q \frac{(-1)^{q+m}}{m^{2q-1}} \left[f^{(2q-2)}(\pi) - \sum_{p=1}^{2P} b_p p^{2q-2} (-1)^{p+q-1} \right]
$$

$$
- \frac{(-1)^q}{m^{2q-1}} \left[f^{(2q-2)}(0) - \sum_{p=1}^{2P} b_p p^{2q-2} (-1)^{q-1} \right]
$$

$$
- \frac{(-1)^Q}{m^{2Q-1}} \int_0^\pi \left[f^{(2Q-1)}(x) - \sum_{p=1}^{2P} b_p \cos^{(2Q-1)} px \right] \cos mx \, dx.
$$

$$(3.106)$$

The first two terms in (3.106) will both vanish if

$$
\sum_{p=1}^{2P} b_p p^{2q-2} (-1)^{p+q-1} = f^{(2q-2)}(\pi) \tag{3.107}
$$

and

$$
\sum_{p=1}^{2P} b_p p^{2q-2} (-1)^{q-1} = f^{(2q-2)}(0), \quad q = 1, 2, \ldots, Q. \tag{3.108}
$$

In order to have a unique and smallest set of coefficients, b_p, we set $Q = P$ in (3.107) and (3.108), or equivalently, in (3.104) and (3.105). The convergence estimate, (3.103), then becomes evident according to Theorem 2.3.

If $2P < n + 1$, the relationship (3.103) can be actually modified to

$$
\lim_{m \to \infty} a_m m^{2P} = 0, \tag{3.109}
$$

which can be further written in a sharper form as

$$
a_m \sim \mathcal{O}(m^{-2P-1}), \text{ for sufficiently large } m. \tag{3.110}
$$

The expansion coefficients, a_m, can be alternatively expressed as

$$a_m = \frac{2}{\pi} \int_0^\pi f(x) \sin mx \, dx - \sum_{p=1}^{2P} \beta_{mp} b_p, \qquad (3.111)$$

where

$$\beta_{mp} = \begin{cases} \dfrac{2m[(-1)^{m+p} - 1]}{\pi(p^2 - m^2)} & \text{for } m \ne p \\ 0 & \text{for } m = p \end{cases}. \qquad (3.112)$$

Actually, referring to (3.59), we have

$$\beta_{mp} = \alpha_{pm}. \qquad (3.113)$$

We can rewrite (3.104) and (3.105) in matrix form as

$$\begin{bmatrix} \mathbf{A}_1 & \mathbf{0} \\ \mathbf{0} & \mathbf{A}_2 \end{bmatrix} \begin{Bmatrix} \mathbf{B}_1 \\ \mathbf{B}_2 \end{Bmatrix} = \begin{Bmatrix} \mathbf{F}_1 \\ \mathbf{F}_2 \end{Bmatrix}, \qquad (3.114)$$

where

$$[\mathbf{A}_1]_{p,q} = (2p - 1)^{2q-2}, \qquad (3.115)$$

$$[\mathbf{A}_2]_{p,q} = (2p)^{2q-2}, \qquad (3.116)$$

$$\mathbf{F}_1 = \{-f_-^{(0)} \quad f_-^{(2)} \quad \cdots \quad (-1)^{q+1} f_-^{(2q-2)} \quad \cdots \quad (-1)^{P+1} f_-^{(2P-2)}\}^{\mathrm{T}}, \qquad (3.117)$$

and

$$\mathbf{F}_2 = \{f_+^{(0)} \quad -f_+^{(2)} \quad \cdots \quad (-1)^q f_+^{(2q-2)} \quad \cdots \quad (-1)^P f_+^{(2P-2)}\}^{\mathrm{T}}. \qquad (3.118)$$

Following the same procedures as described earlier, coefficients b_p can be obtained from

$$b_p = \sum_{k=1}^{P} (-1)^P [f^{(2k-2)}(\pi) + (-1)^P f^{(2k-2)}(0)] \frac{\displaystyle\sum_{\substack{1 \le j_1 < \cdots < j_{P-k} \le P \\ j_1, \cdots, j_{P-k} \ne i}} x_{j_1}^2 \cdots x_{j_{P-k}}^2}{\displaystyle\prod_{\substack{j=1 \\ j \ne i}}^{P} (x_j^2 - x_i^2)}, \qquad (3.119)$$

where

$$p = x_i = \begin{cases} 2i - 1 & \text{if } p \text{ is odd} \\ 2i & \text{if } p \text{ is even} \end{cases} \quad (i = 1, 2, ..., P). \tag{3.120}$$

Using this formula, the first several coefficients are easily determined as:
for $P = 1$,

$$b_1 = -f_-^{(0)}, \tag{3.121}$$

$$b_2 = f_+^{(0)}; \tag{3.122}$$

for $P = 2$,

$$b_1 = -9f_-^{(0)}/8 - f_-^{(2)}/8, \tag{3.123}$$

$$b_2 = 4f_+^{(0)}/3 + f_+^{(2)}/12, \tag{3.124}$$

$$b_3 = f_-^{(0)}/8 + f_-^{(2)}/8, \tag{3.125}$$

$$b_4 = -f_+^{(0)}/3 - f_+^{(2)}/12; \tag{3.126}$$

for $P = 3$,

$$b_1 = -75f_-^{(0)}/64 - 17f_-^{(2)}/96 - f_-^{(4)}/192, \tag{3.127}$$

$$b_2 = 3f_+^{(0)}/2 + 13f_+^{(2)}/96 + f_+^{(4)}/384, \tag{3.128}$$

$$b_3 = 25f_-^{(0)}/128 + 13f_-^{(2)}/64 + f_-^{(4)}/128, \tag{3.129}$$

$$b_4 = -3f_+^{(0)}/5 - f_+^{(2)}/6 - f_+^{(4)}/240, \tag{3.130}$$

$$b_5 = -3f_-^{(0)}/128 - 5f_-^{(2)}/192 - f_-^{(4)}/384, \tag{3.131}$$

$$b_6 = f_+^{(0)}/10 + f_+^{(2)}/32 + f_+^{(4)}/640. \tag{3.132}$$

Example 3.2 Expand the function $f(x) = Ax^2 + Bx + C$ $(0 \le x \le \pi)$ into the generalized Fourier sine series.
    Its conventional sine series expansion is easily determined as

$$f(x) = \sum_{m=1}^{\infty} a_m \sin mx, \tag{3.133}$$

where

$$
a_m = \begin{cases} \dfrac{2m^2(\pi^2 A + \pi B + 2C) - 8A}{m^3 \pi} & \text{if } m \text{ is odd} \\[3mm] -\dfrac{2(\pi A + B)}{m} & \text{if } m \text{ is even} \end{cases} . \qquad (3.134)
$$

Under the current framework, this function can be expressed as:

for $P = 1$,

$$
f(x) = \mathcal{F}_{\infty,2}[f](x) = \sum_{m=1}^{\infty} a_m \sin mx
$$
$$
-\frac{1}{2}(\pi^2 A + \pi B)\cos x + \frac{1}{2}(\pi^2 A + \pi B + 2C)\cos 2x, \quad 0 \le x \le \pi, \qquad (3.135)
$$

where

$$
a_m = \begin{cases} \dfrac{-8m^2[(\pi^2 + 1)A + \pi B + 2C] + 32A}{\pi m^3 (m^2 - 4)} & \text{if } m \text{ is odd} \\[3mm] \dfrac{2(\pi A + B)}{m(m^2 - 1)} & \text{if } m \text{ is even} \end{cases} ; \qquad (3.136)
$$

for $P = 2$,

$$
f(x) = \mathcal{F}_{\infty,4}[f](x) = \sum_{m=1}^{\infty} a_m \sin mx
$$
$$
-\frac{9}{16}(\pi^2 A + \pi B)\cos x + \frac{1}{6}(4\pi^2 A + 4\pi B + 8C + A)\cos 2x
$$
$$
+\frac{1}{16}(\pi^2 A + \pi B)\cos 3x - \frac{1}{6}(\pi^2 A + \pi B + 2C + A)\cos 4x, \quad 0 \le x \le \pi, \qquad (3.137)
$$

where

$$
a_m = \begin{cases} \dfrac{32[m^2(5 + 4\pi^2) - 16]A + 128m^2(\pi B + 2C)}{\pi m^3 (m^2 - 4)(m^2 - 16)} & \text{if } m \text{ is odd} \\[3mm] -\dfrac{18(\pi A + B)}{m(m^2 - 1)(m^2 - 9)} & \text{if } m \text{ is even} \end{cases} . \qquad (3.138)
$$

It is seen that while the conventional sine series decays with $1/m$, the generalized Fourier series expansions, (3.135) and (3.137), converge with much faster rates of $1/m^3$ and $1/m^5$, respectively.

3.4 THE GENERALIZED FOURIER SERIES EXPANSION WITH ACCELERATED CONVERGENCE

Let $f(x)$ be defined on the interval $[-\pi, \pi]$. It can also be expanded into a complete trigonometric series as

$$f(x) = \mathcal{F}_{\infty, 2P, 2Q}[f](x)$$

$$= a_0/2 + \sum_{m=1}^{\infty} [a_m + \bar{a}_{2Q,m} \text{sn}(x)] \cos mx + [b_m + \bar{b}_{2P,m} \text{sn}(x)] \sin mx \quad (3.139)$$

$(\bar{a}_{2Q,m} \equiv 0, \text{ for } m > 2Q; \bar{b}_{2P,m} \equiv 0, \text{ for } m > 2P), \quad -\pi \le x \le \pi,$

where

$$\text{sn}(x) = \begin{cases} 1 & x \ge 0 \\ -1 & x < 0 \end{cases}, \quad (3.140)$$

and a_m and b_m are the Fourier coefficients calculated from

$$a_m = \frac{1}{\pi} \int_{-\pi}^{\pi} f(x) \cos mx \, dx - \frac{2}{\pi} \int_{0}^{\pi} \left(\sum_{p=1}^{2P} \bar{b}_{2P,p} \sin px \right) \cos mx \, dx \quad (3.141)$$

and

$$b_m = \frac{1}{\pi} \int_{-\pi}^{\pi} f(x) \sin mx \, dx - \frac{2}{\pi} \int_{0}^{\pi} \left(\sum_{q=1}^{2Q} \bar{a}_{2Q,q} \cos qx \right) \sin mx \, dx, \quad (3.142)$$

in which the coefficients $\bar{a}_{2Q,m}$ and $\bar{b}_{2P,m}$ are to be determined as described below.

Theorem 3.3 *Let $f(x)$ be of C^{n-1} continuity on the interval $[-\pi, \pi]$ and the n-th derivative is absolutely integrable (the n-th derivative may not exist at certain points). Then the Fourier coefficients of $f(x)$, as defined in (3.141) and (3.142), decay at polynomial rates as*

$$\lim_{m \to \infty} a_m m^{2P} = 0 \quad (2P \le n) \quad (3.143)$$

and

$$\lim_{m \to \infty} b_m m^{2Q-1} = 0 \quad (2Q \le n+1), \quad (3.144)$$

provided that

$$\sum_{p=1}^{P} \overline{b}_{2P,2p}(2p)^{2r-1} = (-1)^{r-1}[f^{(2r-1)}(\pi) - f^{(2r-1)}(-\pi)]/4, \qquad (3.145)$$

$$\sum_{p=1}^{P} \overline{b}_{2P,2p-1}(2p-1)^{2r-1} = (-1)^{r}[f^{(2r-1)}(\pi) - f^{(2r-1)}(-\pi)]/4 \quad (r=1,2,...,P),$$
$$(3.146)$$

$$\sum_{q=1}^{Q} \overline{a}_{2Q,2q}(2q)^{2r-2} = (-1)^{r-1}[f^{(2r-2)}(\pi) - f^{(2r-2)}(-\pi)] / 4 \qquad (3.147)$$

and

$$\sum_{q=1}^{Q} \overline{a}_{2Q,2q-1}(2q-1)^{2r-2} = (-1)^{r}[f^{(2r-2)}(\pi) - f^{(2r-2)}(-\pi)] / 4 \quad (r=1,2,...,Q).$$
$$(3.148)$$

Proof. The function $f(x)$ can be considered as the superposition of an even function, $g(x) = [f(x) + f(-x)]/2$, and an odd function, $h(x) = [f(x) - f(-x)]/2$. According to Theorem 3.1, $g(x)$ can be expanded into the generalized Fourier cosine over $[0, \pi]$ and expression (3.143) holds, if

$$\sum_{p=1}^{P} \overline{b}_{2P,2p}(2p)^{2r-1} = (-1)^{r-1}[g^{(2r-1)}(\pi) + g^{(2r-1)}(0)]/2 \qquad (3.149)$$

and

$$\sum_{p=1}^{P} \overline{b}_{2P,2p-1}(2p-1)^{2r-1} = (-1)^{r}[g^{(2r-1)}(\pi) - g^{(2r-1)}(0)]/2 \quad (r=1,2,...,P).$$
$$(3.150)$$

Since

$$g^{(2r-1)}(0) = [f^{(2r-1)}(x) - f^{(2r-1)}(-x)]/2\big|_{x=0} \equiv 0 \qquad (3.151)$$

and

$$g^{(2r-1)}(\pi) = [f^{(2r-1)}(\pi) - f^{(2r-1)}(-\pi)]/2, \qquad (3.152)$$

(3.149) and (3.150) can be rewritten as (3.145) and (3.146), respectively.

Similarly, expression (3.144) is readily obtained from applying Theorem 3.2 to the odd function $h(x)$ on interval $[0,\pi]$ and by recognizing that

$$h^{(2r-2)}(0) = [f^{(2r-2)}(x) - f^{(2r-2)}(-x)]/2_{x=0} \equiv 0 \qquad (3.153)$$

and

$$h^{(2r-2)}(\pi) = [f^{(2r-2)}(\pi) - f^{(2r-2)}(-\pi)]/2. \qquad (3.154)$$

The Fourier coefficients of $g(x)$ are determined from

$$
\begin{aligned}
a_m &= \frac{2}{\pi} \int_0^\pi \left[\frac{f(x)+f(-x)}{2} - \sum_{p=1}^{2P} \bar{b}_{2P,p} \sin px \right] \cos mx\,dx \\
&= \frac{1}{\pi} \int_0^\pi f(x)\cos mx\,dx + \frac{1}{\pi}\int_{-\pi}^0 f(x)\cos mx\,dx - \frac{2}{\pi}\int_0^\pi \left(\sum_{p=1}^{2P} \bar{b}_{2P,p}\sin px \right) \cos mx\,dx \\
&= \frac{1}{\pi} \int_{-\pi}^\pi f(x)\cos mx\,dx - \frac{2}{\pi}\int_0^\pi \left(\sum_{p=1}^{2P} \bar{b}_{2P,p}\sin px \right) \cos mx\,dx.
\end{aligned}
\qquad (3.155)
$$

Similarly, the expansion coefficients of $h(x)$ are determined from

$$
\begin{aligned}
b_m &= \frac{2}{\pi} \int_0^\pi \left[\frac{f(x)-f(-x)}{2} - \sum_{q=1}^{2Q} \bar{a}_{2Q,q} \cos qx \right] \sin mx\,dx \\
&= \frac{1}{\pi} \int_{-\pi}^\pi f(x)\sin mx\,dx - \frac{2}{\pi}\int_0^\pi \left(\sum_{q=1}^{2Q} \bar{a}_{2Q,q}\cos qx \right) \sin mx\,dx.
\end{aligned}
\qquad (3.156)
$$

Since $g(x) = [f(x)+f(-x)]/2$ is already an even function over $[-\pi,\pi]$, the function $g(x) - \sum_{p=1}^{2P}\bar{b}_{2P,p}\sin px$ defined over $[0,\pi]$ can then be easily even-mapped onto $[-\pi,0)$ by introducing the "sign function" $sn(x)$ as given in (3.140); and the function $h(x) - \sum_{q=1}^{2Q}\bar{a}_{2Q,q}\cos qx$ can be similarly odd-mapped onto $[-\pi,0)$. Thus, expression (3.139) will become evident.

Alternatively, (3,141) and (3.142) can be expressed as

$$a_m = \frac{1}{\pi}\int_{-\pi}^\pi f(x)\cos mx\,dx - \sum_{p=1}^{2P} \alpha_{mp}\bar{b}_{2P,p} \qquad (3.157)$$

and

$$b_m = \frac{1}{\pi}\int_{-\pi}^\pi f(x)\sin mx\,dx - \sum_{q=1}^{2Q} \alpha_{qm}\bar{a}_{2Q,q}, \qquad (3.158)$$

where α_{mp} is given in (3.59).

The coefficients $\bar{b}_{2P,p}$ and $\bar{a}_{2Q,q}$ can then respectively be calculated from (3.78) and (3.119) by letting

$$f_+^{(2r-1)} = f_-^{(2r-1)} = [f^{(2r-1)}(\pi) - f^{(2r-1)}(-\pi)]/4, \quad r = 1,2,...,P \qquad (3.159)$$

and

$$f_+^{(2r-2)} = f_-^{(2r-2)} = [f^{(2r-2)}(\pi) - f^{(2r-2)}(-\pi)]/4, \quad r = 1,2,...,Q. \qquad (3.160)$$

Example 3.3   Expand the function $f(x) = Ax^2 + Bx + C \, (-\pi \le x \le \pi)$ into the generalized Fourier series.

Its conventional Fourier expansion is already known as

$$f(x) = \frac{\pi^2 A}{3} + C + 4A \sum_{m=1}^{\infty} \frac{(-1)^m}{m^2} \cos mx - 2B \sum_{m=1}^{\infty} \frac{(-1)^m}{m} \sin mx, \quad -\pi < x < \pi. \qquad (3.161)$$

By setting $P = 0$ and $Q = 1$ in (3.139), we have

$$\bar{a}_{2Q,1} = -\bar{a}_{2Q,2} = -\frac{f(\pi) - f(-\pi)}{4} = -\frac{\pi B}{2} \qquad (3.162)$$

and

$$f(x) = \frac{\pi^2 A}{3} + C + 4A \sum_{m=1}^{\infty} \frac{(-1)^m}{m^2} \cos mx - \frac{\pi B}{2} \operatorname{sn}(x) \cos x + \frac{\pi B}{2} \operatorname{sn}(x) \cos 2x$$

$$+ \sum_{m=1}^{\infty} b_m \sin mx, \quad -\pi \le x \le \pi, \qquad (3.163)$$

where

$$b_m = \frac{(-1)^{m+1} 2B}{m} - \bar{a}_{2Q,1} \alpha_{1m} - \bar{a}_{2Q,2} \alpha_{2m}$$

$$= \begin{cases} \dfrac{-8B}{m(m^2 - 4)} & \text{if } m \text{ is odd} \\[3mm] \dfrac{2B}{m(m^2 - 1)} & \text{if } m \text{ is even} \end{cases} . \qquad (3.164)$$

It is seen from (3.164) that the sine series now converges at a rate of m^{-3}, which is faster than m^{-2} for the cosine series part. If desired, the convergence of the series expansion in the form of (3.139) can be further accelerated by setting $P = Q = 1$. Accordingly, in addition to (3.162), we need to calculate

$$\bar{b}_{2P,1} = -2\bar{b}_{2P,2} = -\frac{f'(\pi) - f'(-\pi)}{4} = -\pi A \qquad (3.165)$$

and

$$f(x) = \frac{\pi^2 A}{3} + C - 2A - \frac{\pi B}{2}\operatorname{sn}(x)\cos x + \frac{\pi B}{2}\operatorname{sn}(x)\cos 2x - \pi A\operatorname{sn}(x)\sin x$$

$$+ \frac{\pi A}{2}\operatorname{sn}(x)\sin 2x + \sum_{m=1}^{\infty}\left(a_m \cos mx + b_m \sin mx\right), \quad -\pi \leq x \leq \pi,$$

(3.166)

where

$$a_m = \frac{(-1)^m 4A}{m} - \bar{b}_{2P,1}\alpha_{m1} - \bar{b}_{2P,2}\alpha_{m2}$$

$$= \begin{cases} \dfrac{16A}{m^2(m^2 - 4)} & \text{if } m \text{ is odd} \\[3mm] \dfrac{-4A}{m^2(m^2 - 1)} & \text{if } m \text{ is even} \end{cases}$$

(3.167)

The series expansion (3.166) will converge at a rate of m^{-3} in comparison with m^{-2} for series (3.163).

It is seen from the above discussions that the complete Fourier series (3.139) looks slightly more complicated than its sine or cosine counterpart, in particular, with the inclusion of the function $\operatorname{sn}(x)$. In addition, its convergence rate is improved at a relatively slower pace of m^{-1} (as compared with m^{-2} for the sine or cosine series) each time when P or Q is increased by 1. However, this "incremental" way of improving the convergence rate may become necessary in some cases, such as when the smoothness of a (solution) function is preferred not to be artificially boosted for whatever reasons.

REFERENCES

Adcock, B. 2011. Convergence acceleration of modified Fourier series in one or more dimensions. *Mathematics of Computation* 80: 225–261.

Baszenski, G., F. J. Delvos, and M. Tasche. 1995. A united approach to accelerating trigonometric expansions. *Computers & Mathematics with Applications* 30: 33–49.

Eckhoff, K. S. 1998. On a high order numerical method for functions with singularities. *Mathematics of Computation* 67: 1063–1087.

Iserles, A., and S. P. Nørsett. 2008. From high oscillation to rapid approximation I: Modified Fourier expansions. *IMA Journal of Numerical Analysis* 28: 862–887.

Jones, W. B., and G. Hardy. 1970. Accelerating convergence of trigonometric approximations. *Mathematics of Computation* 24: 547–560.

Lanczos, C. 1966. *Discourse on Fourier Series*. New York: Hafner.

4 The Generalized Fourier Series Solutions of the Euler-Bernoulli Beam Equation

Beams represent one of the most common types of structural components used in engineering and design. The formal establishment of basic theory about beam deformations is often attributed to the pioneering works of Bernoulli and Euler in the mid of 18th century, even though the beam problems might be actually attempted earlier by others. At the center of the beam theory is the so-called Euler-Bernoulli equation, a fourth order linear differential equation that, together with the suitable boundary conditions, governs the deflections of elastic beams under various loading conditions. In Chapter 3, different forms of supplementary functions have been used to condition a sufficiently smooth function defined on a compact interval so that the involved Fourier series can have an improved rate of convergence. In what follows, this generalized form of Fourier series expansions is extended from a mathematical scheme for effectively approximating a function into a powerful and general means for solving boundary value problems: specifically, the Euler-Bernoulli beam equation subjected to a set of arbitrary boundary conditions.

4.1 LINEAR DIFFERENTIAL EQUATIONS WITH CONSTANT COEFFICIENTS

Consider a $2r$-th order linear differential equation

$$\mathcal{L}y(x) = f(x), \quad x \in (a, b), \tag{4.1}$$

where the differential operator

$$\mathcal{L} = \sum_{k=0}^{2r} a_k \frac{d^k}{dx^k} \tag{4.2}$$

with r being a positive integer, and a_k, $k = 0, 1, \cdots, 2r$, a set of constants.

In solving a boundary value problem, the solution of the differential equation (4.1) also needs to satisfy the prescribed boundary conditions

$$\mathbf{B}_\alpha y = \mathbf{g}_\alpha \quad (\alpha = 1, 2), \tag{4.3}$$

DOI: 10.1201/9781003194859-4

where the vector of the differential operators is in the form of

$$\mathbf{B}_\alpha = [\mathcal{B}_{\alpha,1} \quad \mathcal{B}_{\alpha,2} \quad \cdots \quad \mathcal{B}_{\alpha,r}]^{\mathrm{T}}, \tag{4.4}$$

and $\mathbf{g}_\alpha(\alpha = 1, 2)$ are the vectors of the prescribed boundary values at the two boundary points.

The differential operators in (4.4) are defined as follows:

$$\mathcal{B}_{\alpha,l} = \sum_{k=0}^{2r-1} b_{\alpha,k}^l \frac{d^k}{dx^k} \quad (l = 1, 2, \cdots, r), \tag{4.5}$$

where $b_{\alpha,k}^l$ are constants.

The system consisting of the differential equation (4.1) and the boundary conditions (4.3) can be used to describe a range of boundary value problems in science and engineering. Specifically, for $r = 1$, the system reduces to

$$a_0 \frac{d^2 y}{dx^2} + a_1 \frac{dy}{dx} + a_2 y = f(x), \quad x \in (a,b) \tag{4.6}$$

and

$$\alpha_1 y'(a) + \beta_1 y(a) = 0, \tag{4.7}$$

and

$$\alpha_2 y'(b) + \beta_2 y(b) = 0. \tag{4.8}$$

In different scientific fields, the solution $y(x)$ of (4.6) typically has its own specific meaning. For instance, in studying the deflection of a string stretched between two points at $x = 0$ and $x = L$, the differential equation (4.6) and the boundary conditions (4.7) and (4.8) accordingly become

$$T \frac{d^2 u}{dx^2} = f(x) \tag{4.9}$$

and

$$u(0) = u(L) = 0, \tag{4.10}$$

where $u(x)$ is the flexural displacement, T is the tension in the string, and $f(x)$ is the transverse force acting on the string.

The solution can be easily obtained as

$$u(x) = \iint f(x)\,dx\,dx + c_1 x + c_2, \tag{4.11}$$

where the integration constants c_1 and c_2 will be determined from the boundary conditions (4.10).

The differential equation (4.9) can also be used to solve for the longitudinal deformation of an elastic rod. In such a case, the tension T should be replaced by EA (E is the Young's modulus of the material and A is the cross-sectional area of the rod) and $f(x)$ should be understood as the distributed force applied along the rod. The boundary condition at an end (e.g., $x = 0$) is accordingly prescribed as:

$$u(0) = 0, \tag{4.12}$$

$$u'(0) = 0, \tag{4.13}$$

or

$$EAu'(0) - k_0 u(0) = 0, \tag{4.14}$$

where k_0 is the stiffness of the elastic restraint at $x = 0$.

Determining the deflection of an elastic beam under a static load requires solving the famous Euler-Bernoulli beam equation

$$D\frac{d^4 w}{dx^4} = f(x), \tag{4.15}$$

where $f(x)$ is the distributed transverse force, and D is the bending rigidity of the beam.

The boundary conditions for a generally supported beam (see Figure 4.1) are given as:

$$Dw'''(0) + k_0 w(0) = 0, \quad Dw''(0) - K_0 w'(0) = 0, \quad \text{at } x = 0 \tag{4.16}$$

and

$$Dw'''(L) - k_1 w(L) = 0, \quad Dw''(L) + K_1 w'(L) = 0, \quad \text{at } x = L, \tag{4.17}$$

where k_0 and k_1 are the linear spring constants, and K_0 and K_1 are the rotational spring constants at $x = 0$ and $x = L$, respectively.

The elastically restrained end conditions mathematically represent a general set of boundary conditions; all the classical homogeneous boundary conditions can be considered as the special cases. For example, the familiar simply supported boundary conditions,

$$w(0) = 0 \text{ and } w''(0) = 0, \text{ at } x = 0, \tag{4.18}$$

FIGURE 4.1　A beam elastically restrained at both ends.

and

$$w(L) = 0 \text{ and } w''(L) = 0, \text{ at } x = L, \tag{4.19}$$

are easily obtained by respectively setting the stiffness constants of the linear and rotational springs at each end to be extremely large and small.

4.2 CHARACTERISTIC SOLUTIONS OF THE BEAM EQUATION

Solving for the static deformations of one-dimensional elastic bodies (such as strings, bars, or beams) will typically involve integrating the loading function a few times and then substituting the results into the boundary conditions to determine integration constants, which may be considered a trivial task. To better illustrate the current solution schemes, we take the vibration of an Euler-Bernoulli beam as example. The beam vibration is governed by the following fourth-order linear differential equation

$$D\frac{\partial^4 w(x,t)}{\partial x^4} + \rho A\frac{\partial^2 w(x,t)}{\partial t^2} = f(x,t), \tag{4.20}$$

where ρ and A are the mass density and the cross-sectional area of the beam, respectively.

In practice, considerable attentions have been paid to the so-called steady-state conditions: the forcing function is sinusoidal with respect to time t, (i.e. $f(x,t) = F(x)e^{i\omega t}$), and the beam will respond it with a harmonic motion, $w(x,t) = W(x)e^{i\omega t}$. Accordingly, the displacement $W(x)$ satisfies the following differential equation

$$D\frac{d^4 W}{dx^4} - \rho A\omega^2 W = F(x). \tag{4.21}$$

Here, we are particularly interested in the characteristic solutions or eigensolutions, which satisfy the homogeneous form of (4.21), namely,

$$\frac{d^4 W}{dx^4} - \mu^4 W = 0, \tag{4.22}$$

where $\mu = \left(\rho A\omega^2 / D\right)^{1/4}$.

Assume $W(x) = e^{\lambda x}$, and substitute it into (4.22)

$$\left(\alpha^4 - \mu^4\right)e^{\alpha x} = 0, \tag{4.23}$$

from which we have

$$\alpha = \pm\mu, \pm i\mu. \tag{4.24}$$

The solution can then be written as

$$W(x) = C_1 e^{\mu x} + C_2 e^{-\mu x} + C_3 e^{i\mu x} + C_4 e^{-i\mu x}, \qquad (4.25)$$

or

$$W(x) = C_1 \sin \mu x + C_2 \cos \mu x + C_3 \sinh \mu x + C_4 \cosh \mu x, \qquad (4.26)$$

where C_i ($i = 1, 2, 3, 4$) are the coefficients to be determined from the boundary conditions.

Take the simply-supported boundary condition at each end as an example. By substituting (4.26) into the boundary conditions (4.18) and (4.19), we have

$$\begin{bmatrix} 0 & 1 & 0 & 1 \\ 0 & -1 & 0 & 1 \\ \sin \mu L & \cos \mu L & \sinh \mu L & \cosh \mu L \\ -\sin \mu L & -\cos \mu L & \sinh \mu L & \cosh \mu L \end{bmatrix} \begin{Bmatrix} C_1 \\ C_2 \\ C_3 \\ C_4 \end{Bmatrix} = \begin{Bmatrix} 0 \\ 0 \\ 0 \\ 0 \end{Bmatrix}. \qquad (4.27)$$

In order to obtain the non-trivial solution of (4.27), the determinant of the coefficient matrix has to vanish, resulting in

$$C_2 = C_3 = C_4 = 0, \mu = \frac{n\pi}{L}. \qquad (4.28)$$

Accordingly, the solution (4.26) reduces to

$$\phi_n(x) = C_1 \sin \mu_n x \quad \left(\mu_n = \frac{n\pi}{L} \right), \qquad (4.29)$$

where the arbitrary constant can be chosen as $C_1 = 1/L$ to make $\int_0^L \phi_n^2(x)dx = 1$. The eigenvalues,

$$\omega_n = \mu_n^2 \sqrt{\frac{D}{\rho A}} = n^2 \sqrt{\frac{\pi^4 D}{\rho A L^4}} \quad (n = 1, 2, 3, \cdots), \qquad (4.30)$$

are often referred to as the eigenfrequencies, characteristic frequencies, modal frequencies, natural frequencies, etc., and the corresponding solutions (4.29) as the eigenfunctions, characteristic solutions, modes, normal modes, etc.

Although the simply supported beam is here used to demonstrate the procedures of determining its eigen-parameters, the whole process equally applies to beams with other boundary conditions, except that the resulting characteristic equations tend to become much more complicated and may have to be solved numerically or approximately.

For a continuous beam there is an infinite number of eigenfunctions. The eigenfunctions for the simply supported beam, as given in (4.29), are evidently orthogonal with each other, and constitute a complete system. This is also true to the eigenfunctions of a beam with any other type of boundary conditions.

Suppose that ϕ_m and ϕ_n are the eigenfunctions corresponding to μ_m and μ_n ($\mu_m \neq \mu_n$) respectively, then

$$\int_0^L (\mu_m^4 - \mu_n^4)\phi_n\phi_m \, dx = \int_0^L \left(\frac{d^4\phi_m}{dx^4}\phi_n - \frac{d^4\phi_n}{dx^4}\phi_m \right) dx$$

$$= \left(\frac{d^3\phi_m}{dx^3}\phi_n - \frac{d^3\phi_n}{dx^3}\phi_m \right)\Bigg|_0^L - \int_0^L \left(\frac{d^3\phi_m}{dx^3}\frac{d\phi_n}{dx} - \frac{d^3\phi_n}{dx^3}\frac{d\phi_m}{dx} \right) dx$$

$$= \left(\frac{d^3\phi_m}{dx^3}\phi_n - \frac{d^3\phi_n}{dx^3}\phi_m \right)\Bigg|_0^L - \left(\frac{d^2\phi_m}{dx^2}\frac{d\phi_n}{dx} - \frac{d^2\phi_n}{dx^2}\frac{d\phi_m}{dx} \right)\Bigg|_0^L \quad (4.31)$$

$$+ \int_0^L \left(\frac{d^2\phi_m}{dx^2}\frac{d^2\phi_n}{dx^2} - \frac{d^2\phi_n}{dx^2}\frac{d^2\phi_m}{dx^2} \right) dx$$

$$= 0.$$

In the last step of (4.31), we have made use of the fact that the first two terms vanish identically due to the boundary conditions (4.16) and (4.17).

In view of the orthogonality and completeness of the eigenfunctions, the solution of (4.21) can be sought as

$$W(x) = \sum_m \xi_m \phi_m(x), \quad (4.32)$$

where ξ_m are the coefficients yet to be determined.

Substituting (4.32) into the beam equation (4.21) leads to

$$D\frac{d^4}{dx^4}\left[\sum_m \xi_m \phi_m(x) \right] - \rho A\omega^2 \sum_m \xi_m \phi_m(x) = F(x). \quad (4.33)$$

Multiplying (4.33) by ϕ_n and making use the orthogonality of the eigenfunctions, we yield

$$\xi_n = \frac{\int_0^L F(x)\phi_n(x)\,dx}{\rho A(\omega_n^2 - \omega^2)}. \quad (4.34)$$

This problem-solving process is often referred to as the eigenfunction method in mathematical physics or the modal superposition method in dynamics.

4.3 FOURIER SERIES SOLUTIONS OF THE BEAM PROBLEMS

In the modal superposition method, the displacement solution is expressed as a linear combination of the eigenfunctions, and its relevant derivatives can be simply obtained from term-by-term differentiations. When the eigenfunctions of a simply supported beam happen to be the sine functions as given in (4.29), the modal superposition method is then equivalent to expanding the solution into the Fourier sine series. Some attempts have been made of extending the Fourier series methods to beams with other boundary conditions (Greif and Mittendorf 1976; Wang and Lin 1996; Maurizi and Robledo 1998). The critical step is to recognize the fact that the Fourier series for the involved derivatives cannot be automatically obtained from the term-by-term differentiations.

To illustrate this point, consider the beam displacement function given in the form of the Fourier sine series

$$w(x) = \sum_{m=1}^{\infty} A_m \sin \lambda_m x, \quad 0 < x < L \quad (\lambda_m = m\pi/L). \tag{4.35}$$

It should be noted that the two end points have been explicitly excluded from (4.35). With reference to Theorem 2.5, its derivative is given by

$$w'(x) = \frac{w(L) - w(0)}{L} + \sum_{m=1}^{\infty} \left(\frac{2}{L} \left[(-1)^m w(L) - w(0) \right] + \lambda_m A_m \right) \cos \lambda_m x. \tag{4.36}$$

The relationship imbedded in (4.36) is sometimes called the Stokes transformation.

To better understand this problem-solving process, let's consider a simply supported beam which is elastically restrained against rotations at both ends, as illustrated in Figure 4.2.

If the displacement is expanded into the Fourier sine series

$$w = \sum_{m=1}^{\infty} A_m \sin \lambda_m x, 0 < x < L, \tag{4.37}$$

its four derivatives can be obtained as

$$\frac{dw}{dx} = \sum_{m=1}^{\infty} \lambda_m A_m \cos \lambda_m x, 0 \le x \le L, \tag{4.38}$$

FIGURE 4.2   A pinned-pinned beam with rotational restraints at both ends.

$$\frac{d^2 w}{dx^2} = -\sum_{m=1}^{\infty} \lambda_m^2 A_m \sin \lambda_m x, \, 0 < x < L, \tag{4.39}$$

$$\frac{d^3 w}{dx^3} = \frac{B_1 - B_0}{L} + \sum_{m=1}^{\infty} \left(\frac{2}{L}(B_1(-1)^m - B_0) - \lambda_m^3 A_m \right) \cos \lambda_m x, \, 0 \le x \le L, \tag{4.40}$$

and

$$\frac{d^4 w}{dx^4} = -\sum_{m=1}^{\infty} \left(\frac{2}{L}(B_1(-1)^m - B_0)\lambda_m - \lambda_m^4 A_m \right) \sin \lambda_m x, \, 0 < x < L, \tag{4.41}$$

where

$$B_0 = w''(0) \tag{4.42}$$

and

$$B_1 = w''(L). \tag{4.43}$$

The first two derivatives can be directly obtained from the term-by-term differentiations of (4.37), as justified by the boundary conditions $w(0) = w(L) = 0$.

Substitution of (4.41) into (4.22) results in

$$\sum_{m=1}^{\infty} \left\{ -\lambda_m \left(\frac{2}{L}(B_1(-1)^m - B_0) - \lambda_m^3 A_m \right) - \rho_D \omega^2 A_m \right\} \sin \lambda_m x = 0 \tag{4.44}$$

or

$$A_m = \frac{\lambda_m}{\lambda_m^4 - \rho_D \omega^2} \left(\frac{2}{L}[B_1(-1)^m - B_0] \right), \tag{4.45}$$

where $\rho_D = \frac{\rho A}{D}$.

Inserting (4.38), (4.42), (4.43), and (4.45) into (4.16b) and (4.17b) will lead to the following homogeneous equations about B_0 and B_1:

$$B_0 - \hat{K}_0 \sum_{m=1}^{\infty} \lambda_m A_m = 0, \, \text{at } x = 0, \tag{4.46}$$

and

$$B_1 + \hat{K}_1 \sum_{m=1}^{\infty} \lambda_m A_m \cos \lambda_m L = 0, \, \text{at } x = L, \tag{4.47}$$

where $\hat{K}_0 = K_0/D$ and $\hat{K}_1 = K_1/D$.

Substituting (4.45) into (4.46) and (4.47) results in

$$B_0 - \hat{K}_0 \sum_{m=1}^{\infty} \frac{\lambda_m^2}{\lambda_m^4 - \rho_D \omega^2} \left(\frac{2}{L} [B_1 (-1)^m - B_0] \right) = 0 \qquad (4.48)$$

and

$$B_1 + \hat{K}_1 \sum_{m=1}^{\infty} \frac{\lambda_m^2}{\lambda_m^4 - \rho_D \omega^2} \left(\frac{2}{L} [B_1 - B_0 (-1)^m] \right) = 0, \qquad (4.49)$$

which can be rewritten in matrix form as

$$\begin{bmatrix} 1 + \hat{K}_0 C_1 & \hat{K}_0 C_2 \\ \hat{K}_1 C_2 & 1 + \hat{K}_1 C_1 \end{bmatrix} \begin{Bmatrix} B_0 \\ B_1 \end{Bmatrix} = \begin{Bmatrix} 0 \\ 0 \end{Bmatrix}, \qquad (4.50)$$

where

$$C_1 = \sum_{m=1}^{\infty} \frac{2\lambda_m^2}{L(\lambda_m^4 - \rho_D \omega^2)} \text{ and } C_2 = \sum_{m=1}^{\infty} \frac{2\lambda_m^2 (-1)^{m+1}}{L(\lambda_m^4 - \rho_D \omega^2)}. \qquad (4.51)$$

Since the elements of the coefficient matrix are explicitly dependent upon the frequency ω, the eigenfrequencies can thus be obtained from (4.50) as a special set of values which make the determinant of the coefficient matrix vanish. Such a process, however, will typically involve solving a nonlinear equation, which is at least a tedious task, if nothing more.

Alternatively, we will adopt a much simpler approach as described below. The boundary values of the second-order derivative, B_0 and B_1, are first determined from the boundary equations (4.46) and (4.47), and then substituted into the governing equation (4.44), leading to

$$\sum_{m=1}^{\infty} \left\{ \lambda_m^4 A_m + \frac{2\lambda_m}{L} \left(\hat{K}_1 \sum_{m'=1}^{\infty} (-1)^{m'+m} \lambda_{m'} A_{m'} + \hat{K}_0 \sum_{m'=1}^{\infty} \lambda_{m'} A_{m'} \right) - \rho_D \omega^2 A_m \right\} \sin \lambda_m x = 0.$$
$$(4.52)$$

Due to the orthogonality of the function system, $\{ \sin \lambda_m x, m = 1, 2, ... \}$, the coefficient for each term in (4.52) must vanish, that is,

$$\lambda_m^4 A_m + \frac{2\lambda_m}{L} \left(\hat{K}_1 \sum_{m'=1}^{\infty} (-1)^{m'+m} \lambda_{m'} A_{m'} + \hat{K}_0 \sum_{m'=1}^{\infty} \lambda_{m'} A_{m'} \right) - \rho_D \omega^2 A_m = 0, \ m = 1, 2, ...,$$
$$(4.53)$$

or, in matrix form

$$[a_{mn}]\{A_n\} - \mu^4 [b_{mn}]\{A_n\} = \{0\}, \tag{4.54}$$

where $b_{mn} = \delta_{mn}$, and

$$a_{mn} = \lambda_m^4 \delta_{mn} + \frac{2\lambda_m \lambda_n}{L}\left(\hat{K}_0 + (-1)^{n+m}\hat{K}_1\right). \tag{4.55}$$

The above equation represents a standard characteristic equation about the matrix $[a_{mn}]$ from which the eigenvalues and the corresponding eigenvectors can be readily determined as adequately discussed in linear algebra.

4.4 THE GENERALIZED FOURIER SERIES SOLUTIONS

In Sec. 4.3, the beam displacement is assumed to vanish identically at each end. As a consequence, its odd-periodic extension is ensured to have C^1 continuity, and its first two derivatives can be directly obtained from term-by-term differentiations. According to Theorem 2.3, the Fourier series (4.35) can be ensured to converge at the speed of $\lambda_m^2 A_m \to 0$ as $m \to \infty$. However, due to the presence of the λ_m^4-like terms in the coefficient matrix, it may not be guaranteed that the characteristic equation (4.54) will lead to the converged solution.

This difficulty, however, can be effectively resolved by using the convergence acceleration schemes described in the previous chapter.

Under the current framework, the solutions of the beam equation, regardless of the boundary conditions, will be invariably expressed as

$$w(x) = \bar{w}(x) + h(x), \tag{4.56}$$

where the supplementary function $h(x)$, as mentioned earlier, is introduced to cope with all the potential discontinuities associated with the displacement and its three derivatives at the ends of the solution domain $[0, L]$ when they are periodically extended. In other words, the "residual" displacement $\bar{w}(x)$ now presents a periodic function and has at least three continuous derivatives. If $\bar{w}(x)$ is expanded into a cosine series

$$\bar{w} = \sum_{m=0}^{\infty} A_m \cos \lambda_m x, \quad 0 \le x \le L, \tag{4.57}$$

then all the required differentiations can be simply carried out term-by-term.

As described in Chapter 3, the supplementary function can be in the form of polynomials, sinusoidal functions, or any other suitable closed-form functions.

In what follows, the supplementary function $h(x)$ is chosen to be a polynomial $p(x)$, and is required to satisfy (refer to (3.39)):

$$p'''(0) = w'''(0) = \alpha_0, \tag{4.58}$$

$$p'''(L) = w'''(L) = \alpha_1, \tag{4.59}$$

$$p'(0) = w'(0) = \beta_0, \tag{4.60}$$

and

$$p'(L) = w'(L) = \beta_1. \tag{4.61}$$

The lowest order polynomial which satisfies (4.58)–(4.61) can be written as

$$p(x) = p_1(x) + p_2(x), \tag{4.62}$$

where

$$p_1'''(x) = \alpha_0(1 - x/L) + \alpha_1 x/L, \tag{4.63}$$

and

$$p_2'(x) = \beta_0(1 - x/L) + \beta_1 x/L. \tag{4.64}$$

Integrating (4.63) three times and choosing the integration constants in such a way that

$$p_1'(0) = 0, \tag{4.65}$$

$$p_1'(L) = 0, \tag{4.66}$$

and

$$\int_0^L p_1(x)dx = 0, \tag{4.67}$$

we yield

$$p_1(x) = \frac{\alpha_1}{24L}(x^4 - 2L^2x^2) - \frac{\alpha_0}{24L}(4L^2x^2 - 4Lx^3 + x^4) + \frac{L^3}{360}(8\alpha_0 + 7\alpha_1). \tag{4.68}$$

Similarly, the second part of the polynomial can be found as

$$p_2(x) = \frac{\beta_1}{6L}(3x^2 - L^2) + \frac{\beta_0}{6L}(6Lx - 2L^2 - 3x^2) \tag{4.69}$$

with

$$\int_0^L p_2(x)dx = 0. \tag{4.70}$$

Substitution of (4.56), (4.57) and (4.62) into (4.22) leads to

$$\frac{\alpha_1 - \alpha_0}{L} + \sum_{n=1}^{\infty} \lambda_n^4 A_n \cos \lambda_n x - \rho_D \omega^2 \left(\sum_{n=0}^{\infty} A_n \cos \lambda_n x + p_1(x) + p_2(x) \right) = 0. \tag{4.71}$$

Multiplying (4.71) with $(2/L)\cos \lambda_m x$ $(m = 0, 1, 2, ...)$ and integrating it from 0 to L result in

$$\lambda_m^4 A_m - \rho_D \omega^2 \left(A_m + P_m \right) = 0, \, m = 1, 2, \ldots, \tag{4.72}$$

and

$$\frac{\alpha_1 - \alpha_0}{L} - \rho_D \omega^2 A_0 = 0, \tag{4.73}$$

where

$$P_m = \frac{2}{L} \int_0^L p(x) \cos \lambda_m x dx = \frac{2}{L} \left(-\frac{\alpha_1(-1)^m - \alpha_0}{\lambda_m^4} + \frac{\beta_1(-1)^m - \beta_0}{\lambda_m^2} \right). \tag{4.74}$$

In (4.74) the values of the first and third derivatives at the ends need to be determined from the boundary conditions (4.16) and (4.17). Using (4.56)–(4.62), (4.68) and (4.69), we are able to obtain:

$$\hat{k}_0 \left(\sum_{m=0}^{\infty} A_m + \frac{8L^3 \alpha_0}{360} + \frac{7L^3 \alpha_1}{360} - \frac{\beta_0 L}{3} - \frac{\beta_1 L}{6} \right) = -\alpha_0, \tag{4.75}$$

$$\hat{k}_1 \left(\sum_{m=0}^{\infty} (-1)^m A_m - \frac{7L^3 \alpha_0}{360} - \frac{8L^3 \alpha_1}{360} + \frac{\beta_0 L}{6} + \frac{\beta_1 L}{3} \right) = \alpha_1, \tag{4.76}$$

$$\hat{K}_0 \beta_0 = \left(-\sum_{m=1}^{\infty} \lambda_m^2 A_m - \frac{\alpha_0 L}{3} - \frac{\alpha_1 L}{6} + \frac{\beta_1}{L} - \frac{\beta_0}{L} \right), \tag{4.77}$$

and

$$\hat{K}_1 \beta_1 = -\left(\sum_{m=1}^{\infty} (-1)^{m+1} \lambda_m^2 A_m + \frac{\alpha_0 L}{6} + \frac{\alpha_1 L}{3} + \frac{\beta_1}{L} - \frac{\beta_0}{L} \right), \qquad (4.78)$$

or in matrix form,

$$\mathbf{H}\bar{\alpha} = \sum_{m=0}^{\infty} \mathbf{Q}_m A_m, \qquad (4.79)$$

where

$$\bar{\alpha} = \left\{ \alpha_0, \alpha_1, \beta_0, \beta_1 \right\}^{\mathrm{T}}, \qquad (4.80)$$

$$\mathbf{H} = \begin{bmatrix} \dfrac{\hat{k}_0 L^3}{45} + 1 & \dfrac{7\hat{k}_0 L^3}{360} & \dfrac{-\hat{k}_0 L}{3} & \dfrac{-\hat{k}_0 L}{6} \\[2ex] \dfrac{7\hat{k}_1 L^3}{360} & \dfrac{\hat{k}_1 L^3}{45} + 1 & \dfrac{-\hat{k}_1 L}{6} & \dfrac{-\hat{k}_1 L}{3} \\[2ex] \dfrac{L}{3} & \dfrac{L}{6} & \hat{K}_0 + \dfrac{1}{L} & \dfrac{-1}{L} \\[2ex] \dfrac{L}{6} & \dfrac{L}{3} & \dfrac{-1}{L} & \hat{K}_1 + \dfrac{1}{L} \end{bmatrix} \qquad (4.81)$$

and

$$\mathbf{Q}_m = \left\{ -\hat{k}_0 \quad (-1)^m \hat{k}_1 \quad -\lambda_m^2 \quad (-1)^m \lambda_m^2 \right\}^{\mathrm{T}}. \qquad (4.82)$$

Combining (4.72), (4.73) and (4.79) gives

$$\lambda_m^4 A_m - \rho_D \omega^2 \left(A_m + \sum_{m'=0}^{\infty} S_{mm'} A_{m'} \right) = 0, \, m = 1, 2, 3, \ldots, \qquad (4.83)$$

and

$$\sum_{m'=0}^{\infty} \mathbf{c}^{\mathrm{T}} \mathbf{H}^{-1} \mathbf{Q}_{m'} A_{m'} - \rho_D \omega^2 A_0 = 0, \qquad (4.84)$$

where

$$\mathbf{c} = \left\{ -1/L \quad 1/L \quad 0 \quad 0 \right\}^{\mathrm{T}},$$ (4.85)

$$S_{mm'} = \mathbf{P}_m^{\mathrm{T}} \mathbf{H}^{-1} \mathbf{Q}_{m'},$$ (4.86)

and

$$\mathbf{P}_m = \frac{2}{L} \left\{ \frac{1}{\lambda_m^4} \quad \frac{(-1)^{m+1}}{\lambda_m^4} \quad \frac{-1}{\lambda_m^2} \quad \frac{(-1)^m}{\lambda_m^2} \right\}^{\mathrm{T}}.$$ (4.87)

If the Fourier series solution is truncated to $m = M$, (4.83) and (4.84) can be combined into

$$\left(\mathbf{K} - \rho_D \omega^2 \mathbf{M} \right) \mathbf{A} = 0,$$ (4.88)

where

$$\mathbf{A} = \left\{ A_0, A_1, ..., A_M \right\}^{\mathrm{T}},$$ (4.89)

$$\mathbf{K} = \begin{bmatrix} \mathbf{c}^{\mathrm{T}}\mathbf{H}^{-1}\mathbf{Q}_0 & \mathbf{c}^{\mathrm{T}}\mathbf{H}^{-1}\mathbf{Q}_1 & \cdots & \mathbf{c}^{\mathrm{T}}\mathbf{H}^{-1}\mathbf{Q}_{m'} & \cdots & \mathbf{c}^{\mathrm{T}}\mathbf{H}^{-1}\mathbf{Q}_M \\ 0 & \lambda_1^4 & \cdots & 0 & \cdots & 0 \\ \vdots & \vdots & \cdots & \vdots & \cdots & \vdots \\ 0 & 0 & \cdots & \lambda_m^4 \delta_{mm'} & \cdots & 0 \\ \vdots & \vdots & \cdots & \vdots & \cdots & \vdots \\ 0 & 0 & \cdots & 0 & \cdots & \lambda_M^4 \end{bmatrix},$$ (4.90)

and

$$\mathbf{M} = \begin{bmatrix} 1 & 0 & \cdots & 0 & \cdots & 0 \\ S_{10} & 1 + S_{11} & \cdots & S_{1m'} & \cdots & S_{1M} \\ \vdots & \vdots & \cdots & \vdots & \cdots & \vdots \\ S_{m0} & S_{m1} & \cdots & \delta_{mm'} + S_{mm'} & \cdots & S_{mM} \\ \vdots & \vdots & \cdots & \vdots & \cdots & \vdots \\ S_{M0} & S_{M1} & \cdots & S_{Mm'} & \cdots & 1 + S_{MM} \end{bmatrix}.$$ (4.91)

The eigenfrequencies and eigenvectors can now be easily determined by solving a standard matrix eigenproblem. The eigenvectors actually contain the expansion coefficients of the Fourier series from which the corresponding eigenfunctions are readily obtained as

$$w(x) = \sum_{m=0}^{M} (\cos \lambda_m x + \mathbf{X}^{\mathrm{T}} \overline{\mathbf{S}}_m) A_m, \tag{4.92}$$

where

$$\mathbf{X} = \left\{ 1 \quad x \quad x^2 \quad x^3 \quad x^4 \right\}^{\mathrm{T}}, \tag{4.93}$$

$$\overline{\mathbf{S}}_m = \mathbf{L} \mathbf{H}^{-1} \mathbf{Q}_m, \tag{4.94}$$

and

$$\mathbf{L} = \begin{bmatrix} 8L^3/360 & 7L^3/360 & -L/3 & -L/6 \\ 0 & 0 & 1 & 0 \\ -L/6 & -L/12 & -1/2L & 1/2L \\ 1/6 & 0 & 0 & 0 \\ -1/24L & 1/24L & 0 & 0 \end{bmatrix}. \tag{4.95}$$

Finally, by making use of

$$\int_0^L w^2(x) dx = L, \tag{4.96}$$

the eigenfunction can be normalized as

$$\phi(x) = \sum_{m=0}^{M} (\cos \lambda_m x + \mathbf{X}^{\mathrm{T}} \overline{\mathbf{S}}_m) \overline{A}_m, \tag{4.97}$$

where

$$\overline{A}_m = A_m / \chi, \ m = 0, 1, 2, ..., M, \tag{4.98}$$

$$\chi = \left(\sum_{m,m'=0} (\delta_{0m} \delta_{0m'} + 1/2 \delta_{mm'} + \overline{\mathbf{S}}_m^{\mathrm{T}} \overline{\mathbf{X}} \overline{\mathbf{S}}_{m'} + 2 \mathbf{C}_m^{\mathrm{T}} \overline{\mathbf{S}}_{m'}) A_m A_{m'} \right)^{1/2}, \tag{4.99}$$

$$\overline{\mathbf{X}} = \frac{1}{L} \int_0^L \mathbf{X} \mathbf{X}^{\mathrm{T}} dx = \left\{ \overline{X}_{ij} \right\} \quad (\overline{X}_{ij} = L^{i+j-2}/(i+j-1)), \tag{4.100}$$

and

$$\mathbf{C}_m = \frac{1}{L} \int_0^L \mathbf{X} \cos \lambda_m x \, dx$$

$$= \begin{cases} \left\{ -\dfrac{1+(-1)^{m+1}}{\lambda_m^2 L} \quad \dfrac{2(-1)^m}{\lambda_m^2} \quad \dfrac{6[1-(-1)^m]+3(-1)^m \lambda_m^2 L^2}{\lambda_m^4 L} \quad \dfrac{[-24+4\lambda_m^2 L^2](-1)^m}{\lambda_m^4} \right\}^{\mathrm{T}}, \\ \qquad m = 1, 2, ..., M. \\ \left\{ 1 \quad \dfrac{L}{2} \quad \dfrac{L^2}{3} \quad \dfrac{L^3}{4} \quad \dfrac{L^4}{5} \right\}^{\mathrm{T}}, \quad m = 0. \end{cases}$$

$$\tag{4.101}$$

The beam solution can also be expanded into a generalized Fourier sine series

$$w(x) = \sum_{m=1}^{\infty} A_m \sin \lambda_m x + p(x), 0 \le x \le L. \tag{4.102}$$

In light of (3.44), the polynomial function $p(x)$ needs to satisfy:

$$p''(0) = w''(0) = \alpha_0, \tag{4.103}$$

$$p''(L) = w''(L) = \alpha_1, \tag{4.104}$$

$$p(0) = w(0) = \beta_0, \tag{4.105}$$

and

$$p(L) = w(L) = \beta_1. \tag{4.106}$$

Such a polynomial can be written as

$$p = \zeta(x)^{\mathrm{T}} \bar{\alpha}, \tag{4.107}$$

where

$$\bar{\alpha} = \left\{ \alpha_0, \alpha_1, \beta_0, \beta_1 \right\}^{\mathrm{T}} \tag{4.108}$$

and

$$\zeta(x)^{\mathrm{T}} = \begin{cases} -(2L^2 x - 3Lx^2 + x^3)/6L \\ (x^3 - L^2 x)/6L \\ (L-x)/L \\ x/L \end{cases}. \tag{4.109}$$

The relationship (4.79) still holds here, except that $\mathbf{H}$ and $\mathbf{Q}_m$ should be accordingly modified to

$$
\mathbf{H} = \begin{bmatrix}
\dfrac{\hat{K}_0 L}{3}+1 & \dfrac{\hat{K}_0 L}{6} & \dfrac{\hat{K}_0}{L} & -\dfrac{\hat{K}_0}{L} \\[2ex]
\dfrac{\hat{K}_1 L}{6} & \dfrac{\hat{K}_1 L}{3}+1 & -\dfrac{\hat{K}_1}{L} & \dfrac{\hat{K}_1}{L} \\[2ex]
-\dfrac{1}{L} & \dfrac{1}{L} & \hat{k}_0 & 0 \\[2ex]
\dfrac{1}{L} & -\dfrac{1}{L} & 0 & \hat{k}_1
\end{bmatrix}
\tag{4.110}
$$

and

$$
\mathbf{Q}_m = \left\{ \hat{K}_0 \lambda_m \quad (-1)^{m+1} \hat{K}_1 \lambda_m \quad \lambda_m^3 \quad (-1)^{m+1} \lambda_m^3 \right\}^{\mathrm{T}}.
\tag{4.111}
$$

Substituting (4.107) into (4.102) leads to

$$
w(x) = \sum_{m=1}^{\infty} A_m \sin \lambda_m x + \zeta(x)^{\mathrm{T}} \alpha,
\tag{4.112}
$$

or

$$
w(x) = \sum_{m=1}^{\infty} A_m \phi_m(x),
\tag{4.113}
$$

where a set of new basis functions is defined as

$$
\phi_m(x) = \sin \lambda_m x + \zeta(x)^{\mathrm{T}} \mathbf{H}^{-1} \mathbf{Q}_m \quad (m = 1, 2, 3, ...).
\tag{4.114}
$$

Evidently, the displacement given in (4.114) has already satisfied all the boundary conditions, and the eigenfrequencies can be simply obtained from the characteristic equations below:

$$
\lambda_m^4 A_m - \rho_D \omega^2 \left(A_m + \sum_{m'=1}^{\infty} S_{mm'} A_{m'} \right) = 0, \, m = 1, 2, 3, ...,
\tag{4.115}
$$

where

$$
S_{mm'} = \mathbf{P}_m^{\mathrm{T}} \mathbf{H}^{-1} \mathbf{Q}_{m'}
\tag{4.116}
$$

and

$$\mathbf{P}_m = \frac{2}{L} \left\{ \frac{-1}{\lambda_m^3} \quad \frac{(-1)^m}{\lambda_m^3} \quad \frac{1}{\lambda_m} \quad \frac{(-1)^{m+1}}{\lambda_m} \right\}^{\mathrm{T}}. \tag{4.117}$$

We can recap the major procedures just depicted for obtaining the generalized Fourier solutions as follows:

1. Selecting a suitable supplementary function of any form;
2. Defining the supplementary function as an explicit function of the unknown boundary constants to ensure a desired degree of continuities for the residual displacement when it is periodically extended;
3. Determining the relationship between the unknown boundary constants and the Fourier coefficients by forcing the generalized Fourier solution to satisfy the boundary conditions;
4. Solving for the Fourier coefficients from the governing differential equation.

However, it must be pointed out that there are plenty of rooms or flexibilities in following these solution steps.

4.5   CONVERGENCE ASSESSMENT

The Fourier series in (4.56) or (4.102) actually represents the residual or conditioned beam solution, whose periodic extension is explicitly required to have at least three continuous derivatives. Thus, it will converge at a faster rate of

$$\lim_{m \to \infty} A_m \lambda_m^4 = 0. \tag{4.118}$$

As a matter of fact, the convergence rate of the generalized Fourier series solution (4.56) can be directly estimated from the governing equation (4.72)

$$A_m = \frac{\rho_D \omega^2 P_m}{\lambda_m^4 - \rho_D \omega^2}, \tag{4.119}$$

where P_m are the Fourier coefficients of the polynomial function $p(x)$. Substituting (4.87) into (4.119) results in

$$A_m = \frac{\rho_D \omega^2 P_m}{\lambda_m^4 - \rho_D \omega^2} = \frac{2\rho_D \omega^2 / \lambda_m^2}{(\lambda_m^4 - \rho_D \omega^2)L} \left(-\frac{\alpha_1(-1)^m - \alpha_0}{\lambda_m^2} + \beta_1(-1)^m - \beta_0 \right). \tag{4.120}$$

If (4.102) is used to expand the beam solution, (4.120) should be accordingly modified to

$$A_m = \frac{\rho_D \omega^2 P_m}{\lambda_m^4 - \rho_D \omega^2} = \frac{2\rho_D \omega^2 / \lambda_m}{(\lambda_m^4 - \rho_D \omega^2)L} \left(\frac{\alpha_1(-1)^m - \alpha_0}{\lambda_m^2} + \beta_1(-1)^{m+1} + \beta_0 \right). \quad (4.121)$$

It must be noted that even though a same set of boundary constants, α_i and β_i, are used in (4.120) and (4.121), they actually have different meanings (refer to (4.63) and (4.64), and (4.103)–(4.106)).

If the displacement is expanded into the conventional Fourier sine series (4.37), (4.45) can be generalized to

$$A_m = \frac{2\lambda_m^3}{(\lambda_m^4 - \rho_D \omega^2)L} \left(\frac{\alpha_1(-1)^m - \alpha_0}{\lambda_m^2} - [\beta_1(-1)^m - \beta_0] \right), \quad (4.122)$$

where the potential discontinuities of displacement at the end points are also taken into account.

If the boundary constants are known *a priori*, then (4.120)–(4.122) can be directly used to assess the convergence rates for the corresponding Fourier series solutions. For instance, if a simply supported beam is elastically restrained by a rotational spring at each end, the constants β_0 and β_1 are then both equal to zero, and the generalized sine series solution will converge according to $A_m \sim O(\lambda_m^{-7})$. In comparison, the generalized Fourier cosine series and the conventional sine series will converge at the rates of $A_m \sim O(\lambda_m^{-6})$ and $A_m \sim O(\lambda_m^{-3})$, respectively. For a generally supported beam, (4.16) and (4.17), the constants β_0 and β_1 will be usually not equal to zero. Therefore, the generalized sine and cosine series solutions will converge according to $A_m \sim O(\lambda_m^{-5})$ and $A_m \sim O(\lambda_m^{-6})$, respectively; in comparison, the conventional Fourier sine series is estimated to converge at a much slower rate of $A_m \sim O(\lambda_m^{-1})$.

4.6   NUMERICAL EXAMPLES

First consider a simply supported beam with rotational constraints at both ends, as shown in Figure 4.2. While the rotational stiffness is set extremely large, such as $\hat{K}_0 L = 10^{10}$, at one end, the rotational restraint at the other end is chosen to have various stiffness values. For the two extreme stiffness values, $\hat{K}_1 L = 0$ and 10^{10}, the general boundary conditions will essentially degenerate into the two familiar cases:

the clamped-pinned supports,

$$w(0) = 0, w'(0) = 0, w(L) = 0, w''(L) = 0; \quad (4.123)$$

the clamped-clamped supports,

$$w(0) = 0, w'(0) = 0, w(L) = 0, w'(L) = 0. \quad (4.124)$$

TABLE 4.1

Frequency Parameters, $\mu_i = L/\pi(\omega_i\sqrt{\rho A/D})^{1/2}$, for the Beam Elastically Restrained with Various Rotational Stiffnesses

	$\mu_i = L/\pi(\omega_i\sqrt{\rho A/D})^{1/2}$				
Mode	$\hat{K}_1 L = 0$	$\hat{K}_1 L = 1$	$\hat{K}_1 L = 10$	$\hat{K}_1 L = 100$	$\hat{K}_1 L = 10^{10}$
1	1.24988	1.28656	1.4102	1.49137	1.50562
2	2.25005	2.27081	2.37138	2.47681	2.49975
3	3.25014	3.26491	3.34927	3.46884	3.50001
4	4.25032	4.26175	4.3337	4.46108	4.5

The frequency parameters, $\mu_i = a/\pi(\omega_i\sqrt{\rho A/D})^{1/2}$, for these two classical boundary conditions are well known as (Blevins 1979): $\mu_i = 1.24988, 2.25, 3.25, 4.25$ and $\mu_i = 1.50562, 2.49975, 3.50001, 4.5$. The related results obtained using the generalized Fourier cosine series are presented in Table 4.1 together with those corresponding to other intermediate stiffness values.

In the above calculations, the generalized Fourier cosine series has been truncated to $M = 20$. To better understand the convergence characteristics of the solution, Table 4.2 compares the first 10 frequency parameters, which are calculated by using different numbers of terms in the generalized cosine series. It is seen that the generalized Fourier series solution converges so fast that only an inclusion of a small number of terms can lead to an excellent prediction. In contrast, as illustrated in Table 4.3, the conventional Fourier series solution based on (4.52) does

TABLE 4.2

Frequency Parameters, $\mu_i = L/\pi(\omega_i\sqrt{\rho A/D})^{1/2}$, for the Clamped-Clamped Beam Obtained by Using Various Numbers of Terms in the Generalized Fourier Series Solution

	$\mu_i = L/\pi(\omega_i\sqrt{\rho A/D})^{1/2}$			
Mode	$M = 5$	$M = 10$	$M = 15$	$M = 20$
1	1.50563	1.50562	1.50562	1.50562
2	2.49985	2.49976	2.49975	2.49975
3	3.50392	3.50003	3.50001	3.50001
4	4.5073	4.5002	4.50001	4.5
5	——	5.50044	5.50005	5.5
6	——	6.50289	6.5001	6.50002
7	——	7.50421	7.50045	7.50004
8	——	8.52423	8.5007	8.50014
9	——	9.52852	9.50251	9.50022
10	——	——	10.5033	10.5007

TABLE 4.3
Frequency Parameters, $\mu_i = L/\pi(\omega_i\sqrt{\rho A/D})^{1/2}$, for the Clamped-Clamped Beam Obtained Using Various Numbers of Terms in the Conventional Fourier Series Solution

	$\mu_i = L/\pi(\omega_i\sqrt{\rho A/D})^{1/2}$			
Mode	$M = 50$	$M = 75$	$M = 100$	$M = 150$
1	1.50984	1.49997	1.53034	1.53753
2	2.52008	2.51498	2.50879	2.51103
3	3.52993	3.51930	3.51461	3.50949
4	4.53881	4.52850	4.52151	4.50753
5	5.54829	5.53089	5.52319	5.51528

not compare as nearly well even though this boundary condition actually represents a favorable scenario: the displacement vanishes simultaneously at both ends.

Once the eigenvalues are found, the corresponding eigenfunctions can be directly determined using (4.97). Plotted in Figures 4.3–4.6 are the first four eigenfunctions for the clamped-clamped beam. The analytical solutions for this case are known as:

$$\phi_i = \cosh\frac{\pi\mu_i x}{L} - \cos\frac{\pi\mu_i x}{L} - \sigma_i\left(\sinh\frac{\pi\mu_i x}{L} - \sin\frac{\pi\mu_i x}{L}\right), \quad (4.125)$$

where

$$\sigma_i = \frac{\cosh\pi\mu_i - \cos\pi\mu_i}{\sinh\pi\mu_i - \sin\pi\mu_i}. \quad (4.126)$$

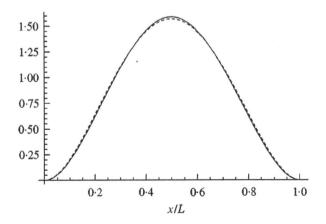

FIGURE 4.3 The first eigenfunction of the clamped-clamped beam: _____, (4.125); _____, (4.97); $M = 1$.

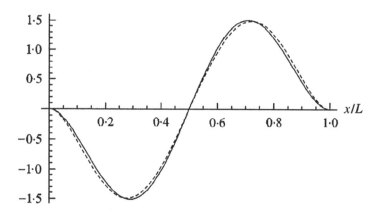

FIGURE 4.4   The second eigenfunction of the clamped-clamped beam: _____, (4.125); _____, (4.97); $M = 2$.

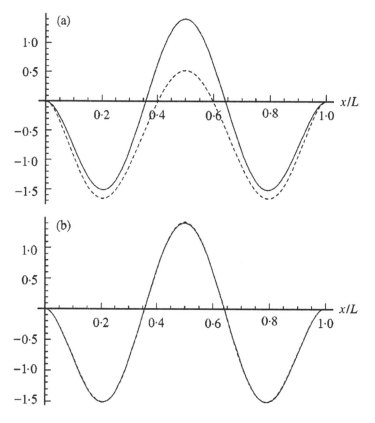

FIGURE 4.5   The third eigenfunction of the clamped-clamped beam: _____, (4.125); _____, (4.97); (a) $M = 3$; (b) $M = 4$.

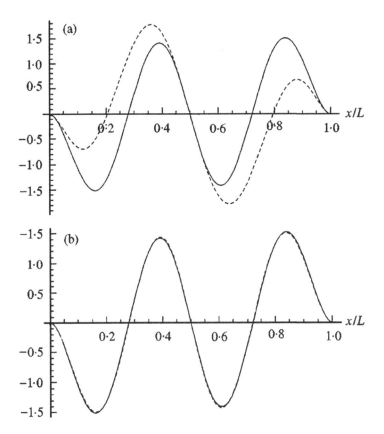

FIGURE 4.6   The fourth eigenfunction of the clamped-clamped beam: _____, (4.125); _____, (4.97); (a) $M = 4$; (b) $M = 5$.

In each of the plots, M denotes the number of cosine terms actually used in the solution; for larger M's, the corresponding curves virtually look identical to their analytical counterparts. All these results have indicated that eigenfunctions can be accurately obtained with only a few terms in the generalized Fourier series. The remarkable convergence characteristic of the current solution is perhaps best shown in Figures 4.5 and 4.6 where adding one more term has really made a big difference.

Let's now consider two examples that involve both rotational and translational restraints. The first one concerns a pinned-free beam with a rotational spring at the pinned end and a translational spring at the free end. Table 4.4 lists the first eigenfrequencies for the various combinations of the spring constants. The results are very close to those previously given in reference (Maurizi et al. 1976).

The second example involves a beam supported at each end by both translational and rotational springs. By assuming $\hat{k}_0 L^3 = \hat{k}_1 L^3 = 1$, the first five frequency parameters, $\mu_i = (a^2 \omega_i \sqrt{\rho A/D})^{1/2}$, are presented in Table 4.5 for the three

TABLE 4.4

The First Eigenfrequency, $\mu_1 = \left(L^2 \omega_1 \sqrt{\rho A/D} \right)^{1/2}$, of a Beam Elastically Restrained with Various Rotational and Translational Stiffnesses

$\hat{K}_0 L$	$\hat{k}_1 L^3$		
	0.01	1	100
0.01	0.4948	1.3134	2.9901
1	1.2520	1.5358	3.1085
100	1.8583	1.9940	3.6134

different stiffness values, $\hat{K}_1 L = \hat{K}_0 L = 0.01, 1, 100$. This problem was previously studied in reference (Rao and Mirza 1989) and an excellent agreement has been observed between the two solutions. The remarkable convergence of the current solution is again demonstrated in Table 4.6 for the elastically restrained beam.

In the above examples, the beam solutions are determined using the generalized Fourier cosine series expansions. It is stated in Sec. 4.5 that the generalized cosine series solution is able to outperform its sine counterpart for beams with general boundary conditions. To illustrate this point, a few more examples will be given below.

Let's start with the clamped-pinned boundary condition, which corresponds to the second column for $\hat{K}_0 L = 0$ in Table 4.1. Shown in Table 4.7 are the four lowest frequency parameters, $\mu = L/\pi (\omega \sqrt{\rho A/D})^{1/2}$, obtained using both generalized sine and cosine series. In this example, while the displacement vanishes at both ends, the first derivative is typically not equal to zero at $x = L$. As mentioned earlier, this case favors the generalized sine series solution.

TABLE 4.5

Frequency Parameters, $\mu_i = \left(L^2 \omega_1 \sqrt{\rho A/D} \right)^{1/2}$, of a Beam Elastically Restrained with Various Rotational Stiffnesses

Mode	$\hat{K}_1 L = \hat{K}_0 L$		
	0.01	1	100
1	1.1843	1.18564	1.1883
2	1.57925	2.23332	3.14418
3	4.75304	5.06326	6.22722
4	7.8607	8.07739	9.33698
5	11.0009	11.1628	12.4499

TABLE 4.6

Frequency Parameters, $\mu_i = \left(L^2 \omega_i \sqrt{\rho A/D}\right)^{1/2}$, Calculated Using Different Numbers of Terms in the Generalized Fourier Cosine Series; $\hat{k}_0 L^3 = \hat{k}_1 L^3 = 1$ and $\hat{K}_1 L = \hat{K}_0 L = 100$

Mode	$M = 5$	$M = 10$	$M = 20$	$M = 40$	Reference[a]
1	1.188301	1.188301	1.188301	1.188301	1.188301
2	3.14418	3.14418	3.14418	3.14418	3.144179
3	6.22726	6.227224	6.227221	6.22722	6.22722
4	9.33717	9.337013	9.336975	9.336971	9.336969
5	12.4514	12.45001	12.44990	12.44988	12.44988

Note:
[a] See reference (Rao and Mirza 1989).

Now, if the stiffness of the rotational spring at $x = L$ also becomes infinitely large (i.e., the clamped-clamped boundary condition), then both the displacement and its derivative will vanish simultaneously at both ends. Even though this case represents a better scenario for both solutions, the generalized cosine series solution is expected to outperform its sine counterpart in terms of convergence (i.e., λ_M^{-4} vs. λ_M^{-3}, according to (4.120) and (4.121)), as roughly confirmed by the results in Table 4.8.

TABLE 4.7

Frequency Parameters, $\mu = L/\pi (\omega \sqrt{\rho A/D})^{1/2}$, for A Clamped-Pinned Beam[a]

$$\mu = L/\pi (\omega \sqrt{\rho A/D})^{1/2}$$

	Cosine Series	Sine Series	Cosine Series	Sine Series	
Mode	$M = 10$	$M = 10$	$M = 20$	$M = 20$	Exact[b]
1	1.24994	1.24988	1.24988	1.24988	1.24988
2	2.25036	2.25002	2.25005	2.25	2.25
3	3.25108	3.25011	3.25014	3.25	3.25
4	4.25293	4.25044	4.25032	4.25002	4.25

Notes:
[a] This table was taken from the *Journal of Sound and Vibration*, 255, W. L. Li, Comparison of Fourier sine and cosine series expansions for beams with arbitrary boundary conditions, 185–194, Copyright Elsevier (2002).
[b] See reference (Blevins 1979).

TABLE 4.8

Frequency Parameters, $\mu = L/\pi(\omega\sqrt{\rho A/D})^{1/2}$, for a Clamped-Clamped Beam[a]

	$\mu = L/\pi(\omega\sqrt{\rho A/D})^{1/2}$				
	Cosine Series	Sine Series	Cosine Series	Sine Series	
Mode	M = 10	M = 10	M = 20	M = 20	Exact[b]
1	1.50562	1.50562	1.50562	1.50562	1.50562
2	2.49976	2.4998	2.49975	2.49975	2.49975
3	3.50003	3.50042	3.50001	3.50002	3.50001
4	4.5002	4.50091	4.5	4.50004	4.5

Notes:

[a] This table was taken from the *Journal of Sound and Vibration*, 255, W. L. Li, Comparison of Fourier sine and cosine series expansions for beams with arbitrary boundary conditions, 185–194, Copyright Elsevier (2002).

[b] See reference (Blevins 1979).

The final example concerns with a general boundary condition in which the beam is elastically restrained by both translational and rotational springs at each end. Assuming $\hat{k}_0 L^3 = \hat{k}_1 L^3 = 1$ and $\hat{K}_1 L = \hat{K}_0 L = 100$, the five lowest frequency parameters $\mu = (L^2\omega\sqrt{\rho A/D})^{1/2}$ are listed in Table 4.9. It is seen that the generalized cosine series solution is far better than the generalized sine series solution in this general case.

TABLE 4.9

Frequency Parameters, $\mu = (L^2\omega\sqrt{\rho A/D})^{1/2}$, for a Beam with General Elastic Restraints, $\hat{K}_1 L = \hat{K}_0 L = 100$ and $\hat{k}_1 L^3 = \hat{k}_0 L^3 = 1$ [a]

	$\mu = (L^2\omega\sqrt{\rho A/D})^{1/2}$						
	Sine	Cosine	Sine	Cosine	Sine	Cosine	
Mode	M = 10	M = 10	M = 20	M = 20	M = 40	M = 40	Reference[b]
1	1.20053	1.188301	1.19436	1.188301	1.19131	1.188301	1.188301
2	3.20606	3.14418	3.17581	3.14418	3.16018	3.14418	3.144179
3	6.37557	6.227224	6.29699	6.227221	6.26121	6.22722	6.22722
4	9.5534	9.337013	9.44123	9.336975	9.38844	9.336971	9.336969
5	12.8067	12.45001	12.6042	12.44990	12.5229	12.44988	12.44988

Notes:

[a] This table was taken from the *Journal of Sound and Vibration*, 255, W. L. Li, Comparison of Fourier sine and cosine series expansions for beams with arbitrary boundary conditions, 185–194, Copyright Elsevier (2002).

[b] See reference (Rao and Mirza 1989).

REFERENCES

Blevins, R. D. 1979. *Formulas for Natural Frequency and Mode Shape*. New York: Van Nostrand Reinhold Company.

Greif, R., and S. C. Mittendorf. 1976. Structural vibrations and Fourier series. *Journal of Sound and Vibration* 48(1): 113–122.

Maurizi, M. J., and G. G. Robledo. 1998. A further note on the "Dynamic analysis of generally supported beams using Fourier series". *Journal of Sound and Vibration* 214(5): 972–973.

Maurizi, M. J., R. E. Rossi, and J. A. Reyes. 1976. Vibration frequencies for a uniform beam with one end spring-hinged and subjected to a translational restraint at the other end. *Journal of Sound and Vibration* 48(4): 565–568.

Rao, C. K., and S. Mirza. 1989. A note on vibration of generally restrained beams. *Journal of Sound and Vibration* 130(3): 453–465.

Wang, J. T.-S., and C.-C. Lin. 1996. Dynamic analysis of generally supported beams using Fourier series. *Journal of Sound and Vibration* 196(3): 285–293.

5 Fourier Series for the Derivatives of One-Dimensional Functions

In Sec. 2.3, the simple integration and differentiation of a Fourier series have been briefly discussed with an important observation that the term-by-term differentiation cannot be automatically applied to the Fourier series of a function to obtain the series representation of its derivative. In this chapter, our attention is directed to the Fourier series for the higher order derivatives of a one-dimensional function. In particular, the general relationships will be established among the Fourier series (actually, the Fourier coefficients) for the derivatives of various orders. Such an interest arises from the need of simultaneously expanding a function and its derivatives into Fourier series over a compact interval.

5.1 INTEGRAL FORMULAS REGARDING THE DERIVATIVES OF FUNCTIONS

Let $u(x)$ be a sufficiently smooth function defined on the interval $[-a, a]$ or $[0, a]$. We first define two specific forms of integrals involved with calculating the Fourier coefficients of higher order derivatives:

$$I_C(p, x_0) = \int_{x_0}^{a} u^{(p)}(x) \cos(\alpha_m x) dx \tag{5.1}$$

and

$$I_S(p, x_0) = \int_{x_0}^{a} u^{(p)}(x) \sin(\alpha_m x) dx, \tag{5.2}$$

where $u^{(p)}(x)$ denotes $\frac{d^p u(x)}{dx^p}$, the p-th order derivative of the function $u(x)$.

Theorem 5.1 Let $\alpha_m = m\pi/a$ and $x_0 = -a$ or 0. Then for a positive integer m and nonnegative integer k, the following equations hold:

$I_C(2k, x_0)$

$$= \sum_{j=1}^{k} (-1)^{k-j} \alpha_m^{2k-2j} [(-1)^m u^{(2j-1)}(a) - \cos(\alpha_m x_0) u^{(2j-1)}(x_0)] \tag{5.3}$$

$$+ (-1)^k \alpha_m^{2k} I_C(0, x_0),$$

DOI: 10.1201/9781003194859-5

$$I_C(2k+1, x_0)$$

$$= \sum_{j=0}^{k} (-1)^{k-j} \alpha_m^{2k-2j} [(-1)^m u^{(2j)}(a) - \cos(\alpha_m x_0) u^{(2j)}(x_0)] \tag{5.4}$$

$$+ (-1)^k \alpha_m^{2k+1} I_S(0, x_0),$$

$$I_S(2k, x_0)$$

$$= \sum_{j=1}^{k} (-1)^{k-j+1} \alpha_m^{2k-2j+1} [(-1)^m u^{(2j-2)}(a) - \cos(\alpha_m x_0) u^{(2j-2)}(x_0)] \tag{5.5}$$

$$+ (-1)^k \alpha_m^{2k} I_S(0, x_0),$$

and

$$I_S(2k+1, x_0)$$

$$= \sum_{j=1}^{k} (-1)^{k-j+1} \alpha_m^{2k-2j+1} [(-1)^m u^{(2j-1)}(a) - \cos(\alpha_m x_0) u^{(2j-1)}(x_0)] \tag{5.6}$$

$$+ (-1)^{k+1} \alpha_m^{2k+1} I_C(0, x_0).$$

Proof. Only (5.3) will be proven as an example.

For $k = 0$, it is obvious that (5.3) holds.

For $k > 0$, we perform integration by parts:

$$I_C(2k, x_0) = \int_{x_0}^{a} u^{(2k)}(x) \cos(\alpha_m x) dx = \int_{x_0}^{a} \cos(\alpha_m x) du^{(2k-1)}(x)$$

$$= \cos(\alpha_m x) u^{(2k-1)}(x) \Big|_{x_0}^{a} - \int_{x_0}^{a} [-\alpha_m \sin(\alpha_m x)] u^{(2k-1)}(x) dx$$

$$= [(-1)^m u^{(2k-1)}(a) - \cos(\alpha_m x_0) u^{(2k-1)}(x_0)] + \alpha_m \int_{x_0}^{a} \sin(\alpha_m x) du^{(2k-2)}(x)$$

$$= [(-1)^m u^{(2k-1)}(a) - \cos(\alpha_m x_0) u^{(2k-1)}(x_0)] + \alpha_m \left[\sin(\alpha_m x) u^{(2k-2)}(x) \Big|_{x_0}^{a} \right.$$

$$\left. - \alpha_m \int_{x_0}^{a} \cos(\alpha_m x) u^{(2k-2)}(x) dx \right]$$

$$= [(-1)^m u^{(2k-1)}(a) - \cos(\alpha_m x_0) u^{(2k-1)}(x_0)] - \alpha_m^2 \int_{x_0}^{a} u^{(2k-2)}(x) \cos(\alpha_m x) dx,$$

that is,

$$I_C(2k, x_0) = [(-1)^m u^{(2k-1)}(a) - \cos(\alpha_m x_0) u^{(2k-1)}(x_0)] - \alpha_m^2 I_C(2k-2, x_0). \tag{5.7}$$

Replacing the variable k in (5.7) with $k-1, k-2, \ldots, 2$ and 1 successively, we then have the following equations:

$$I_C(2k-2,x_0) = [(-1)^m u^{(2k-3)}(a) - \cos(\alpha_m x_0)u^{(2k-3)}(x_0)] - \alpha_m^2 I_C(2k-4,x_0), \quad (5.8)$$

$$I_C(2k-4,x_0) = [(-1)^m u^{(2k-5)}(a) - \cos(\alpha_m x_0)u^{(2k-5)}(x_0)] - \alpha_m^2 I_C(2k-6,x_0), \quad (5.9)$$

$$\ldots\ldots,$$

$$I_C(4,x_0) = [(-1)^m u^{(3)}(a) - \cos(\alpha_m x_0)u^{(3)}(x_0)] - \alpha_m^2 I_C(2,x_0), \quad (5.10)$$

$$I_C(2,x_0) = [(-1)^m u^{(1)}(a) - \cos(\alpha_m x_0)u^{(1)}(x_0)] - \alpha_m^2 I_C(0,x_0). \quad (5.11)$$

Multiplying (5.8) by $-\alpha_m^2$, (5.9) by $(-\alpha_m^2)^2$, $\ldots$, (5.10) by $(-\alpha_m^2)^{k-2}$, and (5.11) by $(-\alpha_m^2)^{k-1}$, and adding the results to (5.7) will lead to

$$\begin{aligned}
I_C(2k,x_0) &= \sum_{j=1}^{k}(-\alpha_m^2)^{j-1}[(-1)^m u^{(2k-2j+1)}(a) - \cos(\alpha_m x_0)u^{(2k-2j+1)}(x_0)] \\
&\quad + (-\alpha_m^2)^k I_C(0,x_0) \\
&= \sum_{j=1}^{k}(-1)^{k-j}\alpha_m^{2k-2j}[(-1)^m u^{(2j-1)}(a) - \cos(\alpha_m x_0)u^{(2j-1)}(x_0)] \\
&\quad + (-1)^k \alpha_m^{2k} I_C(0,x_0).
\end{aligned}$$

In the last step, the terms in the summation are rearranged in a sequence from the lowest to the highest order derivatives of the function $u(x)$.

5.2 FOURIER COEFFICIENTS FOR THE DERIVATIVES OF FUNCTIONS

The function $u(x)$ and its derivatives can typically be expanded into one of the three different forms:

1. The function and its derivatives are expanded into full-range Fourier series over the interval $[-a,a]$;
2. The function is expanded into the half-range cosine series over the interval $[0,a]$, and its derivatives are expanded into the half-range Fourier series;
3. The function is expanded into the half-range sine series over the interval $[0,a]$, and its derivatives are expanded into the half-range Fourier series.

Unlike in the previous discussions, the words "full-range" and "half-range" are adopted here when the simple terms, such as, the complete Fourier series, cosine series, and sine series cannot be adequately used to fully specify the natures of the relevant series representations. However, whenever such clarification is unnecessary, those words will be dropped off for simplicity. Actually, this shall not cause much confusion: "full-range" is often associated with the interval $[-a, a]$ and "half-range" with the interval $[0, a]$.

In the (full-range) Fourier series expansion, the function $u(x)$ and its derivatives, $u^{(2k)}(x)$ and $u^{(2k+1)}(x)$, are expressed as

$$u(x) = \sum_{m=0}^{\infty} \mu_m [U_{1m} \cos(\alpha_m x) + U_{2m} \sin(\alpha_m x)], \qquad (5.12)$$

$$u^{(2k)}(x) = \sum_{m=0}^{\infty} \mu_m [U_{1m}^{(2k)} \cos(\alpha_m x) + U_{2m}^{(2k)} \sin(\alpha_m x)], \qquad (5.13)$$

and

$$u^{(2k+1)}(x) = \sum_{m=0}^{\infty} \mu_m [U_{1m}^{(2k+1)} \cos(\alpha_m x) + U_{2m}^{(2k+1)} \sin(\alpha_m x)], \qquad (5.14)$$

where

$$\mu_m = \begin{cases} 1/2, & m = 0 \\ 1, & m > 0 \end{cases},$$

and U_{1m}, U_{2m}, $U_{1m}^{(2k)}$, $U_{2m}^{(2k)}$, $U_{1m}^{(2k+1)}$, and $U_{2m}^{(2k+1)}$ are the Fourier coefficients of function $u(x)$, its derivatives $u^{(2k)}(x)$ and $u^{(2k+1)}(x)$, respectively.

The full-range Fourier series corresponds to the situation of setting $x_0 = -a$ in (5.7) and the recursion formulas between the Fourier coefficients of the derivatives of different orders are accordingly obtained as

$$U_{1m}^{(2j)} = \frac{1}{a}(-1)^m [u^{(2j-1)}(a) - u^{(2j-1)}(-a)] - \alpha_m^2 U_{1m}^{(2j-2)}, \quad j = 1, 2, \cdots, k. \qquad (5.15)$$

More directly, by substituting $x_0 = -a$ into (5.3)–(5.6), Theorem 5.1 can be readily used to establish the relationships among the Fourier coefficients of the function $u(x)$ and its derivatives $u^{(2k)}(x)$ and $u^{(2k+1)}(x)$:

for $m = 0$ and $k = 0$,

$$U_{10}^{(0)} = U_{10}, \qquad (5.16)$$

and

$$U_{10}^{(1)} = \frac{1}{a}[u(a) - u(-a)]; \qquad (5.17)$$

for $m = 0$ and $k > 0$,

$$U_{10}^{(2k)} = \frac{1}{a}[u^{(2k-1)}(a) - u^{(2k-1)}(-a)], \tag{5.18}$$

and

$$U_{10}^{(2k+1)} = \frac{1}{a}[u^{(2k)}(a) - u^{(2k)}(-a)]; \tag{5.19}$$

for $m > 0$ and $k \geq 0$,

$$U_{1m}^{(2k)} = \frac{1}{a}\sum_{j=1}^{k}(-1)^{m+k-j}\alpha_m^{2k-2j}[u^{(2j-1)}(a) - u^{(2j-1)}(-a)] + (-1)^k \alpha_m^{2k}U_{1m}, \tag{5.20}$$

$$U_{2m}^{(2k)} = \frac{1}{a}\sum_{j=1}^{k}(-1)^{m+k-j+1}\alpha_m^{2k-2j+1}[u^{(2j-2)}(a) - u^{(2j-2)}(-a)] + (-1)^k \alpha_m^{2k}U_{2m}, \tag{5.21}$$

$$U_{1m}^{(2k+1)} = \frac{1}{a}\sum_{j=0}^{k}(-1)^{m+k-j}\alpha_m^{2k-2j}[u^{(2j)}(a) - u^{(2j)}(-a)] + (-1)^k \alpha_m^{2k+1}U_{2m}, \tag{5.22}$$

and

$$U_{2m}^{(2k+1)} = \frac{1}{a}\sum_{j=1}^{k}(-1)^{m+k-j+1}\alpha_m^{2k-2j+1}[u^{(2j-1)}(a) - u^{(2j-1)}(-a)] + (-1)^{k+1}\alpha_m^{2k+1}U_{1m}. \tag{5.23}$$

When the function and its derivatives are expanded into the half-range Fourier series, e.g.,

$$u(x) = \sum_{m=0}^{\infty}\mu_m U_{1m}\cos(\alpha_m x), \tag{5.24}$$

$$u^{(2k)}(x) = \sum_{m=0}^{\infty}\mu_m U_{1m}^{(2k)}\cos(\alpha_m x) \tag{5.25}$$

and

$$u^{(2k+1)}(x) = \sum_{m=1}^{\infty}U_{2m}^{(2k+1)}\sin(\alpha_m x), \tag{5.26}$$

we have:

for $m = 0$ and $k = 0$,

$$U_{10}^{(0)} = U_{10}; \tag{5.27}$$

for $m = 0$ and $k > 0$,

$$U_{10}^{(2k)} = \frac{2}{a}[u^{(2k-1)}(a) - u^{(2k-1)}(0)]; \tag{5.28}$$

for $m > 0$ and $k \geq 0$,

$$U_{1m}^{(2k)} = \frac{2}{a}\sum_{j=1}^{k}(-1)^{k-j}\alpha_m^{2k-2j}[(-1)^m u^{(2j-1)}(a) - u^{(2j-1)}(0)] + (-1)^k \alpha_m^{2k}U_{1m} \tag{5.29}$$

and

$$U_{2m}^{(2k+1)} = \frac{2}{a}\sum_{j=1}^{k}(-1)^{k-j+1}\alpha_m^{2k-2j+1}[(-1)^m u^{(2j-1)}(a) - u^{(2j-1)}(0)] + (-1)^{k+1}\alpha_m^{2k+1}U_{1m}. \tag{5.30}$$

The above equations can be readily obtained from Theorem 5.1 by substituting $x_0 = 0$ into (5.3)–(5.6).

Similarly, if the function and its derivatives are expanded into the half-range Fourier series as:

$$u(x) = \sum_{m=1}^{\infty}U_{2m}\sin(\alpha_m x), \tag{5.31}$$

$$u^{(2k)}(x) = \sum_{m=1}^{\infty}U_{2m}^{(2k)}\sin(\alpha_m x) \tag{5.32}$$

and

$$u^{(2k+1)}(x) = \sum_{m=0}^{\infty}\mu_m U_{1m}^{(2k+1)}\cos(\alpha_m x), \tag{5.33}$$

then we accordingly have:

for $m = 0$ and $k \geq 0$,

$$U_{10}^{(2k+1)} = \frac{2}{a}[u^{(2k)}(a) - u^{(2k)}(0)]; \tag{5.34}$$

for $m > 0$ and $k \geq 0$,

$$U_{2m}^{(2k)} = \frac{2}{a} \sum_{j=1}^{k} (-1)^{k-j+1} \alpha_m^{2k-2j+1} [(-1)^m u^{(2j-2)}(a) - u^{(2j-2)}(0)] + (-1)^k \alpha_m^{2k} U_{2m}$$

(5.35)

and

$$U_{1m}^{(2k+1)} = \frac{2}{a} \sum_{j=0}^{k} (-1)^{k-j} \alpha_m^{2k-2j} [(-1)^m u^{(2j)}(a) - u^{(2j)}(0)] + (-1)^k \alpha_m^{2k+1} U_{2m}.$$ (5.36)

The above results show that the Fourier coefficients for any given order derivative can be calculated directly from those for the function $u(x)$ and the boundary values of the lower order derivatives. It must be pointed out that calculating the boundary values of the lower order derivatives is not equivalent to knowing their Fourier coefficients unless the corresponding Fourier series expansions are known to converge correctly at the boundary points.

5.3 EXAMPLES

In this section, the formulas above derived for calculating the Fourier coefficients of higher order derivatives will be verified by two different sample functions: polynomials and trigonometric functions.

EXAMPLE 5.1   Determine the Fourier series for the first three derivatives of the polynomial function $u(x) = Ax^2 + Bx + C$ ($-\pi < x < \pi$) where A, B, and C are constants.

This function was previously studied in Sec. 2.1, and its full-range Fourier series is known as

$$u(x) = \frac{A\pi^2}{3} + C + 4A \sum_{m=1}^{\infty} \frac{(-1)^m}{m^2} \cos mx - 2B \sum_{m=1}^{\infty} \frac{(-1)^m}{m} \sin mx.$$ (5.37)

As a cross-check, we will take three different approaches.

First, the expressions for the first three derivatives will be given explicitly and each of them is subsequently expanded into the Fourier series directly. The three derivatives of $u(x) = Ax^2 + Bx + C$ are:

$$u^{(1)}(x) = 2Ax + B,$$ (5.38)

$$u^{(2)}(x) = 2A,$$ (5.39)

and

$$u^{(3)}(x) = 0.$$ (5.40)

The second and third derivatives, (5.39) and (5.40), are already in the form of Fourier series. In light of (2.13), the Fourier series of $u^{(1)}(x)$ in (5.38) is readily available

$$u^{(1)}(x) = B - 4A \sum_{m=1}^{\infty} \frac{(-1)^m}{m} \sin mx. \qquad (5.41)$$

In the second approach, we successively apply the formula (2.49) to determine the Fourier series expansions of $u^{(1)}(x)$ and $u^{(2)}(x)$.

According to (5.37), the Fourier coefficients of the function $u(x)$ are

$$U_{10} = \frac{2A\pi^2}{3} + 2C, U_{1m} = \frac{4A(-1)^m}{m^2}, \text{ and } U_{2m} = -\frac{2B(-1)^m}{m}. \qquad (5.42)$$

By making use of the formula (2.49), the Fourier coefficients of the first order derivative $u^{(1)}(x)$ can be calculated as

$$U_{10}^{(1)} = \frac{1}{\pi}[u(\pi) - u(-\pi)] = \frac{1}{\pi}[B\pi - B(-\pi)] = 2B = c, \qquad (5.43)$$

$$U_{1m}^{(1)} = (-1)^m c + mU_{2m} = (-1)^m \cdot (2B) - m \cdot \frac{2B(-1)^m}{m} = 0, \qquad (5.44)$$

and

$$U_{2m}^{(1)} = -mU_{1m} = -m \cdot \frac{4A(-1)^m}{m^2} = -\frac{4A(-1)^m}{m}. \qquad (5.45)$$

Using (2.49) again, the Fourier coefficients of the derivative of $u^{(1)}(x)$, namely $u^{(2)}(x)$, can be obtained as

$$U_{10}^{(2)} = \frac{1}{\pi}[u^{(1)}(\pi) - u^{(1)}(-\pi)] = \frac{1}{\pi}[2A\pi - 2A(-\pi)] = 4A = c, \qquad (5.46)$$

$$U_{1m}^{(2)} = (-1)^m c + mU_{2m}^{(1)} = (-1)^m \cdot (4A) - m \cdot \frac{4A(-1)^m}{m} = 0, \qquad (5.47)$$

and

$$U_{2m}^{(2)} = -mU_{1m}^{(1)} = -m \cdot 0 = 0. \qquad (5.48)$$

Similarly, the Fourier coefficients of the derivative of $u^{(2)}(x)$, namely $u^{(3)}(x)$, can be obtained as

$$U_{10}^{(3)} = \frac{1}{\pi}[u^{(2)}(\pi) - u^{(2)}(-\pi)] = \frac{1}{\pi}[2A - 2A] = 0 = c, \qquad (5.49)$$

$$U_{1m}^{(3)} = (-1)^m c + mU_{2m}^{(2)} = (-1)^m \cdot 0 + m \cdot 0 = 0, \qquad (5.50)$$

and

$$U_{2m}^{(3)} = -mU_{1m}^{(2)} = -m \cdot 0 = 0. \tag{5.51}$$

Consequently, the Fourier series (5.39)–(5.41) are again obtained for the first three derivatives.

Finally, the general formulas, (5.17)–(5.23), will be used to calculate the Fourier coefficients for the first three derivatives of the function $u(x)$:

for $m = 0$,

$$U_{10}^{(1)} = \frac{1}{\pi}[u(\pi) - u(-\pi)] = 2B, \tag{5.52}$$

$$U_{10}^{(2)} = \frac{1}{\pi}[u^{(1)}(\pi) - u^{(1)}(-\pi)] = 4A \tag{5.53}$$

and

$$U_{10}^{(3)} = \frac{1}{\pi}[u^{(2)}(\pi) - u^{(2)}(-\pi)] = 0; \tag{5.54}$$

for $m > 0$,

$$U_{1m}^{(1)} = \frac{1}{\pi}\sum_{j=0}^{0}(-1)^{m-j}m^{-2j}[u^{(2j)}(\pi) - u^{(2j)}(-\pi)] + mU_{2m}$$

$$= \frac{1}{\pi}(-1)^{m}[u(\pi) - u(-\pi)] + mU_{2m} \tag{5.55}$$

$$= 0,$$

$$U_{2m}^{(1)} = \frac{1}{\pi}\sum_{j=1}^{0}(-1)^{m-j+1}m^{-2j+1}[u^{(2j-1)}(\pi) - u^{(2j-1)}(-\pi)] + (-1) \cdot mU_{1m}$$

$$= -mU_{1m} = -\frac{4A(-1)^{m}}{m}, \tag{5.56}$$

$$U_{1m}^{(2)} = \frac{1}{\pi}\sum_{j=1}^{1}(-1)^{m+1-j}m^{2-2j}[u^{(2j-1)}(\pi) - u^{(2j-1)}(-\pi)] + (-1) \cdot m^{2}U_{1m}$$

$$= \frac{1}{\pi}(-1)^{m}[u^{(1)}(\pi) - u^{(1)}(-\pi)] + (-1) \cdot m^{2}U_{1m} \tag{5.57}$$

$$= \frac{1}{\pi}(-1)^{m}[2A\pi - 2A(-\pi)] + (-1) \cdot m^{2} \cdot \frac{4A(-1)^{m}}{m^{2}}$$

$$= 0,$$

$$U_{2m}^{(2)} = \frac{1}{\pi} \sum_{j=1}^{1} (-1)^{m+1-j+1} m^{2-2j+1} [u^{(2j-2)}(\pi) - u^{(2j-2)}(-\pi)] + (-1) \cdot m^2 U_{2m}$$

$$= \frac{1}{\pi} (-1)^{m+1} m[u(\pi) - u(-\pi)] + (-1) \cdot m^2 U_{2m} \qquad (5.58)$$

$$= \frac{1}{\pi} (-1)^{m+1} m[B\pi - B(-\pi)] + m^2 \cdot \frac{2B(-1)^m}{m}$$

$$= 0,$$

$$U_{1m}^{(3)} = \frac{1}{\pi} \sum_{j=0}^{1} (-1)^{m+1-j} m^{2-2j} [u^{(2j)}(\pi) - u^{(2j)}(-\pi)] + (-1) \cdot m^3 U_{2m}$$

$$= \frac{1}{\pi} (-1)^{m+1} m^2 [u(\pi) - u(-\pi)] + \frac{1}{\pi} (-1)^m [u^{(2)}(\pi) - u^{(2)}(-\pi)] + (-1) \cdot m^3 U_{2m} \quad (5.59)$$

$$= \frac{1}{\pi} (-1)^{m+1} m^2 [B\pi - B(-\pi)] + \frac{1}{\pi} (-1)^m [2A - 2A] + m^3 \cdot \frac{2B(-1)^m}{m}$$

$$= 0,$$

and

$$U_{2m}^{(3)} = \frac{1}{\pi} \sum_{j=1}^{1} (-1)^{m+1-j+1} m^{2-2j+1} [u^{(2j-1)}(\pi) - u^{(2j-1)}(-\pi)] + (-1)^2 m^3 U_{1m}$$

$$= \frac{1}{\pi} (-1)^{m+1} m[2A\pi - 2A(-\pi)] + (-1)^2 m^3 U_{1m} \qquad (5.60)$$

$$= \frac{1}{\pi} (-1)^{m+1} m[2A\pi - 2A(-\pi)] + m^3 \cdot \frac{4A(-1)^m}{m^2}$$

$$= 0,$$

Substituting these Fourier coefficients into (5.13) and (5.14) will result in the same series expansions as those just obtained from the other two approaches.

The next two examples will only involve using the general formulas to calculate the Fourier coefficients of the derivatives.

EXAMPLE 5.2   Let $u(x) = \sin(\alpha_0 x)$, $\alpha_0 = \pi/2a$, be defined on the interval $[-a, a]$. Find the full-range Fourier series for its derivatives.

The function $u(x) = \sin(\alpha_0 x)$ is odd on the interval, and its full-range Fourier series is given as

$$u(x) = \sum_{m=0}^{\infty} \mu_m [U_{1m} \cos(\alpha_m x) + U_{2m} \sin(\alpha_m x)], \qquad (5.61)$$

where $\alpha_m = m\pi/a$, $\mu_m = \begin{cases} 1/2, & m = 0 \\ 1, & m > 0 \end{cases}$,

$$U_{1m} \equiv 0, \tag{5.62}$$

and

$$U_{2m} = \frac{2\alpha_m}{a(\alpha_m^2 - \alpha_0^2)}(-1)^{m+1}\sin(\alpha_0 a). \tag{5.63}$$

The derivatives $u^{(2k)}(x)$ and $u^{(2k+1)}(x)$ are assumed to be expanded into the full-range Fourier series as

$$u^{(2k)}(x) = \sum_{m=0}^{\infty} \mu_m[U_{1m}^{(2k)}\cos(\alpha_m x) + U_{2m}^{(2k)}\sin(\alpha_m x)], \tag{5.64}$$

and

$$u^{(2k+1)}(x) = \sum_{m=0}^{\infty} \mu_m[U_{1m}^{(2k+1)}\cos(\alpha_m x) + U_{2m}^{(2k+1)}\sin(\alpha_m x)], \tag{5.65}$$

where k is a nonnegative integer.

For the given function $u(x) = \sin(\alpha_0 x)$, its derivatives are readily obtained as

$$u^{(2k)}(x) = (-1)^k \alpha_0^{2k}\sin(\alpha_0 x) \quad (k = 1, 2, \cdots), \tag{5.66}$$

$$u^{(2k+1)}(x) = (-1)^k \alpha_0^{2k+1}\cos(\alpha_0 x) \quad (k = 0, 1, 2, \cdots) \tag{5.67}$$

and accordingly,

$$u^{(2k)}(a) - u^{(2k)}(-a) = 2 \cdot (-1)^k \alpha_0^{2k}\sin(\alpha_0 a), \tag{5.68}$$

$$u^{(2k+1)}(a) - u^{(2k+1)}(-a) = 0. \tag{5.69}$$

By making use of the general formulas for the Fourier coefficients of the derivatives, we have:

for $k = 1, 2, \cdots$,

$$U_{10}^{(2k)} = \frac{1}{a}[u^{(2k-1)}(a) - u^{(2k-1)}(-a)] = 0, \tag{5.70}$$

$$
\begin{aligned}
U_{1m}^{(2k)} &= \frac{1}{a}\sum_{j=1}^{k}(-1)^{m+k-j}\alpha_m^{2k-2j}[u^{(2j-1)}(a) - u^{(2j-1)}(-a)] + (-1)^k \alpha_m^{2k}U_{1m} \\
&= \frac{1}{a}\sum_{j=1}^{k}(-1)^{m+k-j}\alpha_m^{2k-2j} \cdot 0 + (-1)^k \alpha_m^{2k} \cdot 0 \\
&= 0,
\end{aligned} \tag{5.71}
$$

and

$$U_{2m}^{(2k)} = \frac{1}{a}\sum_{j=1}^{k}(-1)^{m+k-j+1}\alpha_m^{2k-2j+1}[u^{(2j-2)}(a) - u^{(2j-2)}(-a)] + (-1)^k\alpha_m^{2k}U_{2m}$$

$$= \frac{1}{a}\sum_{j=1}^{k}(-1)^{m+k-j+1}\alpha_m^{2k-2j+1}\cdot 2\cdot(-1)^{j-1}\alpha_0^{2j-2}\sin(\alpha_0 a)$$

$$+ (-1)^k\alpha_m^{2k}\cdot\frac{2\alpha_m}{a(\alpha_m^2 - \alpha_0^2)}(-1)^{m+1}\sin(\alpha_0 a)$$

$$= \frac{2}{a}(-1)^{m+k}\alpha_m^{2k-1}\sin(\alpha_0 a)\cdot\sum_{j=1}^{k}\left(\frac{\alpha_0}{\alpha_m}\right)^{2j-2} + (-1)^k\alpha_m^{2k}\cdot\frac{2\alpha_m}{a(\alpha_m^2 - \alpha_0^2)}(-1)^{m+1}\sin(\alpha_0 a)$$

$$= \frac{2}{a}(-1)^{m+k}\sin(\alpha_0 a)\cdot\frac{\alpha_m^{2k} - \alpha_0^{2k}}{\alpha_m^2 - \alpha_0^2}\cdot\alpha_m + (-1)^k\alpha_m^{2k}\cdot\frac{2\alpha_m}{a(\alpha_m^2 - \alpha_0^2)}(-1)^{m+1}\sin(\alpha_0 a)$$

$$= (-1)^{m+k+1}\alpha_0^{2k}\cdot\frac{2\alpha_m}{a(\alpha_m^2 - \alpha_0^2)}\sin(\alpha_0 a);$$

$$(5.72)$$

for $k = 0, 1, 2, \cdots$,

$$U_{10}^{(2k+1)} = \frac{1}{a}[u^{(2k)}(a) - u^{(2k)}(-a)] = \frac{1}{a}[2\cdot(-1)^k\alpha_0^{2k}\sin(\alpha_0 a)] = \frac{2}{a}(-1)^k\alpha_0^{2k}\sin(\alpha_0 a),$$

$$(5.73)$$

$$U_{1m}^{(2k+1)} = \frac{1}{a}\sum_{j=0}^{k}(-1)^{m+k-j}\alpha_m^{2k-2j}[u^{(2j)}(a) - u^{(2j)}(-a)] + (-1)^k\alpha_m^{2k+1}U_{2m}$$

$$= \frac{1}{a}\sum_{j=0}^{k}(-1)^{m+k-j}\alpha_m^{2k-2j}[2\cdot(-1)^j\alpha_0^{2j}\sin(\alpha_0 a)] + (-1)^k\alpha_m^{2k+1}U_{2m}$$

$$= \frac{2}{a}(-1)^{m+k}\alpha_m^{2k}\sin(\alpha_0 a)\cdot\sum_{j=0}^{k}\left(\frac{\alpha_0}{\alpha_m}\right)^{2j} + (-1)^k\alpha_m^{2k+1}\cdot\frac{2\alpha_m}{a(\alpha_m^2 - \alpha_0^2)}(-1)^{m+1}\sin(\alpha_0 a)$$

$$= \frac{2}{a}(-1)^{m+k}\cdot\frac{\alpha_m^{2k+2} - \alpha_0^{2k+2}}{\alpha_m^2 - \alpha_0^2}\sin(\alpha_0 a) + (-1)^{m+k+1}\cdot\frac{2\alpha_m^{2k+2}}{a(\alpha_m^2 - \alpha_0^2)}\sin(\alpha_0 a)$$

$$= \frac{2}{a}(-1)^{m+k+1}\cdot\frac{\alpha_0^{2k+2}}{\alpha_m^2 - \alpha_0^2}\sin(\alpha_0 a),$$

$$(5.74)$$

and

$$U_{2m}^{(2k+1)} = \frac{1}{a}\sum_{j=1}^{k}(-1)^{m+k-j+1}\alpha_m^{2k-2j+1}[u^{(2j-1)}(a) - u^{(2j-1)}(-a)] + (-1)^{k+1}\alpha_m^{2k+1}U_{1m}$$

$$= \frac{1}{a}\sum_{j=1}^{k}(-1)^{m+k-j+1}\alpha_m^{2k-2j+1}\cdot 0 + (-1)^{k+1}\alpha_m^{2k+1}\cdot 0$$

$$(5.75)$$

$$= 0.$$

EXAMPLE 5.3   Let $u(x) = \cos(\alpha_0 x)$, $\alpha_0 = \pi/2a$, be expanded into a half-range sine series on the interval $[0, a]$. Determine the Fourier coefficients for all its derivatives.

Assume that

$$u(x) = \sum_{m=1}^{\infty} U_{2m} \sin(\alpha_m x), \qquad (5.76)$$

$$u^{(2k)}(x) = \sum_{m=1}^{\infty} U_{2m}^{(2k)} \sin(\alpha_m x), \qquad (5.77)$$

and

$$u^{(2k+1)}(x) = \sum_{m=0}^{\infty} \mu_m U_{1m}^{(2k+1)} \cos(\alpha_m x), \qquad (5.78)$$

where k are nonnegative integers, $\alpha_m = m\pi/a$, and $\mu_m = \begin{cases} 1/2, & m = 0 \\ 1, & m > 0 \end{cases}$.

It is useful to note that the Fourier coefficients of the function $u(x)$ are

$$U_{2m} = \frac{2\alpha_m}{a(\alpha_m^2 - \alpha_0^2)}[1 - (-1)^m \cos(\alpha_0 a)]. \qquad (5.79)$$

The derivatives of $u(x)$ are readily written as

$$u^{(2k)}(x) = (-1)^k \alpha_0^{2k} \cos(\alpha_0 x), \qquad (5.80)$$

and

$$u^{(2k+1)}(x) = (-1)^{k+1} \alpha_0^{2k+1} \sin(\alpha_0 x). \qquad (5.81)$$

It follows that

$$(-1)^m u^{(2k)}(a) - u^{(2k)}(0) = (-1)^k \alpha_0^{2k}[(-1)^m \cos(\alpha_0 a) - 1], \quad k = 0, 1, 2, \cdots. \qquad (5.82)$$

By using the general formulas for the Fourier coefficients of derivatives, we have:
for $k = 0, 1, 2, \cdots$,

$$U_{10}^{(2k+1)} = \frac{2}{a}[u^{(2k)}(a) - u^{(2k)}(0)] = \frac{2}{a}(-1)^{k+1} \alpha_0^{2k}[1 - \cos(\alpha_0 a)], \qquad (5.83)$$

and

$$U_{1m}^{(2k+1)} = \frac{2}{a} \sum_{j=0}^{k} (-1)^{k-j} \alpha_m^{2k-2j} [(-1)^m u^{(2j)}(a) - u^{(2j)}(0)] + (-1)^k \alpha_m^{2k+1} U_{2m}$$

$$= \frac{2}{a} \sum_{j=0}^{k} (-1)^{k-j} \alpha_m^{2k-2j} \cdot (-1)^j \alpha_0^{2j} [(-1)^m \cos(\alpha_0 a) - 1] + (-1)^k \alpha_m^{2k+1} U_{2m}$$

$$= \frac{2}{a} (-1)^k \alpha_m^{2k} [(-1)^m \cos(\alpha_0 a) - 1] \sum_{j=0}^{k} \left(\frac{\alpha_0}{\alpha_m}\right)^{2j}$$

$$+ (-1)^k \alpha_m^{2k+1} \cdot \frac{2}{a} \frac{\alpha_m}{\alpha_m^2 - \alpha_0^2} [1 - (-1)^m \cos(\alpha_0 a)]$$

$$= \frac{2}{a} (-1)^k [(-1)^m \cos(\alpha_0 a) - 1] \frac{\alpha_m^{2k+2} - \alpha_0^{2k+2}}{\alpha_m^2 - \alpha_0^2}$$ (5.84)

$$+ (-1)^k \alpha_m^{2k+1} \cdot \frac{2}{a} \frac{\alpha_m}{\alpha_m^2 - \alpha_0^2} [1 - (-1)^m \cos(\alpha_0 a)]$$

$$= \frac{2}{a} (-1)^k [1 - (-1)^m \cos(\alpha_0 a)] \frac{\alpha_0^{2k+2} - \alpha_m^{2k+2}}{\alpha_m^2 - \alpha_0^2}$$

$$+ (-1)^k \alpha_m^{2k+1} \cdot \frac{2}{a} \frac{\alpha_m}{\alpha_m^2 - \alpha_0^2} [1 - (-1)^m \cos(\alpha_0 a)]$$

$$= \frac{2}{a} (-1)^k \frac{\alpha_0^{2k+2}}{\alpha_m^2 - \alpha_0^2} [1 - (-1)^m \cos(\alpha_0 a)];$$

for $k = 1, 2, \cdots,$

$$U_{2m}^{(2k)} = \frac{2}{a} \sum_{j=1}^{k} (-1)^{k-j+1} \alpha_m^{2k-2j+1} [(-1)^m u^{(2j-2)}(a) - u^{(2j-2)}(0)] + (-1)^k \alpha_m^{2k} U_{2m}$$

$$= \frac{2}{a} \sum_{j=1}^{k} (-1)^{k-j+1} \alpha_m^{2k-2j+1} \cdot (-1)^{j-1} \alpha_0^{2j-2} [(-1)^m \cos(\alpha_0 a) - 1]$$

$$+ (-1)^k \alpha_m^{2k} U_{2m}$$

$$= \frac{2}{a} (-1)^k \alpha_m^{2k-1} [(-1)^m \cos(\alpha_0 a) - 1] \sum_{j=1}^{k} \left(\frac{\alpha_0}{\alpha_m}\right)^{2j-2}$$

$$+ (-1)^k \alpha_m^{2k} \cdot \frac{2}{a} \frac{\alpha_m}{\alpha_m^2 - \alpha_0^2} [1 - (-1)^m \cos(\alpha_0 a)]$$

$$= \frac{2}{a} (-1)^k \alpha_m [(-1)^m \cos(\alpha_0 a) - 1] \frac{\alpha_m^{2k} - \alpha_0^{2k}}{\alpha_m^2 - \alpha_0^2}$$ (5.85)

$$+ (-1)^k \alpha_m^{2k} \cdot \frac{2}{a} \frac{\alpha_m}{\alpha_m^2 - \alpha_0^2} [1 - (-1)^m \cos(\alpha_0 a)]$$

$$= \frac{2}{a} (-1)^k \alpha_m [1 - (-1)^m \cos(\alpha_0 a)] \frac{\alpha_0^{2k} - \alpha_m^{2k}}{\alpha_m^2 - \alpha_0^2}$$

$$+ (-1)^k \alpha_m^{2k} \cdot \frac{2}{a} \frac{\alpha_m}{\alpha_m^2 - \alpha_0^2} [1 - (-1)^m \cos(\alpha_0 a)]$$

$$= \frac{2}{a} (-1)^k \frac{\alpha_m \alpha_0^{2k}}{\alpha_m^2 - \alpha_0^2} [1 - (-1)^m \cos(\alpha_0 a)].$$

5.4 SUFFICIENT CONDITIONS FOR THE TERM-BY-TERM DIFFERENTIATIONS OF FOURIER SERIES

The above discussions have described a framework for directly determining the Fourier series for the higher order derivatives of a function. In the meantime, it also becomes evident that the applications of these general formulas require the specific knowledge about the values of the function and the relevant lower order derivatives at the boundary points of the interval over which the function is defined. This may not actually be a trivial task in the context of, such as, solving general differential equations or boundary value problems.

By now, it should have become clear that the difficulties associated with determining the Fourier coefficients for higher order derivatives are primarily caused by those terms which require the explicit calculations, if possible at all, of the boundary values of the lower order derivatives. Instead of taking a head-on approach to attack the problem, alternative options will be pursued here, and their feasibilities are based on the following theorems.

Theorem 5.2 Let the function $u(x) \in C^{2r-1}([-a,a])$ and its $2r$-th derivative be absolutely integrable. If $u(x)$ is given in the form of the Fourier series over the interval $[-a,a]$, then for any integer $p = 1,2,\cdots,2r$, the sufficient condition for this series to be termwise differentiable for the corresponding series of $u^{(p)}(x)$ is that

$$u^{(k)}(a) = u^{(k)}(-a), k = 0,1,\cdots,2r-1. \tag{5.86}$$

Proof. If (5.86) are simultaneously satisfied, the terms in (5.17)–(5.23) which involve the boundary values of lower order derivatives will vanish all together, which is mathematically equivalent to the applicability of the term-by-term differential operations.

In light of (5.28)–(5.30) and (5.34)–(5.36), similar theorems exist regarding the half-range Fourier series expansions.

Theorem 5.3 Let the function $u(x) \in C^{2r-1}([0,a])$ and its $2r$-th derivative be absolutely integrable. If $u(x)$ is given in the form of the Fourier cosine series over the interval $[0,a]$, then for any integer $p = 1,2,\cdots,2r$, the sufficient condition for this series to be termwise differentiable for the corresponding series of $u^{(p)}(x)$ is that

$$u^{(2k+1)}(a) = 0, u^{(2k+1)}(0) = 0, k = 0,1,\cdots,r-1. \tag{5.87}$$

Theorem 5.4 Let the function $u(x) \in C^{2r-1}([0,a])$ and its $2r$-th derivative be absolutely integrable. If $u(x)$ is given in the form of the Fourier sine series over the interval $[0,a]$, then for any integer $p = 1,2,\cdots,2r$, the sufficient condition for this series to be termwise differentiable for the corresponding series of $u^{(p)}(x)$ is that

$$u^{(2k)}(a) = 0, u^{(2k)}(0) = 0, k = 0, 1, \cdots, r - 1. \tag{5.88}$$

The implications and realizations of these theorems will be discussed later in applying the Fourier series methods to solving linear differential equations and boundary value problems.

6 Fourier Series for the Partial Derivatives of Two-Dimensional Functions

As a continuation to the previous discussions, we now turn to the Fourier series for the partial derivatives of two-dimensional functions. In comparison with the one-dimensional cases, the general formulas for calculating the Fourier coefficients of the partial derivatives will here become more complicated and tedious, but they are still quite manageable due to the similarities to the previous ones. More importantly, these formulas will be useful in establishing the sufficient conditions for the term-by-term differentiations of the corresponding Fourier series for the partial derivatives of two-dimensional functions.

6.1 INTEGRAL FORMULAS REGARDING HIGHER ORDER PARTIAL DERIVATIVES OF TWO-DIMENSIONAL FUNCTIONS

First, define four basic integrals:

$$I_{CC}(p_1, p_2, x_0, y_0) = \int_{x_0}^{a}\int_{y_0}^{b} u^{(p_1, p_2)}(x, y)\cos(\alpha_m x)\cos(\beta_n y)dxdy, \tag{6.1}$$

$$I_{SC}(p_1, p_2, x_0, y_0) = \int_{x_0}^{a}\int_{y_0}^{b} u^{(p_1, p_2)}(x, y)\sin(\alpha_m x)\cos(\beta_n y)dxdy, \tag{6.2}$$

$$I_{CS}(p_1, p_2, x_0, y_0) = \int_{x_0}^{a}\int_{y_0}^{b} u^{(p_1, p_2)}(x, y)\cos(\alpha_m x)\sin(\beta_n y)dxdy, \tag{6.3}$$

and

$$I_{SS}(p_1, p_2, x_0, y_0) = \int_{x_0}^{a}\int_{y_0}^{b} u^{(p_1, p_2)}(x, y)\sin(\alpha_m x)\sin(\beta_n y)dxdy, \tag{6.4}$$

DOI: 10.1201/9781003194859-6

where $u^{(p_1,p_2)}(x,y)$ denotes $\frac{\partial^{p_1+p_2} u(x,y)}{\partial x^{p_1} \partial y^{p_2}}$, the (p_1+p_2)-th order partial derivative of the function $u(x,y)$.

Theorem 6.1 *Let m and n be positive integers, and k_1 and k_2 nonnegative integers. Then the following equation holds*

$$I_{CC}(2k_1, 2k_2, x_0, y_0)$$

$$= \sum_{j=1}^{k_1}\sum_{l=1}^{k_2} (-1)^{k_1+k_2-j-l} \alpha_m^{2k_1-2j} \beta_n^{2k_2-2l} \cdot [(-1)^{m+n} u^{(2j-1,2l-1)}(a,b)$$

$$- (-1)^m \cos(\beta_n y_0) u^{(2j-1,2l-1)}(a,y_0) - (-1)^n \cos(\alpha_m x_0) u^{(2j-1,2l-1)}(x_0,b)$$

$$+ \cos(\alpha_m x_0)\cos(\beta_n y_0) u^{(2j-1,2l-1)}(x_0,y_0)]$$

$$+ \sum_{j=1}^{k_1} (-1)^{k_1+k_2-j} \alpha_m^{2k_1-2j} \beta_n^{2k_2} \int_{y_0}^{b} [(-1)^m u^{(2j-1,0)}(a,y)$$

$$- \cos(\alpha_m x_0) u^{(2j-1,0)}(x_0,y)]\cos(\beta_n y)dy$$

$$+ \sum_{l=1}^{k_2} (-1)^{k_1+k_2-l} \alpha_m^{2k_1} \beta_n^{2k_2-2l} \int_{x_0}^{a} [(-1)^n u^{(0,2l-1)}(x,b)$$

$$- \cos(\beta_n y_0) u^{(0,2l-1)}(x,y_0)]\cos(\alpha_m x)dx + (-1)^{k_1+k_2} \alpha_m^{2k_1} \beta_n^{2k_2} I_{CC}(0,0,x_0,y_0),$$

$$(6.5)$$

where $x_0 = -a$ or 0, $y_0 = -b$ or 0, $\alpha_m = m\pi/a$, and $\beta_n = n\pi/b$.

Proof. For $k_1 = 0$ and $k_2 = 0$, (6.5) is obviously true.

For $k_1 > 0$ and $k_2 = 0$, making use of (5.3) leads to

$$I_{CC}(2k_1, 0, x_0, y_0) = \int_{x_0}^{a}\int_{y_0}^{b} u^{(2k_1,0)}(x,y)\cos(\alpha_m x)\cos(\beta_n y)dxdy$$

$$= \int_{y_0}^{b} \cos(\beta_n y)dy \int_{x_0}^{a} u^{(2k_1,0)}(x,y)\cos(\alpha_m x)dx$$

$$= \int_{y_0}^{b} \cos(\beta_n y)dy \left\{ \sum_{j=1}^{k_1} (-1)^{k_1-j} \alpha_m^{2k_1-2j} [(-1)^m u^{(2j-1,0)}(a,y) \right.$$

$$\left. - \cos(\alpha_m x_0) u^{(2j-1,0)}(x_0,y)] + (-1)^{k_1} \alpha_m^{2k_1} \int_{x_0}^{a} u(x,y)\cos(\alpha_m x)dx \right\},$$

$$(6.6)$$

which is the special case of (6.5).

Similarly, (6.5) applies to the cases of $k_1 = 0$ and $k_2 > 0$.

For $k_1 > 0$ and $k_2 > 0$, the repetitive applications of (5.3) to both x and y integrations will result in

$$I_{CC}(2k_1, 2k_2, x_0, y_0)$$

$$= \int_{x_0}^{a} \int_{y_0}^{b} u^{(2k_1, 2k_2)}(x, y) \cos(\alpha_m x) \cos(\beta_n y) \, dx \, dy$$

$$= \int_{x_0}^{a} \cos(\alpha_m x) \, dx \left\{ \sum_{l=1}^{k_2} (-1)^{k_2 - l} \beta_n^{2k_2 - 2l} [(-1)^n u^{(2k_1, 2l-1)}(x, b) \right.$$

$$\left. - \cos(\beta_n y_0) u^{(2k_1, 2l-1)}(x, y_0)] + (-1)^{k_2} \beta_n^{2k_2} \int_{y_0}^{b} u^{(2k_1, 0)}(x, y) \cos(\beta_n y) dy \right\}$$

$$= \sum_{l=1}^{k_2} (-1)^{k_2 - l} \beta_n^{2k_2 - 2l} \left\{ \sum_{j=1}^{k_1} (-1)^{k_1 - j} \alpha_m^{2k_1 - 2j} [(-1)^{n+m} u^{(2j-1, 2l-1)}(a, b) \right.$$

$$+ (-1)^{m+1} \cos(\beta_n y_0) u^{(2j-1, 2l-1)}(a, y_0) + (-1)^{n+1} \cos(\alpha_m x_0) u^{(2j-1, 2l-1)}(x_0, b)$$

$$+ \cos(\alpha_m x_0) \cos(\beta_n y_0) u^{(2j-1, 2l-1)}(x_0, y_0)]$$

$$\left. + (-1)^{k_1} \alpha_m^{2k_1} \int_{x_0}^{a} [(-1)^n u^{(0, 2l-1)}(x, b) - \cos(\beta_n y_0) u^{(0, 2l-1)}(x, y_0)] \cos(\alpha_m x) dx \right\}$$

$$+ (-1)^{k_2} \beta_n^{2k_2} \int_{y_0}^{b} \cos(\beta_n y) \, dy \left\{ \sum_{j=1}^{k_1} (-1)^{k_1 - j} \alpha_m^{2k_1 - 2j} [(-1)^m u^{(2j-1, 0)}(a, y) \right.$$

$$\left. - \cos(\alpha_m x_0) u^{(2j-1, 0)}(x_0, y)] + (-1)^{k_1} \alpha_m^{2k_1} \int_{x_0}^{a} u(x, y) \cos(\alpha_m x) \, dx \right\}, \qquad (6.7)$$

which can be easily rearranged into (6.5).

In (6.5), the dual-orders, (p_1, p_2), of the partial derivatives are both assumed to be even numbers. Obviously, in parallel with (6.5) there exist three other possible parity combinations which call for different expressions. For conciseness, the corresponding results, together with those for the other integrals in (6.2)–(6.4), are all given in Appendix A placed at the end of this chapter.

6.2 FOURIER COEFFICIENTS FOR PARTIAL DERIVATIVES OF TWO-DIMENSIONAL FUNCTIONS

Using the integral formula (6.5) and the likes in Appendix A, the Fourier coefficients for the higher order partial derivatives of a two-dimensional function can be directly calculated in a way similar to what was described in Chapter 5 for one-dimensional functions.

The determinations of the Fourier coefficients for a two-dimensional function and its higher order partial derivatives are here specifically divided into the following five categories:

1. The function and its partial derivatives are expanded into full-range or complete Fourier series over the domain $[-a,a] \times [-b,b]$;
2. The function is expanded into a half-range cosine-cosine series over the domain $[0,a] \times [0,b]$, and its partial derivatives are expanded into half-range Fourier series over the domain $[0,a] \times [0,b]$;
3. The function is expanded into a half-range sine-cosine series over the domain $[0,a] \times [0,b]$, and its partial derivatives are expanded into half-range Fourier series over the domain $[0,a] \times [0,b]$;
4. The function is expanded into a half-range cosine-sine series over the domain $[0,a] \times [0,b]$, and its partial derivatives are expanded into half-range Fourier series over the domain $[0,a] \times [0,b]$;
5. The function is expanded into a half-range sine-sine series over the domain $[0,a] \times [0,b]$, and its partial derivatives are expanded into half-range Fourier series over the domain $[0,a] \times [0,b]$.

In the above, the term "half-range Fourier series" is generically used to refer to one of the four specific forms of series expansions: the half-range Fourier cosine-cosine series, the half-range Fourier sine-cosine series, the half-range Fourier cosine-sine series, and the half-range Fourier sine-sine series. In what follows, our discussions will only be focused on two of the five different categories.

6.2.1 THE FULL-RANGE FOURIER SERIES FOR THE PARTIAL DERIVATIVES

Let $u(x,y)$ be a sufficiently smooth function defined on $[-a,a] \times [-b,b]$. Suppose the function and its partial derivative $u^{(k_1,k_2)}(x,y)$ are expanded into full-range Fourier series as

$$u(x,y) = \sum_{m=0}^{\infty} \sum_{n=0}^{\infty} \lambda_{mn} [U_{1mn} \cos(\alpha_m x)\cos(\beta_n y) + U_{2mn} \sin(\alpha_m x)\cos(\beta_n y)$$

$$+ U_{3mn} \cos(\alpha_m x)\sin(\beta_n y) + U_{4mn} \sin(\alpha_m x)\sin(\beta_n y)],$$

(6.8)

and

$$u^{(k_1,k_2)}(x,y) = \sum_{m=0}^{\infty}\sum_{n=0}^{\infty}\lambda_{mn}[U_{1mn}^{(k_1,k_2)}\cos(\alpha_m x)\cos(\beta_n y) + U_{2mn}^{(k_1,k_2)}\sin(\alpha_m x)\cos(\beta_n y)$$

$$+ U_{3mn}^{(k_1,k_2)}\cos(\alpha_m x)\sin(\beta_n y) + U_{4mn}^{(k_1,k_2)}\sin(\alpha_m x)\sin(\beta_n y)], \qquad (6.9)$$

where k_1 and k_2 are nonnegative integers; $\alpha_m = m\pi/a$, $\beta_n = n\pi/b$, and

$$\lambda_{mn} = \begin{cases} 1/4 & m=0, n=0 \\ 1/2 & m=1,2,\cdots, n=0 \\ 1/2 & m=0, n=1,2,\cdots \\ 1 & m,n=1,2,\cdots \end{cases};$$

U_{1mn}, U_{2mn}, U_{3mn}, U_{4mn} and $U_{1mn}^{(k_1,k_2)}$, $U_{2mn}^{(k_1,k_2)}$, $U_{3mn}^{(k_1,k_2)}$, $U_{4mn}^{(k_1,k_2)}$ are the Fourier coefficients for the function $u(x,y)$ and its partial derivative $u^{(k_1,k_2)}(x,y)$, respectively. In addition, let

$$u^{(j,0)}(a,y) = \sum_{n=0}^{\infty}\mu_n[c_{1n}^{(j)}\cos(\beta_n y) + c_{2n}^{(j)}\sin(\beta_n y)], j = 0,1,\cdots,2k_1, \qquad (6.10)$$

$$u^{(j,0)}(-a,y) = \sum_{n=0}^{\infty}\mu_n[d_{1n}^{(j)}\cos(\beta_n y) + d_{2n}^{(j)}\sin(\beta_n y)], j = 0,1,\cdots,2k_1, \qquad (6.11)$$

$$u^{(0,l)}(x,b) = \sum_{m=0}^{\infty}\mu_m[e_{1m}^{(l)}\cos(\alpha_m x) + e_{2m}^{(l)}\sin(\alpha_m x)], l = 0,1,\cdots,2k_2, \qquad (6.12)$$

and

$$u^{(0,l)}(x,-b) = \sum_{m=0}^{\infty}\mu_m[f_{1m}^{(l)}\cos(\alpha_m x) + f_{2m}^{(l)}\sin(\alpha_m x)], l = 0,1,\cdots,2k_2, \qquad (6.13)$$

where $\mu_n = \begin{cases} 1/2 & n=0 \\ 1 & n>0 \end{cases}$, and $c_{1n}^{(j)}$, $c_{2n}^{(j)}$, $d_{1n}^{(j)}$, $d_{2n}^{(j)}$, $e_{1m}^{(l)}$, $e_{2m}^{(l)}$, $f_{1m}^{(l)}$, and $f_{2m}^{(l)}$ are the

(boundary) Fourier coefficients of $u^{(j,0)}(a,y)$, $u^{(j,0)}(-a,y)$, $u^{(0,l)}(x,b)$, and $u^{(0,l)}(x,-b)$, respectively.

By substituting $x_0 = -a$ and $y_0 = -b$ into (6.5) and (A.1)–(A.15), the Fourier coefficients for the partial derivative $u^{(k_1,k_2)}(x,y)$ are determined as below.

Based on the parity of the dual orders of the derivatives, there are four different scenarios which need to be treated separately.

Let's first consider the Fourier coefficients

$$U_{1mn}^{(2k_1,2k_2)} = \frac{1}{ab} \int\limits_{-a}^{a}\int\limits_{-b}^{b} u^{(2k_1,2k_2)}(x,y)\cos(\alpha_m x)\cos(\beta_n y)dxdy. \tag{6.14}$$

If $m = 0$ and $n = 0$, then

$$\begin{aligned}
U_{100}^{(0,2k_2)} &= \frac{1}{ab} \int\limits_{-a}^{a}\int\limits_{-b}^{b} u^{(0,2k_2)}(x,y)dxdy = \frac{1}{ab}\int\limits_{-a}^{a}dx \int\limits_{-b}^{b} u^{(0,2k_2)}(x,y)dy \\
&= \frac{1}{ab}\int\limits_{-a}^{a}[u^{(0,2k_2-1)}(x,b) - u^{(0,2k_2-1)}(x,-b)]dx \\
&= \frac{1}{b}[e_{10}^{(2k_2-1)} - f_{10}^{(2k_2-1)}] \quad (k_1 = 0 \text{ and } k_2 > 0),
\end{aligned} \tag{6.15}$$

$$\begin{aligned}
U_{100}^{(2k_1,0)} &= \frac{1}{ab} \int\limits_{-a}^{a}\int\limits_{-b}^{b} u^{(2k_1,0)}(x,y)dxdy = \frac{1}{ab}\int\limits_{-b}^{b}dy \int\limits_{-a}^{a} u^{(2k_1,0)}(x,y)dx \\
&= \frac{1}{ab}\int\limits_{-b}^{b}[u^{(2k_1-1,0)}(a,y) - u^{(2k_1-1,0)}(-a,y)]dy \\
&= \frac{1}{a}[c_{10}^{(2k_1-1)} - d_{10}^{(2k_1-1)}] \quad (k_1 > 0 \text{ and } k_2 = 0),
\end{aligned} \tag{6.16}$$

$$\begin{aligned}
U_{100}^{(2k_1,2k_2)} &= \frac{1}{ab} \int\limits_{-a}^{a}\int\limits_{-b}^{b} u^{(2k_1,2k_2)}(x,y)dxdy = \frac{1}{ab}\int\limits_{-b}^{b}dy \int\limits_{-a}^{a} u^{(2k_1,2k_2)}(x,y)dx \\
&= \frac{1}{ab}\int\limits_{-b}^{b}[u^{(2k_1-1,2k_2)}(a,y) - u^{(2k_1-1,2k_2)}(-a,y)]dy
\end{aligned} \tag{6.17}$$

or

$$\begin{aligned}
U_{100}^{(2k_1,2k_2)} &= \frac{1}{ab}[u^{(2k_1-1,2k_2-1)}(a,b) - u^{(2k_1-1,2k_2-1)}(a,-b) \\
&\quad - u^{(2k_1-1,2k_2-1)}(-a,b) + u^{(2k_1-1,2k_2-1)}(-a,-b)] \quad (k_1 > 0 \text{ and } k_2 > 0).
\end{aligned} \tag{6.18}$$

The formulas for the Fourier coefficients corresponding to other possible combinations of indexes m, n, k_1, and k_2 can be similarly obtained as:

for $m = 0$, $n > 0$,

$$U_{10n}^{(0,2k_2)} = \frac{1}{b} \sum_{l=1}^{k_2} (-1)^{n+k_2-l} \beta_n^{2k_2-2l} [e_{10}^{(2l-1)} - f_{10}^{(2l-1)}] + (-1)^{k_2} \beta_n^{2k_2} U_{10n}$$

(6.19)

$$(k_1 = 0 \text{ and } k_2 > 0),$$

$$U_{10n}^{(2k_1,0)} = \frac{1}{a} [c_{1n}^{(2k_1-1)} - d_{1n}^{(2k_1-1)}] \quad (k_1 > 0 \text{ and } k_2 = 0),$$

(6.20)

and

$$U_{10n}^{(2k_1,2k_2)} = \frac{1}{ab} \sum_{l=1}^{k_2} (-1)^{n+k_2-l} \beta_n^{2k_2-2l} [u^{(2k_1-1,2l-1)}(a,b) - u^{(2k_1-1,2l-1)}(a,-b)$$

$$- u^{(2k_1-1,2l-1)}(-a,b) + u^{(2k_1-1,2l-1)}(-a,-b)]$$

(6.21)

$$+ \frac{1}{a} (-1)^{k_2} \beta_n^{2k_2} [c_{1n}^{(2k_1-1)} - d_{1n}^{(2k_1-1)}] \quad (k_1 > 0 \text{ and } k_2 > 0);$$

for $m > 0$ and $n = 0$,

$$U_{1m0}^{(0,2k_2)} = \frac{1}{b} [e_{1m}^{(2k_2-1)} - f_{1m}^{(2k_2-1)}] \quad (k_1 = 0 \text{ and } k_2 > 0),$$

(6.22)

$$U_{1m0}^{(2k_1,0)} = \frac{1}{a} \sum_{j=1}^{k_1} (-1)^{m+k_1-j} \alpha_m^{2k_1-2j} [c_{10}^{(2j-1)} - d_{10}^{(2j-1)}] + (-1)^{k_1} \alpha_m^{2k_1} U_{1m0}$$

(6.23)

$$(k_1 > 0 \text{ and } k_2 = 0),$$

and

$$U_{1m0}^{(2k_1,2k_2)} = \frac{1}{ab} \sum_{j=1}^{k_1} (-1)^{m+k_1-j} \alpha_m^{2k_1-2j} [u^{(2j-1,2k_2-1)}(a,b) - u^{(2j-1,2k_2-1)}(a,-b)$$

$$- u^{(2j-1,2k_2-1)}(-a,b) + u^{(2j-1,2k_2-1)}(-a,-b)]$$

(6.24)

$$+ \frac{1}{b} (-1)^{k_1} \alpha_m^{2k_1} [e_{1m}^{(2k_2-1)} - f_{1m}^{(2k_2-1)}] \quad (k_1 > 0 \text{ and } k_2 > 0);$$

for $m > 0$ and $n > 0$,

$$U_{1mn}^{(2k_1,2k_2)} = \frac{1}{ab} \sum_{j=1}^{k_1} \sum_{l=1}^{k_2} (-1)^{m+n+k_1+k_2-j-l} \alpha_m^{2k_1-2j} \beta_n^{2k_2-2l} [u^{(2j-1,2l-1)}(a,b)$$

$$- u^{(2j-1,2l-1)}(a,-b) - u^{(2j-1,2l-1)}(-a,b) + u^{(2j-1,2l-1)}(-a,-b)]$$

$$+ \frac{1}{a} \sum_{j=1}^{k_1} (-1)^{m+k_1+k_2-j} \alpha_m^{2k_1-2j} \beta_n^{2k_2} [c_{1n}^{(2j-1)} - d_{1n}^{(2j-1)}]$$

$$+ \frac{1}{b} \sum_{l=1}^{k_2} (-1)^{n+k_1+k_2-l} \alpha_m^{2k_1} \beta_n^{2k_2-2l} [e_{1m}^{(2l-1)} - f_{1m}^{(2l-1)}] + (-1)^{k_1+k_2} \alpha_m^{2k_1} \beta_n^{2k_2} U_{1mn}.$$

$$(6.25)$$

The formulas for calculating the Fourier coefficients for other three terms in (6.9) can be derived in a similar manner. To avoid lengthy and tedious intermediate details, only the final expressions are presented below.

    a. Fourier coefficients $U_{2mn}^{(2k_1,2k_2)}$
       for $m > 0$ and $n = 0$,

$$U_{2m0}^{(0,2k_2)} = \frac{1}{b} [e_{2m}^{(2k_2-1)} - f_{2m}^{(2k_2-1)}] \quad (k_1 = 0 \text{ and } k_2 > 0), \qquad (6.26)$$

$$U_{2m0}^{(2k_1,0)} = \frac{1}{a} \sum_{j=1}^{k_1} (-1)^{m+k_1-j+1} \alpha_m^{2k_1-2j+1} [c_{10}^{(2j-2)} - d_{10}^{(2j-2)}]$$

$$+ (-1)^{k_1} \alpha_m^{2k_1} U_{2m0} \quad (k_1 > 0 \text{ and } k_2 = 0), \qquad (6.27)$$

and

$$U_{2m0}^{(2k_1,2k_2)} = \frac{1}{ab} \sum_{j=1}^{k_1} (-1)^{m+k_1-j+1} \alpha_m^{2k_1-2j+1} [u^{(2j-2,2k_2-1)}(a,b) - u^{(2j-2,2k_2-1)}(a,-b)$$

$$- u^{(2j-2,2k_2-1)}(-a,b) + u^{(2j-2,2k_2-1)}(-a,-b)]$$

$$+ \frac{1}{b} (-1)^{k_1} \alpha_m^{2k_1} [e_{2m}^{(2k_2-1)} - f_{2m}^{(2k_2-1)}] \quad (k_1 > 0 \text{ and } k_2 > 0); \qquad (6.28)$$

for $m > 0$ and $n > 0$,

$$U_{2mn}^{(2k_1,2k_2)} = \frac{1}{ab}\sum_{j=1}^{k_1}\sum_{l=1}^{k_2}(-1)^{m+n+k_1+k_2-j-l+1}\alpha_m^{2k_1-2j+1}\beta_n^{2k_2-2l}[u^{(2j-2,2l-1)}(a,b)$$

$$-u^{(2j-2,2l-1)}(a,-b)-u^{(2j-2,2l-1)}(-a,b)+u^{(2j-2,2l-1)}(-a,-b)]$$

$$+\frac{1}{a}\sum_{j=1}^{k_1}(-1)^{m+k_1+k_2-j+1}\alpha_m^{2k_1-2j+1}\beta_n^{2k_2}[c_{1n}^{(2j-2)}-d_{1n}^{(2j-2)}]$$

$$+\frac{1}{b}\sum_{l=1}^{k_2}(-1)^{n+k_1+k_2-l}\alpha_m^{2k_1}\beta_n^{2k_2-2l}[e_{2m}^{(2l-1)}-f_{2m}^{(2l-1)}]$$

$$+(-1)^{k_1+k_2}\alpha_m^{2k_1}\beta_n^{2k_2}U_{2mn}. \tag{6.29}$$

b. Fourier coefficients $U_{3mn}^{(2k_1,2k_2)}$
   for $m = 0$ and $n > 0$,

$$U_{30n}^{(0,2k_2)} = \frac{1}{b}\sum_{l=1}^{k_2}(-1)^{n+k_2-l+1}\beta_n^{2k_2-2l+1}[e_{10}^{(2l-2)}-f_{10}^{(2l-2)}]$$

$$+(-1)^{k_2}\beta_n^{2k_2}U_{30n} \quad (k_1 = 0 \text{ and } k_2 > 0), \tag{6.30}$$

$$U_{30n}^{(2k_1,0)} = \frac{1}{a}[c_{2n}^{(2k_1-1)}-d_{2n}^{(2k_1-1)}] \quad (k_1 > 0 \text{ and } k_2 = 0), \tag{6.31}$$

and

$$U_{30n}^{(2k_1,2k_2)} = \frac{1}{ab}\sum_{l=1}^{k_2}(-1)^{n+k_2-l+1}\beta_n^{2k_2-2l+1}[u^{(2k_1-1,2l-2)}(a,b)$$

$$-u^{(2k_1-1,2l-2)}(a,-b)-u^{(2k_1-1,2l-2)}(-a,b)+u^{(2k_1-1,2l-2)}(-a,-b)]$$

$$+\frac{1}{a}(-1)^{k_2}\beta_n^{2k_2}[c_{2n}^{(2k_1-1)}-d_{2n}^{(2k_1-1)}] \quad (k_1 > 0 \text{ and } k_2 > 0); \tag{6.32}$$

for $m > 0$ and $n > 0$,

$$
\begin{aligned}
U_{3mn}^{(2k_1,2k_2)} =\ & \frac{1}{ab} \sum_{j=1}^{k_1} \sum_{l=1}^{k_2} (-1)^{m+n+k_1+k_2-j-l+1} \alpha_m^{2k_1-2j} \beta_n^{2k_2-2l+1} [u^{(2j-1,2l-2)}(a,b) \\
& - u^{(2j-1,2l-2)}(a,-b) - u^{(2j-1,2l-2)}(-a,b) + u^{(2j-1,2l-2)}(-a,-b)] \\
& + \frac{1}{a} \sum_{j=1}^{k_1} (-1)^{m+k_1+k_2-j} \alpha_m^{2k_1-2j} \beta_n^{2k_2} [c_{2n}^{(2j-1)} - d_{2n}^{(2j-1)}] \\
& + \frac{1}{b} \sum_{l=1}^{k_2} (-1)^{n+k_1+k_2-l+1} \alpha_m^{2k_1} \beta_n^{2k_2-2l+1} [e_{1m}^{(2l-2)} - f_{1m}^{(2l-2)}] \\
& + (-1)^{k_1+k_2} \alpha_m^{2k_1} \beta_n^{2k_2} U_{3mn}.
\end{aligned}
\tag{6.33}
$$

c. Fourier coefficients $U_{4mn}^{(2k_1,2k_2)}$
for $m > 0$ and $n > 0$,

$$
\begin{aligned}
U_{4mn}^{(2k_1,2k_2)} =\ & \frac{1}{ab} \sum_{j=1}^{k_1} \sum_{l=1}^{k_2} (-1)^{m+n+k_1+k_2-j-l+2} \alpha_m^{2k_1-2j+1} \beta_n^{2k_2-2l+1} [u^{(2j-2,2l-2)}(a,b) \\
& - u^{(2j-2,2l-2)}(a,-b) - u^{(2j-2,2l-2)}(-a,b) + u^{(2j-2,2l-2)}(-a,-b)] \\
& + \frac{1}{a} \sum_{j=1}^{k_1} (-1)^{m+k_1+k_2-j+1} \alpha_m^{2k_1-2j+1} \beta_n^{2k_2} [c_{2n}^{(2j-2)} - d_{2n}^{(2j-2)}] \\
& + \frac{1}{b} \sum_{l=1}^{k_2} (-1)^{n+k_1+k_2-l+1} \alpha_m^{2k_1} \beta_n^{2k_2-2l+1} [e_{2m}^{(2l-2)} - f_{2m}^{(2l-2)}] \\
& + (-1)^{k_1+k_2} \alpha_m^{2k_1} \beta_n^{2k_2} U_{4mn}.
\end{aligned}
\tag{6.34}
$$

It should be noted that the above formulas are specifically used to calculate the Fourier coefficients for the derivatives of even orders in both x and y directions. The corresponding results for other different order parities can be easily obtained because converting a derivative order from even into odd, or *vice versa*, will simply involve one step of integration-by-part. Thus, only the final expressions are directly given below for the sake of completeness.

The Fourier coefficients for the full-range Fourier series of $u^{(2k_1+1,2k_2)}(x,y)$ can be calculated from:

a. Fourier coefficients $U_{1mn}^{(2k_1+1,2k_2)}$
for $m = 0$ and $n = 0$,

$$
U_{100}^{(2k_1+1,0)} = \frac{1}{a} [c_{10}^{(2k_1)} - d_{10}^{(2k_1)}] \quad (k_2 = 0),
\tag{6.35}
$$

and

$$U_{100}^{(2k_1+1,2k_2)} = \frac{1}{ab}[u^{(2k_1,2k_2-1)}(a,b) - u^{(2k_1,2k_2-1)}(a,-b) - u^{(2k_1,2k_2-1)}(-a,b)$$
$$+ u^{(2k_1,2k_2-1)}(-a,-b)] \quad (k_2 > 0); \tag{6.36}$$

for $m = 0$ and $n > 0$,

$$U_{10n}^{(2k_1+1,0)} = \frac{1}{a}[c_{1n}^{(2k_1)} - d_{1n}^{(2k_1)}] \quad (k_2 = 0), \tag{6.37}$$

and

$$U_{10n}^{(2k_1+1,2k_2)} = \frac{1}{ab}\sum_{l=1}^{k_2}(-1)^{n+k_2-l}\beta_n^{2k_2-2l}[u^{(2k_1,2l-1)}(a,b) - u^{(2k_1,2l-1)}(a,-b)$$
$$- u^{(2k_1,2l-1)}(-a,b) + u^{(2k_1,2l-1)}(-a,-b)]$$
$$+ \frac{1}{a}(-1)^{k_2}\beta_n^{2k_2}[c_{1n}^{(2k_1)} - d_{1n}^{(2k_1)}] \quad (k_2 > 0); \tag{6.38}$$

for $m > 0$ and $n = 0$,

$$U_{1m0}^{(2k_1+1,0)} = \frac{1}{a}\sum_{j=0}^{k_1}(-1)^{m+k_1-j}\alpha_m^{2k_1-2j}[c_{10}^{(2j)} - d_{10}^{(2j)}]$$
$$+ (-1)^{k_1}\alpha_m^{2k_1+1}U_{2m0} \quad (k_2 = 0), \tag{6.39}$$

and

$$U_{1m0}^{(2k_1+1,2k_2)} = \frac{1}{ab}\sum_{j=0}^{k_1}(-1)^{m+k_1-j}\alpha_m^{2k_1-2j}[u^{(2j,2k_2-1)}(a,b) - u^{(2j,2k_2-1)}(a,-b)$$
$$- u^{(2j,2k_2-1)}(-a,b) + u^{(2j,2k_2-1)}(-a,-b)]$$
$$+ \frac{1}{b}(-1)^{k_1}\alpha_m^{2k_1+1}[e_{2m}^{(2k_2-1)} - f_{2m}^{(2k_2-1)}] \quad (k_2 > 0); \tag{6.40}$$

for $m > 0$ and $n > 0$,

$$U_{1mn}^{(2k_1+1,2k_2)} = \frac{1}{ab}\sum_{j=0}^{k_1}\sum_{l=1}^{k_2}(-1)^{m+n+k_1+k_2-j-l}\alpha_m^{2k_1-2j}\beta_n^{2k_2-2l}[u^{(2j,2l-1)}(a,b)$$

$$-u^{(2j,2l-1)}(a,-b)-u^{(2j,2l-1)}(-a,b)+u^{(2j,2l-1)}(-a,-b)]$$

$$+\frac{1}{a}\sum_{j=0}^{k_1}(-1)^{m+k_1+k_2-j}\alpha_m^{2k_1-2j}\beta_n^{2k_2}[c_{1n}^{(2j)}-d_{1n}^{(2j)}] \qquad (6.41)$$

$$+\frac{1}{b}\sum_{l=1}^{k_2}(-1)^{n+k_1+k_2-l}\alpha_m^{2k_1+1}\beta_n^{2k_2-2l}[e_{2m}^{(2l-1)}-f_{2m}^{(2l-1)}]$$

$$+(-1)^{k_1+k_2}\alpha_m^{2k_1+1}\beta_n^{2k_2}U_{2mn}.$$

b. Fourier coefficients $U_{2mn}^{(2k_1+1,2k_2)}$

for $m > 0$ and $n = 0$,

$$U_{2m0}^{(2k_1+1,0)} = \frac{1}{a}\sum_{j=1}^{k_1}(-1)^{m+k_1-j+1}\alpha_m^{2k_1-2j+1}[c_{10}^{(2j-1)}-d_{10}^{(2j-1)}]$$

$$+(-1)^{k_1+1}\alpha_m^{2k_1+1}U_{1m0} \quad (k_2=0), \qquad (6.42)$$

and

$$U_{2m0}^{(2k_1+1,2k_2)} = \frac{1}{ab}\sum_{j=1}^{k_1}(-1)^{m+k_1-j+1}\alpha_m^{2k_1-2j+1}[u^{(2j-1,2k_2-1)}(a,b)-u^{(2j-1,2k_2-1)}(a,-b)$$

$$-u^{(2j-1,2k_2-1)}(-a,b)+u^{(2j-1,2k_2-1)}(-a,-b)]$$

$$+\frac{1}{b}(-1)^{k_1+1}\alpha_m^{2k_1+1}[e_{1m}^{(2k_2-1)}-f_{1m}^{(2k_2-1)}] \quad (k_2>0); \qquad (6.43)$$

for $m > 0$ and $n > 0$,

$$U_{2mn}^{(2k_1+1,2k_2)} = \frac{1}{ab}\sum_{j=1}^{k_1}\sum_{l=1}^{k_2}(-1)^{m+n+k_1+k_2-j-l+1}\alpha_m^{2k_1-2j+1}\beta_n^{2k_2-2l}[u^{(2j-1,2l-1)}(a,b)$$

$$-u^{(2j-1,2l-1)}(a,-b)-u^{(2j-1,2l-1)}(-a,b)+u^{(2j-1,2l-1)}(-a,-b)]$$

$$+\frac{1}{a}\sum_{j=1}^{k_1}(-1)^{m+k_1+k_2-j+1}\alpha_m^{2k_1-2j+1}\beta_n^{2k_2}[c_{1n}^{(2j-1)}-d_{1n}^{(2j-1)}]$$

$$+\frac{1}{b}\sum_{l=1}^{k_2}(-1)^{n+k_1+k_2-l+1}\alpha_m^{2k_1+1}\beta_n^{2k_2-2l}[e_{1m}^{(2l-1)}-f_{1m}^{(2l-1)}]$$

$$+(-1)^{k_1+k_2+1}\alpha_m^{2k_1+1}\beta_n^{2k_2}U_{1mn}. \qquad (6.44)$$

c. Fourier coefficients $U_{3mn}^{(2k_1+1,2k_2)}$

for $m = 0$ and $n > 0$,

$$U_{30n}^{(2k_1+1,0)} = \frac{1}{a}[c_{2n}^{(2k_1)} - d_{2n}^{(2k_1)}] \quad (k_2 = 0), \tag{6.45}$$

and

$$U_{30n}^{(2k_1+1,2k_2)} = \frac{1}{ab}\sum_{l=1}^{k_2}(-1)^{n+k_2-l+1}\beta_n^{2k_2-2l+1}[u^{(2k_1,2l-2)}(a,b) - u^{(2k_1,2l-2)}(a,-b)$$

$$- u^{(2k_1,2l-2)}(-a,b) + u^{(2k_1,2l-2)}(-a,-b)]$$

$$+ \frac{1}{a}(-1)^{k_2}\beta_n^{2k_2}[c_{2n}^{(2k_1)} - d_{2n}^{(2k_1)}] \quad (k_2 > 0); \tag{6.46}$$

for $m > 0$ and $n > 0$,

$$U_{3mn}^{(2k_1+1,2k_2)} = \frac{1}{ab}\sum_{j=0}^{k_1}\sum_{l=1}^{k_2}(-1)^{m+n+k_1+k_2-j-l+1}\alpha_m^{2k_1-2j}\beta_n^{2k_2-2l+1}[u^{(2j,2l-2)}(a,b)$$

$$- u^{(2j,2l-2)}(a,-b) - u^{(2j,2l-2)}(-a,b) + u^{(2j,2l-2)}(-a,-b)]$$

$$+ \frac{1}{a}\sum_{j=0}^{k_1}(-1)^{m+k_1+k_2-j}\alpha_m^{2k_1-2j}\beta_n^{2k_2}[c_{2n}^{(2j)} - d_{2n}^{(2j)}]$$

$$+ \frac{1}{b}\sum_{l=1}^{k_2}(-1)^{n+k_1+k_2-l+1}\alpha_m^{2k_1+1}\beta_n^{2k_2-2l+1}[e_{2m}^{(2l-2)} - f_{2m}^{(2l-2)}]$$

$$+ (-1)^{k_1+k_2}\alpha_m^{2k_1+1}\beta_n^{2k_2}U_{4mn}. \tag{6.47}$$

d. Fourier coefficients $U_{4mn}^{(2k_1+1,2k_2)}$

for $m > 0$ and $n > 0$,

$$U_{4mn}^{(2k_1+1,2k_2)} = \frac{1}{ab}\sum_{j=1}^{k_1}\sum_{l=1}^{k_2}(-1)^{m+n+k_1+k_2-j-l+2}\alpha_m^{2k_1-2j+1}\beta_n^{2k_2-2l+1}[u^{(2j-1,2l-2)}(a,b)$$

$$- u^{(2j-1,2l-2)}(a,-b) - u^{(2j-1,2l-2)}(-a,b) + u^{(2j-1,2l-2)}(-a,-b)]$$

$$+ \frac{1}{a}\sum_{j=1}^{k_1}(-1)^{m+k_1+k_2-j+1}\alpha_m^{2k_1-2j+1}\beta_n^{2k_2}[c_{2n}^{(2j-1)} - d_{2n}^{(2j-1)}]$$

$$+ \frac{1}{b}\sum_{l=1}^{k_2}(-1)^{n+k_1+k_2-l+2}\alpha_m^{2k_1+1}\beta_n^{2k_2-2l+1}[e_{1m}^{(2l-2)} - f_{1m}^{(2l-2)}]$$

$$+ (-1)^{k_1+k_2+1}\alpha_m^{2k_1+1}\beta_n^{2k_2}U_{3mn}. \tag{6.48}$$

The Fourier coefficients for the full-range Fourier series of $u^{(2k_1,2k_2+1)}(x_1,x_2)$ can be calculated from:

a. Fourier coefficients $U_{1mn}^{(2k_1,2k_2+1)}$

for $m = 0$ and $n = 0$,

$$U_{100}^{(0,2k_2+1)} = \frac{1}{b}[e_{10}^{(2k_2)} - f_{10}^{(2k_2)}] \quad (k_1 = 0), \tag{6.49}$$

and

$$U_{100}^{(2k_1,2k_2+1)} = \frac{1}{ab}[u^{(2k_1-1,2k_2)}(a,b) - u^{(2k_1-1,2k_2)}(a,-b) - u^{(2k_1-1,2k_2)}(-a,b)$$

$$+ u^{(2k_1-1,2k_2)}(-a,-b)] \quad (k_1 > 0); \tag{6.50}$$

for $m = 0$ and $n > 0$,

$$U_{10n}^{(0,2k_2+1)} = \frac{1}{b}\sum_{l=0}^{k_2}(-1)^{n+k_2-l}\beta_n^{2k_2-2l}[e_{10}^{(2l)} - f_{10}^{(2l)}] + (-1)^{k_2}\beta_n^{2k_2+1}U_{30n} \quad (k_1 = 0), \tag{6.51}$$

and

$$U_{10n}^{(2k_1,2k_2+1)} = \frac{1}{ab}\sum_{l=0}^{k_2}(-1)^{n+k_2-l}\beta_n^{2k_2-2l}[u^{(2k_1-1,2l)}(a,b) - u^{(2k_1-1,2l)}(a,-b)$$

$$- u^{(2k_1-1,2l)}(-a,b) + u^{(2k_1-1,2l)}(-a,-b)]$$

$$+ \frac{1}{a}(-1)^{k_2}\beta_n^{2k_2+1}[c_{2n}^{(2k_1-1)} - d_{2n}^{(2k_1-1)}] \quad (k_1 > 0); \tag{6.52}$$

for $m > 0$ and $n = 0$,

$$U_{1m0}^{(0,2k_2+1)} = \frac{1}{b}[e_{1m}^{(2k_2)} - f_{1m}^{(2k_2)}] \quad (k_1 = 0), \tag{6.53}$$

and

$$U_{1m0}^{(2k_1,2k_2+1)} = \frac{1}{ab}\sum_{j=1}^{k_1}(-1)^{m+k_1-j}\alpha_m^{2k_1-2j}[u^{(2j-1,2k_2)}(a,b) - u^{(2j-1,2k_2)}(a,-b)$$

$$- u^{(2j-1,2k_2)}(-a,b) + u^{(2j-1,2k_2)}(-a,-b)]$$

$$+ \frac{1}{b}(-1)^{k_1}\alpha_m^{2k_1}[e_{1m}^{(2k_2)} - f_{1m}^{(2k_2)}] \quad (k_1 > 0); \tag{6.54}$$

for $m > 0$ and $n > 0$,

$$
U_{1mn}^{(2k_1,2k_2+1)} = \frac{1}{ab} \sum_{j=1}^{k_1} \sum_{l=0}^{k_2} (-1)^{m+n+k_1+k_2-j-l} \alpha_m^{2k_1-2j} \beta_n^{2k_2-2l} [u^{(2j-1,2l)}(a,b)
$$
$$
- u^{(2j-1,2l)}(a,-b) - u^{(2j-1,2l)}(-a,b) + u^{(2j-1,2l)}(-a,-b)]
$$
$$
+ \frac{1}{a} \sum_{j=1}^{k_1} (-1)^{m+k_1+k_2-j} \alpha_m^{2k_1-2j} \beta_n^{2k_2+1} [c_{2n}^{(2j-1)} - d_{2n}^{(2j-1)}]
$$
$$
+ \frac{1}{b} \sum_{l=0}^{k_2} (-1)^{n+k_1+k_2-l} \alpha_m^{2k_1} \beta_n^{2k_2-2l} [e_{1m}^{(2l)} - f_{1m}^{(2l)}]
$$
$$
+ (-1)^{k_1+k_2} \alpha_m^{2k_1} \beta_n^{2k_2+1} U_{3mn}. \tag{6.55}
$$

b. Fourier coefficients $U_{2mn}^{(2k_1,2k_2+1)}$

for $m > 0$ and $n = 0$,

$$
U_{2m0}^{(0,2k_2+1)} = \frac{1}{b} [e_{2m}^{(2k_2)} - f_{2m}^{(2k_2)}] \quad (k_1 = 0), \tag{6.56}
$$

and

$$
U_{2m0}^{(2k_1,2k_2+1)} = \frac{1}{ab} \sum_{j=1}^{k_1} (-1)^{m+k_1-j+1} \alpha_m^{2k_1-2j+1} [u^{(2j-2,2k_2)}(a,b) - u^{(2j-2,2k_2)}(a,-b)
$$
$$
- u^{(2j-2,2k_2)}(-a,b) + u^{(2j-2,2k_2)}(-a,-b)]
$$
$$
+ \frac{1}{b} (-1)^{k_1} \alpha_m^{2k_1} [e_{2m}^{(2k_2)} - f_{2m}^{(2k_2)}] \quad (k_1 > 0); \tag{6.57}
$$

for $m > 0$ and $n > 0$,

$$
U_{2mn}^{(2k_1,2k_2+1)} = \frac{1}{ab} \sum_{j=1}^{k_1} \sum_{l=0}^{k_2} (-1)^{m+n+k_1+k_2-j-l+1} \alpha_m^{2k_1-2j+1} \beta_n^{2k_2-2l} [u^{(2j-2,2l)}(a,b)
$$
$$
- u^{(2j-2,2l)}(a,-b) - u^{(2j-2,2l)}(-a,b) + u^{(2j-2,2l)}(-a,-b)]
$$
$$
+ \frac{1}{a} \sum_{j=1}^{k_1} (-1)^{m+k_1+k_2-j+1} \alpha_m^{2k_1-2j+1} \beta_n^{2k_2+1} [c_{2n}^{(2j-2)} - d_{2n}^{(2j-2)}]
$$
$$
+ \frac{1}{b} \sum_{l=0}^{k_2} (-1)^{n+k_1+k_2-l} \alpha_m^{2k_1} \beta_n^{2k_2-2l} [e_{2m}^{(2l)} - f_{2m}^{(2l)}]
$$
$$
+ (-1)^{k_1+k_2} \alpha_m^{2k_1} \beta_n^{2k_2+1} U_{4mn}. \tag{6.58}
$$

c. Fourier coefficients $U_{3mn}^{(2k_1,2k_2+1)}$

for $m = 0$ and $n > 0$,

$$U_{30n}^{(0,2k_2+1)} = \frac{1}{b} \sum_{l=1}^{k_2} (-1)^{n+k_2-l+1} \beta_n^{2k_2-2l+1} [e_{10}^{(2l-1)} - f_{10}^{(2l-1)}]$$

$$+ (-1)^{k_2+1} \beta_n^{2k_2+1} U_{10n} \quad (k_1 = 0),$$

(6.59)

and

$$U_{30n}^{(2k_1,2k_2+1)} = \frac{1}{ab} \sum_{l=1}^{k_2} (-1)^{n+k_2-l+1} \beta_n^{2k_2-2l+1} [u^{(2k_1-1,2l-1)}(a,b) - u^{(2k_1-1,2l-1)}(a,-b)$$

$$- u^{(2k_1-1,2l-1)}(-a,b) + u^{(2k_1-1,2l-1)}(-a,-b)]$$

$$+ \frac{1}{a} (-1)^{k_2+1} \beta_n^{2k_2+1} [c_{1n}^{(2k_1-1)} - d_{1n}^{(2k_1-1)}] \quad (k_1 > 0);$$

(6.60)

for $m > 0$ and $n > 0$,

$$U_{3mn}^{(2k_1,2k_2+1)} = \frac{1}{ab} \sum_{j=1}^{k_1} \sum_{l=1}^{k_2} (-1)^{m+n+k_1+k_2-j-l+1} \alpha_m^{2k_1-2j} \beta_n^{2k_2-2l+1} [u^{(2j-1,2l-1)}(a,b)$$

$$- u^{(2j-1,2l-1)}(a,-b) - u^{(2j-1,2l-1)}(-a,b) + u^{(2j-1,2l-1)}(-a,-b)]$$

$$+ \frac{1}{a} \sum_{j=1}^{k_1} (-1)^{m+k_1+k_2-j+1} \alpha_m^{2k_1-2j} \beta_n^{2k_2+1} [c_{1n}^{(2j-1)} - d_{1n}^{(2j-1)}]$$

$$+ \frac{1}{b} \sum_{l=1}^{k_2} (-1)^{n+k_1+k_2-l+1} \alpha_m^{2k_1} \beta_n^{2k_2-2l+1} [e_{1m}^{(2l-1)} - f_{1m}^{(2l-1)}]$$

$$+ (-1)^{k_1+k_2+1} \alpha_m^{2k_1} \beta_n^{2k_2+1} U_{1mn}.$$

(6.61)

d. Fourier coefficients $U_{4mn}^{(2k_1,2k_2+1)}$

For $m > 0$ and $n > 0$,

$$U_{4mn}^{(2k_1,2k_2+1)} = \frac{1}{ab} \sum_{j=1}^{k_1} \sum_{l=1}^{k_2} (-1)^{m+n+k_1+k_2-j-l+2} \alpha_m^{2k_1-2j+1} \beta_n^{2k_2-2l+1} [u^{(2j-2,2l-1)}(a,b)$$

$$- u^{(2j-2,2l-1)}(a,-b) - u^{(2j-2,2l-1)}(-a,b) + u^{(2j-2,2l-1)}(-a,-b)]$$

$$+ \frac{1}{a} \sum_{j=1}^{k_1} (-1)^{m+k_1+k_2-j+2} \alpha_m^{2k_1-2j+1} \beta_n^{2k_2+1} [c_{1n}^{(2j-2)} - d_{1n}^{(2j-2)}]$$

$$+ \frac{1}{b} \sum_{l=1}^{k_2} (-1)^{n+k_1+k_2-l+1} \alpha_m^{2k_1} \beta_n^{2k_2-2l+1} [e_{2m}^{(2l-1)} - f_{2m}^{(2l-1)}]$$

$$+ (-1)^{k_1+k_2+1} \alpha_m^{2k_1} \beta_n^{2k_2+1} U_{2mn}. \tag{6.62}$$

The Fourier coefficients for the full-range Fourier series of $u^{(2k_1+1,2k_2+1)}(x,y)$ can be obtained from:

a. Fourier coefficients $U_{1mn}^{(2k_1+1,2k_2+1)}$

for $m = 0$ and $n = 0$,

$$U_{100}^{(2k_1+1,2k_2+1)} = \frac{1}{ab} [u^{(2k_1,2k_2)}(a,b) - u^{(2k_1,2k_2)}(a,-b)$$

$$- u^{(2k_1,2k_2)}(-a,b) + u^{(2k_1,2k_2)}(-a,-b)]; \tag{6.63}$$

for $m = 0$ and $n > 0$,

$$U_{10n}^{(2k_1+1,2k_2+1)} = \frac{1}{ab} \sum_{l=0}^{k_2} (-1)^{n+k_2-l} \beta_n^{2k_2-2l} [u^{(2k_1,2l)}(a,b) - u^{(2k_1,2l)}(a,-b)$$

$$- u^{(2k_1,2l)}(-a,b) + u^{(2k_1,2l)}(-a,-b)] \tag{6.64}$$

$$+ \frac{1}{a} (-1)^{k_2} \beta_n^{2k_2+1} [c_{2n}^{(2k_1)} - d_{2n}^{(2k_1)}];$$

for $m > 0$ and $n = 0$,

$$U_{1m0}^{(2k_1+1,2k_2+1)} = \frac{1}{ab} \sum_{j=0}^{k_1} (-1)^{m+k_1-j} \alpha_m^{2k_1-2j} [u^{(2j,2k_2)}(a,b) - u^{(2j,2k_2)}(a,-b)$$

$$- u^{(2j,2k_2)}(-a,b) + u^{(2j,2k_2)}(-a,-b)] \tag{6.65}$$

$$+ \frac{1}{b}(-1)^{k_1} \alpha_m^{2k_1+1} [e_{2m}^{(2k_2)} - f_{2m}^{(2k_2)}];$$

for $m > 0$ and $n > 0$,

$$U_{1mn}^{(2k_1+1,2k_2+1)} = \frac{1}{ab} \sum_{j=0}^{k_1} \sum_{l=0}^{k_2} (-1)^{m+n+k_1+k_2-j-l} \alpha_m^{2k_1-2j} \beta_n^{2k_2-2l} [u^{(2j,2l)}(a,b)$$

$$- u^{(2j,2l)}(a,-b) - u^{(2j,2l)}(-a,b) + u^{(2j,2l)}(-a,-b)]$$

$$+ \frac{1}{a} \sum_{j=0}^{k_1} (-1)^{m+k_1+k_2-j} \alpha_m^{2k_1-2j} \beta_n^{2k_2+1} [c_{2n}^{(2j)} - d_{2n}^{(2j)}]$$

$$+ \frac{1}{b} \sum_{l=0}^{k_2} (-1)^{n+k_1+k_2-l} \alpha_m^{2k_1+1} \beta_n^{2k_2-2l} [e_{2m}^{(2l)} - f_{2m}^{(2l)}]$$

$$+ (-1)^{k_1+k_2} \alpha_m^{2k_1+1} \beta_n^{2k_2+1} U_{4mn}. \tag{6.66}$$

b. Fourier coefficients $U_{2mn}^{(2k_1+1,2k_2+1)}$

for $m > 0$ and $n = 0$,

$$U_{2m0}^{(2k_1+1,2k_2+1)} = \frac{1}{ab} \sum_{j=1}^{k_1} (-1)^{m+k_1-j+1} \alpha_m^{2k_1-2j+1} [u^{(2j-1,2k_2)}(a,b) - u^{(2j-1,2k_2)}(a,-b)$$

$$- u^{(2j-1,2k_2)}(-a,b) + u^{(2j-1,2k_2)}(-a,-b)]$$

$$+ \frac{1}{b}(-1)^{k_1+1} \alpha_m^{2k_1+1} [e_{1m}^{(2k_2)} - f_{1m}^{(2k_2)}]; \tag{6.67}$$

for $m > 0$ and $n > 0$,

$$U_{2mn}^{(2k_1+1,2k_2+1)} = \frac{1}{ab}\sum_{j=1}^{k_1}\sum_{l=0}^{k_2}(-1)^{m+n+k_1+k_2-j-l+1}\alpha_m^{2k_1-2j+1}\beta_n^{2k_2-2l}[u^{(2j-1,2l)}(a,b)$$

$$-u^{(2j-1,2l)}(a,-b)-u^{(2j-1,2l)}(-a,b)+u^{(2j-1,2l)}(-a,-b)]$$

$$+\frac{1}{a}\sum_{j=1}^{k_1}(-1)^{m+k_1+k_2-j+1}\alpha_m^{2k_1-2j+1}\beta_n^{2k_2+1}[c_{2n}^{(2j-1)}-d_{2n}^{(2j-1)}]$$

$$+\frac{1}{b}\sum_{l=0}^{k_2}(-1)^{n+k_1+k_2-l+1}\alpha_m^{2k_1+1}\beta_n^{2k_2-2l}[e_{1m}^{(2l)}-f_{1m}^{(2l)}]$$

$$+(-1)^{k_1+k_2+1}\alpha_m^{2k_1+1}\beta_n^{2k_2+1}U_{3mn}. \tag{6.68}$$

c. Fourier coefficients $U_{3mn}^{(2k_1+1,2k_2+1)}$

for $m = 0$ and $n > 0$,

$$U_{30n}^{(2k_1+1,2k_2+1)} = \frac{1}{ab}\sum_{l=1}^{k_2}(-1)^{n+k_2-l+1}\beta_n^{2k_2-2l+1}[u^{(2k_1,2l-1)}(a,b)-u^{(2k_1,2l-1)}(a,-b)$$

$$-u^{(2k_1,2l-1)}(-a,b)+u^{(2k_1,2l-1)}(-a,-b)]$$

$$+\frac{1}{a}(-1)^{k_2+1}\beta_n^{2k_2+1}[c_{1n}^{(2k_1)}-d_{1n}^{(2k_1)}]; \tag{6.69}$$

for $m > 0$ and $n > 0$,

$$U_{3mn}^{(2k_1+1,2k_2+1)} = \frac{1}{ab}\sum_{j=0}^{k_1}\sum_{l=1}^{k_2}(-1)^{m+n+k_1+k_2-j-l+1}\alpha_m^{2k_1-2j}\beta_n^{2k_2-2l+1}[u^{(2j,2l-1)}(a,b)$$

$$-u^{(2j,2l-1)}(a,-b)-u^{(2j,2l-1)}(-a,b)+u^{(2j,2l-1)}(-a,-b)]$$

$$+\frac{1}{a}\sum_{j=0}^{k_1}(-1)^{m+k_1+k_2-j+1}\alpha_m^{2k_1-2j}\beta_n^{2k_2+1}[c_{1n}^{(2j)}-d_{1n}^{(2j)}]$$

$$+\frac{1}{b}\sum_{l=1}^{k_2}(-1)^{n+k_1+k_2-l+1}\alpha_m^{2k_1+1}\beta_n^{2k_2-2l+1}[e_{2m}^{(2l-1)}-f_{2m}^{(2l-1)}]$$

$$+(-1)^{k_1+k_2+1}\alpha_m^{2k_1+1}\beta_n^{2k_2+1}U_{2mn}. \tag{6.70}$$

d. Fourier coefficients $U_{4mn}^{(2k_1+1,2k_2+1)}$

for $m > 0$ and $n > 0$,

$$
\begin{aligned}
U_{4mn}^{(2k_1+1,2k_2+1)} = &\frac{1}{ab}\sum_{j=1}^{k_1}\sum_{l=1}^{k_2}(-1)^{m+n+k_1+k_2-j-l+2}\alpha_m^{2k_1-2j+1}\beta_n^{2k_2-2l+1}[u^{(2j-1,2l-1)}(a,b)\\
&-u^{(2j-1,2l-1)}(a,-b)-u^{(2j-1,2l-1)}(-a,b)+u^{(2j-1,2l-1)}(-a,-b)]\\
&+\frac{1}{a}\sum_{j=1}^{k_1}(-1)^{m+k_1+k_2-j+2}\alpha_m^{2k_1-2j+1}\beta_n^{2k_2+1}[c_{1n}^{(2j-1)}-d_{1n}^{(2j-1)}]\\
&+\frac{1}{b}\sum_{l=1}^{k_2}(-1)^{n+k_1+k_2-l+2}\alpha_m^{2k_1+1}\beta_n^{2k_2-2l+1}[e_{1m}^{(2l-1)}-f_{1m}^{(2l-1)}]\\
&+(-1)^{k_1+k_2+2}\alpha_m^{2k_1+1}\beta_n^{2k_2+1}U_{1mn}.
\end{aligned}
\tag{6.71}
$$

The significance of these formulas lies in the fact that the Fourier coefficients for the derivatives can be calculated from those for the lower order derivatives if their boundary values are also known explicitly. In the meantime, these formulas also show that the Fourier series for the partial derivatives of a function cannot be generally obtained from the term-by-term differentiations of the Fourier series for their "root" functions unless certain conditions are satisfied as clarified later.

6.2.2 THE HALF-RANGE FOURIER SERIES FOR THE PARTIAL DERIVATIVES

It is obvious from the previous section that expanding the partial derivatives into the full-range Fourier series may easily become a tedious task in view of the fact that each of the four types of terms can have multiple expressions corresponding to different index combinations, (m, n, k_1, k_2). In this regard, the half-range Fourier (sine-sine, cosine-cosine, sine-cosine, or cosine-sine) series may become more attractive because the number of terms involved is greatly reduced.

For instance, suppose that the two-dimensional function $u(x,y)$ is expanded into a (half-range) Fourier sine-sine series on the domain $[0,a]\times[0,b]$ as

$$
u(x,y)=\sum_{m=1}^{\infty}\sum_{n=1}^{\infty}U_{4mn}\sin(\alpha_m x)\sin(\beta_n y),
\tag{6.72}
$$

and its partial derivatives can be accordingly expressed as

$$
u^{(2k_1,2k_2)}(x,y)=\sum_{m=1}^{\infty}\sum_{n=1}^{\infty}U_{4mn}^{(2k_1,2k_2)}\sin(\alpha_m x)\sin(\beta_n y),
\tag{6.73}
$$

$$u^{(2k_1+1,2k_2)}(x,y) = \sum_{m=0}^{\infty}\sum_{n=1}^{\infty}\lambda_{mn}U_{3mn}^{(2k_1+1,2k_2)}\cos(\alpha_m x)\sin(\beta_n y), \qquad (6.74)$$

$$u^{(2k_1,2k_2+1)}(x,y) = \sum_{m=1}^{\infty}\sum_{n=0}^{\infty}\lambda_{mn}U_{2mn}^{(2k_1,2k_2+1)}\sin(\alpha_m x)\cos(\beta_n y), \qquad (6.75)$$

and

$$u^{(2k_1+1,2k_2+1)}(x,y) = \sum_{m=0}^{\infty}\sum_{n=0}^{\infty}\lambda_{mn}U_{1mn}^{(2k_1+1,2k_2+1)}\cos(\alpha_m x)\cos(\beta_n y), \qquad (6.76)$$

where k_1 and k_2 are nonnegative integers, $\alpha_m = m\pi/a$, $\beta_n = n\pi/b$ and

$$\lambda_{mn} = \begin{cases} 1/4 & m=0, n=0 \\ 1/2 & m=1,2,\cdots, n=0 \\ 1/2 & m=0, n=1,2,\cdots \\ 1 & m,n=1,2,\cdots \end{cases} ;$$

U_{4mn}, $U_{4mn}^{(2k_1,2k_2)}$, $U_{3mn}^{(2k_1+1,2k_2)}$, $U_{2mn}^{(2k_1,2k_2+1)}$, and $U_{1mn}^{(2k_1+1,2k_2+1)}$ are the Fourier coefficients of the functions $u(x,y)$, $u^{(2k_1,2k_2)}(x,y)$, $u^{(2k_1+1,2k_2)}(x,y)$, $u^{(2k_1,2k_2+1)}(x,y)$, and $u^{(2k_1+1,2k_2+1)}(x,y)$, respectively.

In addition, let

$$u^{(2j,0)}(a,y) = \sum_{n=1}^{\infty}c_{2n}^{(2j)}\sin(\beta_n y), j=0,1,\cdots,k_1, \qquad (6.77)$$

$$u^{(2j,0)}(0,y) = \sum_{n=1}^{\infty}d_{2n}^{(2j)}\sin(\beta_n y), j=0,1,\cdots,k_1, \qquad (6.78)$$

$$u^{(0,2l)}(x,b) = \sum_{m=1}^{\infty}e_{2m}^{(2l)}\sin(\alpha_m x), l=0,1,\cdots,k_2, \qquad (6.79)$$

$$u^{(0,2l)}(x,0) = \sum_{m=1}^{\infty}f_{2m}^{(2l)}\sin(\alpha_m x), l=0,1,\cdots,k_2, \qquad (6.80)$$

where $c_{2n}^{(2j)}, d_{2n}^{(2j)}, e_{2m}^{(2l)}$, and $f_{2m}^{(2l)}$ are the Fourier coefficients of $u^{(2j,0)}(a,y), u^{(2j,0)}(0,y)$, $u^{(0,2l)}(x,b)$, and $u^{(0,2l)}(x,0)$, respectively; or they can be referred to as the boundary Fourier coefficients of the corresponding function and partial derivatives.

The Fourier coefficients for the half-range Fourier series of $u^{(2k_1,2k_2)}(x,y)$ are obtained from:

$$U_{4mn}^{(2k_1,2k_2)} = \frac{4}{ab} \sum_{j=1}^{k_1} \sum_{l=1}^{k_2} (-1)^{k_1+k_2-j-l+2} \alpha_m^{2k_1-2j+1} \beta_n^{2k_2-2l+1} [(-1)^{m+n} u^{(2j-2,2l-2)}(a,b)$$

$$- (-1)^m u^{(2j-2,2l-2)}(a,0) - (-1)^n u^{(2j-2,2l-2)}(0,b) + u^{(2j-2,2l-2)}(0,0)]$$

$$+ \frac{2}{a} \sum_{j=1}^{k_1} (-1)^{k_1+k_2-j+1} \alpha_m^{2k_1-2j+1} \beta_n^{2k_2} [(-1)^m c_{2n}^{(2j-2)} - d_{2n}^{(2j-2)}]$$

$$+ \frac{2}{b} \sum_{l=1}^{k_2} (-1)^{k_1+k_2-l+1} \alpha_m^{2k_1} \beta_n^{2k_2-2l+1} [(-1)^n e_{2m}^{(2l-2)} - f_{2m}^{(2l-2)}]$$

$$+ (-1)^{k_1+k_2} \alpha_m^{2k_1} \beta_n^{2k_2} U_{4mn}. \tag{6.81}$$

The Fourier coefficients for the half-range Fourier series of $u^{(2k_1+1,2k_2)}(x,y)$ are calculated from:

for $m = 0$ and $n > 0$,

$$U_{30n}^{(2k_1+1,0)} = \frac{2}{a} [c_{2n}^{(2k_1)} - d_{2n}^{(2k_1)}] \quad (k_2 = 0), \tag{6.82}$$

and

$$U_{30n}^{(2k_1+1,2k_2)} = \frac{4}{ab} \sum_{l=1}^{k_2} (-1)^{k_2-l+1} \beta_n^{2k_2-2l+1} [(-1)^n u^{(2k_1,2l-2)}(a,b) - u^{(2k_1,2l-2)}(a,0)$$

$$- (-1)^n u^{(2k_1,2l-2)}(0,b) + u^{(2k_1,2l-2)}(0,0)] + \frac{2}{a} (-1)^{k_2} \beta_n^{2k_2} [c_{2n}^{(2k_1)} - d_{2n}^{(2k_1)}]$$

$$(k_2 > 0); \tag{6.83}$$

for $m > 0$ and $n > 0$,

$$U_{3mn}^{(2k_1+1,2k_2)} = \frac{4}{ab} \sum_{j=0}^{k_1} \sum_{l=1}^{k_2} (-1)^{k_1+k_2-j-l+1} \alpha_m^{2k_1-2j} \beta_n^{2k_2-2l+1} [(-1)^{m+n} u^{(2j,2l-2)}(a,b)$$

$$- (-1)^m u^{(2j,2l-2)}(a,0) - (-1)^n u^{(2j,2l-2)}(0,b) + u^{(2j,2l-2)}(0,0)]$$

$$+ \frac{2}{a} \sum_{j=0}^{k_1} (-1)^{k_1+k_2-j} \alpha_m^{2k_1-2j} \beta_n^{2k_2} [(-1)^m c_{2n}^{(2j)} - d_{2n}^{(2j)}] \tag{6.84}$$

$$+ \frac{2}{b} \sum_{l=1}^{k_2} (-1)^{k_1+k_2-l+1} \alpha_m^{2k_1+1} \beta_n^{2k_2-2l+1} [(-1)^n e_{2m}^{(2l-2)} - f_{2m}^{(2l-2)}]$$

$$+ (-1)^{k_1+k_2} \alpha_m^{2k_1+1} \beta_n^{2k_2} U_{4mn}.$$

The Fourier coefficients for the half-range Fourier series of $u^{(2k_1,2k_2+1)}(x,y)$ are determined from:

for $m > 0$ and $n = 0$,

$$U_{2m0}^{(0,2k_2+1)} = \frac{2}{b}[e_{2m}^{(2k_2)} - f_{2m}^{(2k_2)}] \quad (k_1 = 0), \tag{6.85}$$

and

$$\begin{aligned}
U_{2m0}^{(2k_1,2k_2+1)} &= \frac{4}{ab}\sum_{j=1}^{k_1}(-1)^{k_1-j+1}\alpha_m^{2k_1-2j+1}[(-1)^m u^{(2j-2,2k_2)}(a,b) \\
&\quad -(-1)^m u^{(2j-2,2k_2)}(a,0) - u^{(2j-2,2k_2)}(0,b) + u^{(2j-2,2k_2)}(0,0)] \\
&\quad +\frac{2}{b}(-1)^{k_1}\alpha_m^{2k_1}[e_{2m}^{(2k_2)} - f_{2m}^{(2k_2)}] \quad (k_1 > 0);
\end{aligned} \tag{6.86}$$

for $m > 0$ and $n > 0$,

$$\begin{aligned}
U_{2mn}^{(2k_1,2k_2+1)} &= \frac{4}{ab}\sum_{j=1}^{k_1}\sum_{l=0}^{k_2}(-1)^{k_1+k_2-j-l+1}\alpha_m^{2k_1-2j+1}\beta_n^{2k_2-2l}[(-1)^{m+n}u^{(2j-2,2l)}(a,b) \\
&\quad -(-1)^m u^{(2j-2,2l)}(a,0) - (-1)^n u^{(2j-2,2l)}(0,b) + u^{(2j-2,2l)}(0,0)] \\
&\quad +\frac{2}{a}\sum_{j=1}^{k_1}(-1)^{k_1+k_2-j+1}\alpha_m^{2k_1-2j+1}\beta_n^{2k_2+1}[(-1)^m c_{2n}^{(2j-2)} - d_{2n}^{(2j-2)}] \\
&\quad +\frac{2}{b}\sum_{l=0}^{k_2}(-1)^{k_1+k_2-l}\alpha_m^{2k_1}\beta_n^{2k_2-2l}[(-1)^n e_{2m}^{(2l)} - f_{2m}^{(2l)}] \\
&\quad +(-1)^{k_1+k_2}\alpha_m^{2k_1}\beta_n^{2k_2+1}U_{4mn}.
\end{aligned} \tag{6.87}$$

The Fourier coefficients for the half-range Fourier series of $u^{(2k_1+1,2k_2+1)}(x,y)$ are determined from:

for $m = 0$ and $n = 0$,

$$\begin{aligned}
U_{100}^{(2k_1+1,2k_2+1)} &= \frac{4}{ab}[u^{(2k_1,2k_2)}(a,b) - u^{(2k_1,2k_2)}(a,0) \\
&\quad - u^{(2k_1,2k_2)}(0,b) + u^{(2k_1,2k_2)}(0,0)];
\end{aligned} \tag{6.88}$$

for $m = 0$ and $n > 0$,

$$U_{10n}^{(2k_1+1,2k_2+1)} = \frac{4}{ab} \sum_{l=0}^{k_2} (-1)^{k_2-l} \beta_n^{2k_2-2l} [(-1)^n u^{(2k_1,2l)}(a,b) - u^{(2k_1,2l)}(a,0)$$

$$- (-1)^n u^{(2k_1,2l)}(0,b) + u^{(2k_1,2l)}(0,0)] + \frac{2}{a}(-1)^{k_2} \beta_n^{2k_2+1} [c_{2n}^{(2k_1)} - d_{2n}^{(2k_1)}];$$

$$(6.89)$$

for $m > 0$ and $n = 0$,

$$U_{1m0}^{(2k_1+1,2k_2+1)} = \frac{4}{ab} \sum_{j=0}^{k_1} (-1)^{k_1-j} \alpha_m^{2k_1-2j} [(-1)^m u^{(2j,2k_2)}(a,b) - (-1)^m u^{(2j,2k_2)}(a,0)$$

$$- u^{(2j,2k_2)}(0,b) + u^{(2j,2k_2)}(0,0)] + \frac{2}{b}(-1)^{k_1} \alpha_m^{2k_1+1} [e_{2m}^{(2k_2)} - f_{2m}^{(2k_2)}];$$

$$(6.90)$$

for $m > 0$ and $n > 0$,

$$U_{1mn}^{(2k_1+1,2k_2+1)} = \frac{4}{ab} \sum_{j=0}^{k_1} \sum_{l=0}^{k_2} (-1)^{k_1+k_2-j-l} \alpha_m^{2k_1-2j} \beta_n^{2k_2-2l} [(-1)^{m+n} u^{(2j,2l)}(a,b)$$

$$- (-1)^m u^{(2j,2l)}(a,0) - (-1)^n u^{(2j,2l)}(0,b) + u^{(2j,2l)}(0,0)]$$

$$+ \frac{2}{a} \sum_{j=0}^{k_1} (-1)^{k_1+k_2-j} \alpha_m^{2k_1-2j} \beta_n^{2k_2+1} [(-1)^m c_{2n}^{(2j)} - d_{2n}^{(2j)}] \qquad (6.91)$$

$$+ \frac{2}{b} \sum_{l=0}^{k_2} (-1)^{k_1+k_2-l} \alpha_m^{2k_1+1} \beta_n^{2k_2-2l} [(-1)^n e_{2m}^{(2l)} - f_{2m}^{(2l)}]$$

$$+ (-1)^{k_1+k_2} \alpha_m^{2k_1+1} \beta_n^{2k_2+1} U_{4mn}.$$

These formulas look like their counterparts in the full-range Fourier series expansions except that a factor of 1/2 (or 1/4) shall be applied to the boundary (or corner) terms, and the lower order derivatives are now calculated at $x = 0$ (and/or $y = 0$), rather than at $x = -a$ (and/or $y = -b$).

A two-dimensional function $u(x,y)$ can also be expanded into a half-range cosine-cosine series, sine-cosine series, or cosine-sine series on the domain $[0,a] \times [0,b]$. The resulting formulas are readily available by comparing the like terms in the full- and half-range series expansions and making some necessary modifications as done above for the Fourier sine-sine series.

6.3 EXAMPLES

In this section, the above formulas derived for calculating the Fourier coefficients of the partial derivatives will be validated by using the polynomials and trigonometric functions given in Table 6.1.

For the given functions, their partial derivatives of any orders are readily obtained, and so are the corresponding Fourier coefficients, including the boundary ones, regardless which particular form of Fourier series is chosen. The Fourier coefficients such calculated will be considered accurate and used as reference values for checking the accuracies of the Fourier coefficients determined using the formulas given in the previous sections.

For this purpose, some error indexes (or, norms) need to be first defined to measure the accuracies of the calculated Fourier coefficients for the partial derivatives of orders up to $2r$ (r is a positive integer). Take the Fourier series expansion over the domain $[-a,a]\times[-b,b]$ for example. For the sake of numerical calculations, we truncate all the series expansions to $m=M$ and $n=N$, and define the error indexes as

$$e^{\|2r\|}(u) = \frac{1}{S_r} \sum_{\substack{k_1\geq0,k_2\geq0 \\ k_1+k_2\leq2r}} \frac{1}{F_0 \cdot \bar{U}_{max}^{(k_1,k_2)}} \left[\sum_{m=0}^{M}\sum_{n=0}^{N}\left|U_{1mn}^{(k_1,k_2)}-\bar{U}_{1mn}^{(k_1,k_2)}\right| + \sum_{m=1}^{M}\sum_{n=0}^{N}\left|U_{2mn}^{(k_1,k_2)}-\bar{U}_{2mn}^{(k_1,k_2)}\right| \right.$$

$$\left. + \sum_{m=0}^{M}\sum_{n=1}^{N}\left|U_{3mn}^{(k_1,k_2)}-\bar{U}_{3mn}^{(k_1,k_2)}\right| + \sum_{m=1}^{M}\sum_{n=1}^{N}\left|U_{4mn}^{(k_1,k_2)}-\bar{U}_{4mn}^{(k_1,k_2)}\right| \right], \tag{6.92}$$

TABLE 6.1
Sample Functions with Different Fourier Series Expansions[a]

No.	Sample Function	Fourier Series Expansion
1	$u(x,y) = \left[\frac{1}{2}-\frac{x}{a}-\frac{1}{2}\left(\frac{x}{a}\right)^2+\left(\frac{x}{a}\right)^3\right]\left[\frac{1}{2}-\frac{y}{b}-\frac{1}{2}\left(\frac{y}{b}\right)^2+\left(\frac{y}{b}\right)^3\right]$, $(x,y)\in[-a,a]\times[-b,b]$	Full-range Fourier series over the domain $[-a,a]\times[-b,b]$
2	$u(x,y) = \sin(\alpha_0 x)\cos(\beta_0 y)$, $\alpha_0=\pi/2a, \beta_0=\pi/2b, (x,y)\in[-a,a]\times[-b,b]$	Full-range Fourier series over the domain $[-a,a]\times[-b,b]$
3	$u(x,y) = \left[\frac{1}{2}-\frac{x}{a}-\frac{1}{2}\left(\frac{x}{a}\right)^2+\left(\frac{x}{a}\right)^3\right]\left[\frac{1}{2}-\frac{y}{b}-\frac{1}{2}\left(\frac{y}{b}\right)^2+\left(\frac{y}{b}\right)^3\right]$, $(x,y)\in[0,a]\times[0,b]$	Half-range sine-sine series over the domain $[0,a]\times[0,b]$
4	$u(x,y) = \sin(\alpha_0 x)\cos(\beta_0 y)$, $\alpha_0=\pi/2a, \beta_0=\pi/2b, (x,y)\in[0,a]\times[0,b]$	Half-range sine-sine series over the domain $[0,a]\times[0,b]$

Note:

[a] This table was taken from Mathematical Methods in the Applied Sciences, 40, W. M. Sun and Z. M. Zhang, On Fourier series for higher order (partial) derivatives of functions, 2197–2218, Copyright John Wiley & Sons, Ltd (2017).

TABLE 6.2

Error Indexes of the Fourier Coefficients for the Partial Derivatives of the Sample Functions[a]

No. of Sample	$e^{\|2r\|}(u)$		
Functions	$2r = 2$	$2r = 4$	$2r = 6$
1	7.391266E–18	5.215461E–14	3.130863E–10
2	1.475631E–17	1.982656E–14	4.810975E–11
3	4.024085E–17	6.004995E–14	4.519101E–10
4	9.975485E–16	1.530610E–12	3.733972E–09

Note:

[a] This table was taken from Mathematical Methods in the Applied Sciences, 40, W. M. Sun and Z. M. Zhang, On Fourier series for higher order (partial) derivatives of functions, 2197–2218, Copyright John Wiley & Sons, Ltd (2017).

where $S_r = (2r+1)(r+1)$ is the total number of the functions involved; $F_0 = 4MN + 2M + 2N + 1$ is the total number of the Fourier coefficients; $U_{1mn}^{(k_1,k_2)}, U_{2mn}^{(k_1,k_2)}, U_{3mn}^{(k_1,k_2)}$, $U_{4mn}^{(k_1,k_2)}$ and $\bar{U}_{1mn}^{(k_1,k_2)}, \bar{U}_{2mn}^{(k_1,k_2)}, \bar{U}_{3mn}^{(k_1,k_2)}, \bar{U}_{4mn}^{(k_1,k_2)}$ are the calculated and analytical (or exact) Fourier coefficients, respectively; $\bar{U}_{max}^{(k_1,k_2)}$ is the maximum value of the analytical Fourier coefficients (in case of $\bar{U}_{max}^{(k_1,k_2)} = 0$, $\bar{U}_{max}^{(k_1,k_2)} = 1$ will be used instead).

By setting $a/b = 1.0$ and $M = N = 40$, presented in Table 6.2 are the error indexes of the Fourier coefficients for the various orders of partial derivatives of the sample functions given in Table 6.1.

It can be seen from Table 6.2 that while the calculated Fourier coefficients are generally very accurate, the error indexes tend to deteriorate as the orders of derivatives increase. Instead of viewing it as an indication of the unreliability of the formulas involved, we believe that this actually results from the increased truncation errors for the higher order derivatives; or more explicitly, each time a function is differentiated, the convergence rate of the corresponding Fourier series is roughly slowed down by a factor of α_m or β_n.

6.4 SUFFICIENT CONDITIONS FOR TERM-BY-TERM DIFFERENTIATIONS OF THE FOURIER SERIES OF TWO-DIMENSIONAL FUNCTIONS

We now establish the sufficient conditions for the term-by-term differentiations of the Fourier series of a two-dimensional function.

Theorem 6.2 *Let the function* $u(x,y) \in C^{2r-1}([-a,a] \times [-b,b])$ *and its $2r$-th order partial derivatives be absolutely integrable. If $u(x,y)$ is given in the form of the full-range Fourier series over the domain $[-a,a] \times [-b,b]$, then for any integers*

$p_1, p_2 = 0,1,2,\cdots$ and $1 \le p_1 + p_2 \le 2r$, the sufficient condition for this series to be termwise differentiable for the corresponding series of $u^{(p_1,p_2)}(x,y)$ is that

$$u^{(k_1,0)}(a,y) = u^{(k_1,0)}(-a,y), \quad y \in (-b,b), \quad k_1 = 0,1,\cdots,2r-1; \qquad (6.93)$$

$$u^{(0,k_2)}(x,b) = u^{(0,k_2)}(x,-b), \quad x \in (-a,a), \quad k_2 = 0,1,\cdots,2r-1; \qquad (6.94)$$

$$u^{(k_1,k_2)}(a,b) - u^{(k_1,k_2)}(a,-b) - u^{(k_1,k_2)}(-a,b) + u^{(k_1,k_2)}(-a,-b) = 0$$
$$k_1,k_2 = 0,1,2,\cdots,k_1 + k_2 \le 2r-2. \qquad (6.95)$$

Proof. Although the complete proof of this theorem is quite lengthy and tedious, the results are readily verified by using the formulas in Sec. 6.2 regarding the calculations of the Fourier coefficients of the partial derivatives. As an illustration, we here only consider the Fourier coefficients $U_{2mn}^{(2k_1,2k_2+1)}$ in the Fourier series:

$$u^{(2k_1,2k_2+1)}(x,y) = \sum_{m=0}^{\infty}\sum_{n=0}^{\infty}\lambda_{mn}[U_{1mn}^{(2k_1,2k_2+1)}\cos(\alpha_m x)\cos(\beta_n y)$$

$$+ U_{2mn}^{(2k_1,2k_2+1)}\sin(\alpha_m x)\cos(\beta_n y) + U_{3mn}^{(2k_1,2k_2+1)}\cos(\alpha_m x)\sin(\beta_n y)$$

$$+ U_{4mn}^{(2k_1,2k_2+1)}\sin(\alpha_m x)\sin(\beta_n y)], \qquad (6.96)$$

where $k_1, k_2 = 0, 1, 2, \cdots$ and $1 \le 2k_1 + 2k_2 + 1 \le 2r$.

In light of expressions (6.93) and (6.94), we have:

for $j = 0,1,\cdots,2r-1$,

$$c_{1n}^{(j)} = d_{1n}^{(j)} \quad (n \ge 0), \qquad (6.97)$$

$$c_{2n}^{(j)} = d_{2n}^{(j)} \quad (n > 0); \qquad (6.98)$$

for $l = 0,1,\cdots,2r-1$,

$$e_{1m}^{(l)} = f_{1m}^{(l)} \quad (m \ge 0), \qquad (6.99)$$

$$e_{2m}^{(l)} = f_{2m}^{(l)} \quad (m > 0). \qquad (6.100)$$

Since $1 \le 2k_1 + 2k_2 + 1 \le 2r$, then we have

$$0 \le 2k_2 \le 2r-2, \qquad (6.101)$$

and

$$e_{2m}^{(2k_2)} = f_{2m}^{(2k_2)} \quad (m > 0). \qquad (6.102)$$

Meanwhile, for $j = 1, 2, \cdots, k_1$,

$$0 \leq 2j - 2 \leq 2j - 2 + 2k_2 \leq 2k_1 + 2k_2 - 2 \leq 2r - 2. \qquad (6.103)$$

Therefore, we have

$$u^{(2j-2,2k_2)}(a,b) - u^{(2j-2,2k_2)}(a,-b) - u^{(2j-2,2k_2)}(-a,b) + u^{(2j-2,2k_2)}(-a,-b) = 0. \qquad (6.104)$$

According to (6.56) and (6.57), it is immediately clear that
for $m > 0$ and $n = 0$,

$$U_{2m0}^{(0,2k_2+1)} = \frac{1}{b}[e_{2m}^{(2k_2)} - f_{2m}^{(2k_2)}] = 0 \quad (k_1 = 0), \qquad (6.105)$$

and

$$U_{2m0}^{(2k_1,2k_2+1)} = \frac{1}{ab}\sum_{j=1}^{k_1}(-1)^{m+k_1-j+1}\alpha_m^{2k_1-2j+1}[u^{(2j-2,2k_2)}(a,b) - u^{(2j-2,2k_2)}(a,-b)$$

$$- u^{(2j-2,2k_2)}(-a,b) + u^{(2j-2,2k_2)}(-a,-b)] \qquad (6.106)$$

$$+ \frac{1}{b}(-1)^{k_1}\alpha_m^{2k_1}[e_{2m}^{(2k_2)} - f_{2m}^{(2k_2)}] = 0 \quad (k_1 > 0).$$

Similarly, we observe that
for $j = 1, 2, \cdots, k_1$, and $l = 0, 1, \cdots, k_2$,

$$0 \leq 2j - 2 + 2l \leq 2k_1 + 2k_2 - 2 \leq 2r - 2, \qquad (6.107)$$

$$0 \leq 2j - 2 \leq 2k_1 - 2 \leq 2r - 2, \qquad (6.108)$$

and

$$0 \leq 2l \leq 2k_2 \leq 2r - 2. \qquad (6.109)$$

Thus, the following equations hold:

$$u^{(2j-2,2l)}(a,b) - u^{(2j-2,2l)}(a,-b) - u^{(2j-2,2l)}(-a,b) + u^{(2j-2,2l)}(-a,-b) = 0, \qquad (6.110)$$

$$c_{2n}^{(2j-2)} - d_{2n}^{(2j-2)} = 0 \quad (n > 0), \qquad (6.111)$$

$$e_{2m}^{(2l)} - f_{2m}^{(2l)} = 0 \quad (m > 0). \qquad (6.112)$$

Finally, for $m > 0$ and $n > 0$,

$$U_{2mn}^{(2k_1,2k_2+1)} = \frac{1}{ab} \sum_{j=1}^{k_1} \sum_{l=0}^{k_2} (-1)^{m+n+k_1+k_2-j-l+1} \alpha_m^{2k_1-2j+1} \beta_n^{2k_2-2l} [u^{(2j-2,2l)}(a,b)$$

$$- u^{(2j-2,2l)}(a,-b) - u^{(2j-2,2l)}(-a,b) + u^{(2j-2,2l)}(-a,-b)]$$

$$+ \frac{1}{a} \sum_{j=1}^{k_1} (-1)^{m+k_1+k_2-j+1} \alpha_m^{2k_1-2j+1} \beta_n^{2k_2+1} [c_{2n}^{(2j-2)} - d_{2n}^{(2j-2)}] \qquad (6.113)$$

$$+ \frac{1}{b} \sum_{l=0}^{k_2} (-1)^{n+k_1+k_2-l} \alpha_m^{2k_1} \beta_n^{2k_2-2l} [e_{2m}^{(2l)} - f_{2m}^{(2l)}]$$

$$+ (-1)^{k_1+k_2} \alpha_m^{2k_1} \beta_n^{2k_2+1} U_{4mn} = (-1)^{k_1+k_2} \alpha_m^{2k_1} \beta_n^{2k_2+1} U_{4mn},$$

which is equivalent to the term-by-term differentiations of the full-range Fourier series, (6.8), of the function $u(x,y)$.

The similar expressions can be derived for other terms to complete the proof of this theorem.

The sufficient conditions for term-by-term differentiations can also be established for the various forms of half-range Fourier series expansions, as stated in the theorems below.

Theorem 6.3 *Let the function* $u(x,y) \in C^{2r-1}([0,a] \times [0,b])$ *and its* $2r$-*th order partial derivatives be absolutely integrable. If the function* $u(x,y)$ *is expanded into half-range Fourier sine-sine series over the domain* $[0,a] \times [0,b]$, *then for any integers* $p_1, p_2 = 0,1,2,\cdots$ *and* $1 \le p_1 + p_2 \le 2r$, *the sufficient condition for this series to be termwise differentiable for the corresponding series of* $u^{(p_1,p_2)}(x,y)$ *is that*

$$u^{(2k_1,0)}(a,y) = 0, u^{(2k_1,0)}(0,y) = 0, y \in (0,b), k_1 = 0,1,\cdots,r-1; \qquad (6.114)$$

$$u^{(0,2k_2)}(x,b) = 0, u^{(0,2k_2)}(x,0) = 0, x \in (0,a), k_2 = 0,1,\cdots,r-1; \qquad (6.115)$$

$$u^{(2k_1,2k_2)}(a,b) = 0, u^{(2k_1,2k_2)}(a,0) = 0, u^{(2k_1,2k_2)}(0,b) = 0, u^{(2k_1,2k_2)}(0,0) = 0,$$
$$k_1, k_2 = 0,1,2,\cdots,k_1 + k_2 \le r-1. \qquad (6.116)$$

It should be noted that the appropriate forms of half-range Fourier series are dictated by the parities of the derivative orders in x- and y-direction, as specifically given in (6.73)–(6.76).

Theorem 6.4 *Let the function* $u(x, y) \in C^{2r-1}([0,a] \times [0,b])$ *and its 2r-th order partial derivatives be absolutely integrable. If the function* $u(x, y)$ *is expanded into half-range Fourier cosine-cosine series over the domain* $[0,a] \times [0,b]$, *then for any integers* $p_1, p_2 = 0,1,2,\cdots$ *and* $1 \le p_1 + p_2 \le 2r$, *the sufficient condition for this series to be termwise differentiable for the corresponding series of* $u^{(p_1,p_2)}(x, y)$ *is that*

$$u^{(2k_1+1,0)}(a, y) = 0, u^{(2k_1+1,0)}(0, y) = 0, y \in (0,b), k_1 = 0,1,\cdots,r-1; \quad (6.117)$$

$$u^{(0,2k_2+1)}(x, b) = 0, u^{(0,2k_2+1)}(x, 0) = 0, x \in (0,a), k_2 = 0,1,\cdots,r-1; \quad (6.118)$$

and for $r > 1$,

$$u^{(2k_1+1,2k_2+1)}(a,b)=0, u^{(2k_1+1,2k_2+1)}(a,0)=0, u^{(2k_1+1,2k_2+1)}(0,b)=0, u^{(2k_1+1,2k_2+1)}(0,0)=0,$$
$$k_1, k_2 = 0,1,2,\cdots,k_1 + k_2 \le r - 2. \quad (6.119)$$

Theorem 6.5 *Let the function* $u(x, y) \in C^{2r-1}([0,a] \times [0,b])$ *and its 2r-th order partial derivatives be absolutely integrable. If the function* $u(x, y)$ *is expanded into half-range Fourier sine-cosine series over the domain* $[0,a] \times [0,b]$, *then for any integers* $p_1, p_2 = 0,1,2,\cdots$ *and* $1 \le p_1 + p_2 \le 2r$, *the sufficient condition for this series to be termwise differentiable for the corresponding series of* $u^{(p_1,p_2)}(x, y)$ *is that*

$$u^{(2k_1,0)}(a, y) = 0, u^{(2k_1,0)}(0, y) = 0, y \in (0,b), k_1 = 0,1,\cdots,r-1; \quad (6.120)$$

$$u^{(0,2k_2+1)}(x, b) = 0, u^{(0,2k_2+1)}(x, 0) = 0, x \in (0,a), k_2 = 0,1,\cdots,r-1; \quad (6.121)$$

and for $r > 1$,

$$u^{(2k_1,2k_2+1)}(a,b) = 0, u^{(2k_1,2k_2+1)}(a,0) = 0, u^{(2k_1,2k_2+1)}(0,b) = 0, u^{(2k_1,2k_2+1)}(0,0) = 0,$$
$$k_1, k_2 = 0,1,2,\cdots,k_1 + k_2 \le r - 2. \quad (6.122)$$

Theorem 6.6 *Let the function* $u(x, y) \in C^{2r-1}([0,a] \times [0,b])$ *and its 2r-th order partial derivatives be absolutely integrable. If the function* $u(x, y)$ *is expanded into half-range Fourier cosine-sine series over the domain* $[0,a] \times [0,b]$, *then for any integers* $p_1, p_2 = 0,1,2,\cdots$ *and* $1 \le p_1 + p_2 \le 2r$, *the sufficient condition for this series to be termwise differentiable for the corresponding series of* $u^{(p_1,p_2)}(x, y)$

is that

$$u^{(2k_1+1,0)}(a,y)=0, u^{(2k_1+1,0)}(0,y)=0, y\in(0,b), k_1=0,1,\cdots,r-1; \quad (6.123)$$

$$u^{(0,2k_2)}(x,b)=0, u^{(0,2k_2)}(x,0)=0, x\in(0,a), k_2=0,1,\cdots,r-1; \quad (6.124)$$

and for $r>1$,

$$u^{(2k_1+1,2k_2)}(a,b)=0, u^{(2k_1+1,2k_2)}(a,0)=0, u^{(2k_1+1,2k_2)}(0,b)=0, u^{(2k_1+1,2k_2)}(0,0)=0,$$
$$k_1,k_2=0,1,2,\cdots,k_1+k_2\le r-2. \quad (6.125)$$

The face values of these theorems lie in the fact that they specifically regulate under what conditions the Fourier series of the partial derivatives can be directly obtained from the term-by-term differentiations of the series for the root functions. More importantly, however, they actually point to a new direction for solving differential equations or boundary values problems as discussed later.

Appendix A: Additional Integral Formulas

For the completeness, other relevant integral formulas are simply given below without proofs:

$$I_{SC}(2k_1,2k_2,x_0,y_0)=\sum_{j=1}^{k_1}\sum_{l=1}^{k_2}(-1)^{k_1+k_2-j-l+1}\alpha_m^{2k_1-2j+1}\beta_n^{2k_2-2l}$$
$$\cdot[(-1)^{m+n}u^{(2j-2,2l-1)}(a,b)-(-1)^m\cos(\beta_n y_0)u^{(2j-2,2l-1)}(a,y_0)$$
$$-(-1)^n\cos(\alpha_m x_0)u^{(2j-2,2l-1)}(x_0,b)$$
$$+\cos(\alpha_m x_0)\cos(\beta_n y_0)u^{(2j-2,2l-1)}(x_0,y_0)]$$
$$+\sum_{j=1}^{k_1}(-1)^{k_1+k_2-j+1}\alpha_m^{2k_1-2j+1}\beta_n^{2k_2}\int_{y_0}^{b}[(-1)^m u^{(2j-2,0)}(a,y)$$
$$-\cos(\alpha_m x_0)u^{(2j-2,0)}(x_0,y)]\cos(\beta_n y)dy$$
$$+\sum_{l=1}^{k_2}(-1)^{k_1+k_2-l}\alpha_m^{2k_1}\beta_n^{2k_2-2l}\int_{x_0}^{a}[(-1)^n u^{(0,2l-1)}(x,b)$$
$$-\cos(\beta_n y_0)u^{(0,2l-1)}(x,y_0)]\sin(\alpha_m x)dx$$
$$+(-1)^{k_1+k_2}\alpha_m^{2k_1}\beta_n^{2k_2}I_{SC}(0,0,x_0,y_0), \quad (A.1)$$

$$I_{CS}(2k_1, 2k_2, x_0, y_0) = \sum_{j=1}^{k_1}\sum_{l=1}^{k_2}(-1)^{k_1+k_2-j-l+1}\alpha_m^{2k_1-2j}\beta_n^{2k_2-2l+1}$$

$$\cdot[(-1)^{m+n}u^{(2j-1,2l-2)}(a,b) - (-1)^m\cos(\beta_n y_0)u^{(2j-1,2l-2)}(a,y_0)$$

$$-(-1)^n\cos(\alpha_m x_0)u^{(2j-1,2l-2)}(x_0,b)$$

$$+\cos(\alpha_m x_0)\cos(\beta_n y_0)u^{(2j-1,2l-2)}(x_0,y_0)]$$

$$+\sum_{j=1}^{k_1}(-1)^{k_1+k_2-j}\alpha_m^{2k_1-2j}\beta_n^{2k_2}\int_{y_0}^{b}[(-1)^m u^{(2j-1,0)}(a,y)$$

$$-\cos(\alpha_m x_0)u^{(2j-1,0)}(x_0,y)]\sin(\beta_n y)dy$$

$$+\sum_{l=1}^{k_2}(-1)^{k_1+k_2-l+1}\alpha_m^{2k_1}\beta_n^{2k_2-2l+1}\int_{x_0}^{a}[(-1)^n u^{(0,2l-2)}(x,b)$$

$$-\cos(\beta_n y_0)u^{(0,2l-2)}(x,y_0)]\cos(\alpha_m x)dx$$

$$+(-1)^{k_1+k_2}\alpha_m^{2k_1}\beta_n^{2k_2}I_{CS}(0,0,x_0,y_0), \tag{A.2}$$

$$I_{SS}(2k_1, 2k_2, x_0, y_0) = \sum_{j=1}^{k_1}\sum_{l=1}^{k_2}(-1)^{k_1+k_2-j-l+2}\alpha_m^{2k_1-2j+1}\beta_n^{2k_2-2l+1}$$

$$\cdot[(-1)^{m+n}u^{(2j-2,2l-2)}(a,b) - (-1)^m\cos(\beta_n y_0)u^{(2j-2,2l-2)}(a,y_0)$$

$$-(-1)^n\cos(\alpha_m x_0)u^{(2j-2,2l-2)}(x_0,b)$$

$$+\cos(\alpha_m x_0)\cos(\beta_n y_0)u^{(2j-2,2l-2)}(x_0,y_0)]$$

$$+\sum_{j=1}^{k_1}(-1)^{k_1+k_2-j+1}\alpha_m^{2k_1-2j+1}\beta_n^{2k_2}\int_{y_0}^{b}[(-1)^m u^{(2j-2,0)}(a,y)$$

$$-\cos(\alpha_m x_0)u^{(2j-2,0)}(x_0,y)]\sin(\beta_n y)dy$$

$$+\sum_{l=1}^{k_2}(-1)^{k_1+k_2-l+1}\alpha_m^{2k_1}\beta_n^{2k_2-2l+1}\int_{x_0}^{a}[(-1)^n u^{(0,2l-2)}(x,b)$$

$$-\cos(\beta_n y_0)u^{(0,2l-2)}(x,y_0)]\sin(\alpha_m x)dx$$

$$+(-1)^{k_1+k_2}\alpha_m^{2k_1}\beta_n^{2k_2}I_{SS}(0,0,x_0,y_0), \tag{A.3}$$

$$I_{CC}(2k_1+1,2k_2,x_0,y_0) = \sum_{j=0}^{k_1}\sum_{l=1}^{k_2}(-1)^{k_1+k_2-j-l}\alpha_m^{2k_1-2j}\beta_n^{2k_2-2l}$$

$$\cdot[(-1)^{m+n}u^{(2j,2l-1)}(a,b)-(-1)^m\cos(\beta_n y_0)u^{(2j,2l-1)}(a,y_0)$$

$$-(-1)^n\cos(\alpha_m x_0)u^{(2j,2l-1)}(x_0,b)$$

$$+\cos(\alpha_m x_0)\cos(\beta_n y_0)u^{(2j,2l-1)}(x_0,y_0)]$$

$$+\sum_{j=0}^{k_1}(-1)^{k_1+k_2-j}\alpha_m^{2k_1-2j}\beta_n^{2k_2}\int_{y_0}^{b}[(-1)^m u^{(2j,0)}(a,y)$$

$$-\cos(\alpha_m x_0)u^{(2j,0)}(x_0,y)]\cos(\beta_n y)dy$$

$$+\sum_{l=1}^{k_2}(-1)^{k_1+k_2-l}\alpha_m^{2k_1+1}\beta_n^{2k_2-2l}\int_{x_0}^{a}[(-1)^n u^{(0,2l-1)}(x,b)$$

$$-\cos(\beta_n y_0)u^{(0,2l-1)}(x,y_0)]\sin(\alpha_m x)dx$$

$$+(-1)^{k_1+k_2}\alpha_m^{2k_1+1}\beta_n^{2k_2}I_{SC}(0,0,x_0,y_0), \tag{A.4}$$

$$I_{SC}(2k_1+1,2k_2,x_0,y_0) = \sum_{j=1}^{k_1}\sum_{l=1}^{k_2}(-1)^{k_1+k_2-j-l+1}\alpha_m^{2k_1-2j+1}\beta_n^{2k_2-2l}$$

$$\cdot[(-1)^{m+n}u^{(2j-1,2l-1)}(a,b)-(-1)^m\cos(\beta_n y_0)u^{(2j-1,2l-1)}(a,y_0)$$

$$-(-1)^n\cos(\alpha_m x_0)u^{(2j-1,2l-1)}(x_0,b)$$

$$+\cos(\alpha_m x_0)\cos(\beta_n y_0)u^{(2j-1,2l-1)}(x_0,y_0)]$$

$$+\sum_{j=1}^{k_1}(-1)^{k_1+k_2-j+1}\alpha_m^{2k_1-2j+1}\beta_n^{2k_2}\int_{y_0}^{b}[(-1)^m u^{(2j-1,0)}(a,y)$$

$$-\cos(\alpha_m x_0)u^{(2j-1,0)}(x_0,y)]\cos(\beta_n y)dy$$

$$+\sum_{l=1}^{k_2}(-1)^{k_1+k_2-l+1}\alpha_m^{2k_1+1}\beta_n^{2k_2-2l}\int_{x_0}^{a}[(-1)^n u^{(0,2l-1)}(x,b)$$

$$-\cos(\beta_n y_0)u^{(0,2l-1)}(x,y_0)]\cos(\alpha_m x)dx$$

$$+(-1)^{k_1+k_2+1}\alpha_m^{2k_1+1}\beta_n^{2k_2}I_{CC}(0,0,x_0,y_0), \tag{A.5}$$

$$I_{CS}(2k_1+1,2k_2,x_0,y_0) = \sum_{j=0}^{k_1}\sum_{l=1}^{k_2}(-1)^{k_1+k_2-j-l+1}\alpha_m^{2k_1-2j}\beta_n^{2k_2-2l+1}$$

$$\cdot[(-1)^{m+n}u^{(2j,2l-2)}(a,b)-(-1)^m\cos(\beta_n y_0)u^{(2j,2l-2)}(a,y_0)$$

$$-(-1)^n\cos(\alpha_m x_0)u^{(2j,2l-2)}(x_0,b)$$

$$+\cos(\alpha_m x_0)\cos(\beta_n y_0)u^{(2j,2l-2)}(x_0,y_0)]$$

$$+\sum_{j=0}^{k_1}(-1)^{k_1+k_2-j}\alpha_m^{2k_1-2j}\beta_n^{2k_2}\int_{y_0}^{b}[(-1)^m u^{(2j,0)}(a,y)$$

$$-\cos(\alpha_m x_0)u^{(2j,0)}(x_0,y)]\sin(\beta_n y)dy$$

$$+\sum_{l=1}^{k_2}(-1)^{k_1+k_2-l+1}\alpha_m^{2k_1+1}\beta_n^{2k_2-2l+1}\int_{x_0}^{a}[(-1)^n u^{(0,2l-2)}(x,b)$$

$$-\cos(\beta_n y_0)u^{(0,2l-2)}(x,y_0)]\sin(\alpha_m x)dx$$

$$+(-1)^{k_1+k_2}\alpha_m^{2k_1+1}\beta_n^{2k_2}I_{SS}(0,0,x_0,y_0),\qquad\text{(A.6)}$$

$$I_{SS}(2k_1+1,2k_2,x_0,y_0) = \sum_{j=1}^{k_1}\sum_{l=1}^{k_2}(-1)^{k_1+k_2-j-l+2}\alpha_m^{2k_1-2j+1}\beta_n^{2k_2-2l+1}$$

$$\cdot[(-1)^{m+n}u^{(2j-1,2l-2)}(a,b)-(-1)^m\cos(\beta_n y_0)u^{(2j-1,2l-2)}(a,y_0)$$

$$-(-1)^n\cos(\alpha_m x_0)u^{(2j-1,2l-2)}(x_0,b)$$

$$+\cos(\alpha_m x_0)\cos(\beta_n y_0)u^{(2j-1,2l-2)}(x_0,y_0)]$$

$$+\sum_{j=1}^{k_1}(-1)^{k_1+k_2-j+1}\alpha_m^{2k_1-2j+1}\beta_n^{2k_2}\int_{y_0}^{b}[(-1)^m u^{(2j-1,0)}(a,y)$$

$$-\cos(\alpha_m x_0)u^{(2j-1,0)}(x_0,y)]\sin(\beta_n y)dy$$

$$+\sum_{l=1}^{k_2}(-1)^{k_1+k_2-l+2}\alpha_m^{2k_1+1}\beta_n^{2k_2-2l+1}\int_{x_0}^{a}[(-1)^n u^{(0,2l-2)}(x,b)$$

$$-\cos(\beta_n y_0)u^{(0,2l-2)}(x,y_0)]\cos(\alpha_m x)dx$$

$$+(-1)^{k_1+k_2+1}\alpha_m^{2k_1+1}\beta_n^{2k_2}I_{CS}(0,0,x_0,y_0),\qquad\text{(A.7)}$$

$$I_{CC}(2k_1, 2k_2 + 1, x_0, y_0) = \sum_{j=1}^{k_1} \sum_{l=0}^{k_2} (-1)^{k_1+k_2-j-l} \alpha_m^{2k_1-2j} \beta_n^{2k_2-2l}$$

$$\cdot [(-1)^{m+n} u^{(2j-1,2l)}(a,b) - (-1)^m \cos(\beta_n y_0) u^{(2j-1,2l)}(a,y_0)$$

$$- (-1)^n \cos(\alpha_m x_0) u^{(2j-1,2l)}(x_0,b)$$

$$+ \cos(\alpha_m x_0) \cos(\beta_n y_0) u^{(2j-1,2l)}(x_0,y_0)]$$

$$+ \sum_{j=1}^{k_1} (-1)^{k_1+k_2-j} \alpha_m^{2k_1-2j} \beta_n^{2k_2+1} \int_{y_0}^{b} [(-1)^m u^{(2j-1,0)}(a,y)$$

$$- \cos(\alpha_m x_0) u^{(2j-1,0)}(x_0,y)] \sin(\beta_n y) dy$$

$$+ \sum_{l=0}^{k_2} (-1)^{k_1+k_2-l} \alpha_m^{2k_1} \beta_n^{2k_2-2l} \int_{x_0}^{a} [(-1)^n u^{(0,2l)}(x,b)$$

$$- \cos(\beta_n y_0) u^{(0,2l)}(x,y_0)] \cos(\alpha_m x) dx$$

$$+ (-1)^{k_1+k_2} \alpha_m^{2k_1} \beta_n^{2k_2+1} I_{CS}(0,0,x_0,y_0), \tag{A.8}$$

$$I_{SC}(2k_1, 2k_2 + 1, x_0, y_0) = \sum_{j=1}^{k_1} \sum_{l=0}^{k_2} (-1)^{k_1+k_2-j-l+1} \alpha_m^{2k_1-2j+1} \beta_n^{2k_2-2l}$$

$$\cdot [(-1)^{m+n} u^{(2j-2,2l)}(a,b) - (-1)^m \cos(\beta_n y_0) u^{(2j-2,2l)}(a,y_0)$$

$$- (-1)^n \cos(\alpha_m x_0) u^{(2j-2,2l)}(x_0,b)$$

$$+ \cos(\alpha_m x_0) \cos(\beta_n y_0) u^{(2j-2,2l)}(x_0,y_0)]$$

$$+ \sum_{j=1}^{k_1} (-1)^{k_1+k_2-j+1} \alpha_m^{2k_1-2j+1} \beta_n^{2k_2+1} \int_{y_0}^{b} [(-1)^m u^{(2j-2,0)}(a,y)$$

$$- \cos(\alpha_m x_0) u^{(2j-2,0)}(x_0,y)] \sin(\beta_n y) dy$$

$$+ \sum_{l=0}^{k_2} (-1)^{k_1+k_2-l} \alpha_m^{2k_1} \beta_n^{2k_2-2l} \int_{x_0}^{a} [(-1)^n u^{(0,2l)}(x,b)$$

$$- \cos(\beta_n y_0) u^{(0,2l)}(x,y_0)] \sin(\alpha_m x) dx$$

$$+ (-1)^{k_1+k_2} \alpha_m^{2k_1} \beta_n^{2k_2+1} I_{SS}(0,0,x_0,y_0), \tag{A.9}$$

$$I_{CS}(2k_1, 2k_2+1, x_0, y_0) = \sum_{j=1}^{k_1}\sum_{l=1}^{k_2}(-1)^{k_1+k_2-j-l+1}\alpha_m^{2k_1-2j}\beta_n^{2k_2-2l+1}$$

$$\cdot[(-1)^{m+n}u^{(2j-1,2l-1)}(a,b)-(-1)^m\cos(\beta_n y_0)u^{(2j-1,2l-1)}(a,y_0)$$

$$-(-1)^n\cos(\alpha_m x_0)u^{(2j-1,2l-1)}(x_0,b)$$

$$+\cos(\alpha_m x_0)\cos(\beta_n y_0)u^{(2j-1,2l-1)}(x_0,y_0)]$$

$$+\sum_{j=1}^{k_1}(-1)^{k_1+k_2-j+1}\alpha_m^{2k_1-2j}\beta_n^{2k_2+1}\int_{y_0}^b[(-1)^m u^{(2j-1,0)}(a,y)$$

$$-\cos(\alpha_m x_0)u^{(2j-1,0)}(x_0,y)]\cos(\beta_n y)dy$$

$$+\sum_{l=1}^{k_2}(-1)^{k_1+k_2-l+1}\alpha_m^{2k_1}\beta_n^{2k_2-2l+1}\int_{x_0}^a[(-1)^n u^{(0,2l-1)}(x,b)$$

$$-\cos(\beta_n y_0)u^{(0,2l-1)}(x,y_0)]\cos(\alpha_m x)dx$$

$$+(-1)^{k_1+k_2+1}\alpha_m^{2k_1}\beta_n^{2k_2+1}I_{CC}(0,0,x_0,y_0), \tag{A.10}$$

$$I_{SS}(2k_1, 2k_2+1, x_0, y_0) = \sum_{j=1}^{k_1}\sum_{l=1}^{k_2}(-1)^{k_1+k_2-j-l+2}\alpha_m^{2k_1-2j+1}\beta_n^{2k_2-2l+1}$$

$$\cdot[(-1)^{m+n}u^{(2j-2,2l-1)}(a,b)-(-1)^m\cos(\beta_n y_0)u^{(2j-2,2l-1)}(a,y_0)$$

$$-(-1)^n\cos(\alpha_m x_0)u^{(2j-2,2l-1)}(x_0,b)$$

$$+\cos(\alpha_m x_0)\cos(\beta_n y_0)u^{(2j-2,2l-1)}(x_0,y_0)]$$

$$+\sum_{j=1}^{k_1}(-1)^{k_1+k_2-j+2}\alpha_m^{2k_1-2j+1}\beta_n^{2k_2+1}\int_{y_0}^b[(-1)^m u^{(2j-2,0)}(a,y)$$

$$-\cos(\alpha_m x_0)u^{(2j-2,0)}(x_0,y)]\cos(\beta_n y)dy$$

$$+\sum_{l=1}^{k_2}(-1)^{k_1+k_2-l+1}\alpha_m^{2k_1}\beta_n^{2k_2-2l+1}\int_{x_0}^a[(-1)^n u^{(0,2l-1)}(x,b)$$

$$-\cos(\beta_n y_0)u^{(0,2l-1)}(x,y_0)]\sin(\alpha_m x)dx$$

$$+(-1)^{k_1+k_2+1}\alpha_m^{2k_1}\beta_n^{2k_2+1}I_{SC}(0,0,x_0,y_0), \tag{A.11}$$

$$I_{CC}(2k_1+1,2k_2+1,x_0,y_0) = \sum_{j=0}^{k_1}\sum_{l=0}^{k_2}(-1)^{k_1+k_2-j-l}\alpha_m^{2k_1-2j}\beta_n^{2k_2-2l}$$

$$\cdot[(-1)^{m+n}u^{(2j,2l)}(a,b)-(-1)^m\cos(\beta_n y_0)u^{(2j,2l)}(a,y_0)$$

$$-(-1)^n\cos(\alpha_m x_0)u^{(2j,2l)}(x_0,b)$$

$$+\cos(\alpha_m x_0)\cos(\beta_n y_0)u^{(2j,2l)}(x_0,y_0)]$$

$$+\sum_{j=0}^{k_1}(-1)^{k_1+k_2-j}\alpha_m^{2k_1-2j}\beta_n^{2k_2+1}\int_{y_0}^{b}[(-1)^m u^{(2j,0)}(a,y)$$

$$-\cos(\alpha_m x_0)u^{(2j,0)}(x_0,y)]\sin(\beta_n y)dy$$

$$+\sum_{l=0}^{k_2}(-1)^{k_1+k_2-l}\alpha_m^{2k_1+1}\beta_n^{2k_2-2l}\int_{x_0}^{a}[(-1)^n u^{(0,2l)}(x,b)$$

$$-\cos(\beta_n y_0)u^{(0,2l)}(x,y_0)]\sin(\alpha_m x)dx$$

$$+(-1)^{k_1+k_2}\alpha_m^{2k_1+1}\beta_n^{2k_2+1}I_{SS}(0,0,x_0,y_0), \qquad (A.12)$$

$$I_{SC}(2k_1+1,2k_2+1,x_0,y_0) = \sum_{j=1}^{k_1}\sum_{l=0}^{k_2}(-1)^{k_1+k_2-j-l+1}\alpha_m^{2k_1-2j+1}\beta_n^{2k_2-2l}$$

$$\cdot[(-1)^{m+n}u^{(2j-1,2l)}(a,b)-(-1)^m\cos(\beta_n y_0)u^{(2j-1,2l)}(a,y_0)$$

$$-(-1)^n\cos(\alpha_m x_0)u^{(2j-1,2l)}(x_0,b)$$

$$+\cos(\alpha_m x_0)\cos(\beta_n y_0)u^{(2j-1,2l)}(x_0,y_0)]$$

$$+\sum_{j=1}^{k_1}(-1)^{k_1+k_2-j+1}\alpha_m^{2k_1-2j+1}\beta_n^{2k_2+1}\int_{y_0}^{b}[(-1)^m u^{(2j-1,0)}(a,y)$$

$$-\cos(\alpha_m x_0)u^{(2j-1,0)}(x_0,y)]\sin(\beta_n y)dy$$

$$+\sum_{l=0}^{k_2}(-1)^{k_1+k_2-l+1}\alpha_m^{2k_1+1}\beta_n^{2k_2-2l}\int_{x_0}^{a}[(-1)^n u^{(0,2l)}(x,b)$$

$$-\cos(\beta_n y_0)u^{(0,2l)}(x,y_0)]\cos(\alpha_m x)dx$$

$$+(-1)^{k_1+k_2+1}\alpha_m^{2k_1+1}\beta_n^{2k_2+1}I_{CS}(0,0,x_0,y_0), \qquad (A.13)$$

$$I_{CS}(2k_1+1,2k_2+1,x_0,y_0) = \sum_{j=0}^{k_1}\sum_{l=1}^{k_2}(-1)^{k_1+k_2-j-l+1}\alpha_m^{2k_1-2j}\beta_n^{2k_2-2l+1}$$

$$\cdot[(-1)^{m+n}u^{(2j,2l-1)}(a,b)-(-1)^m\cos(\beta_n y_0)u^{(2j,2l-1)}(a,y_0)$$

$$-(-1)^n\cos(\alpha_m x_0)u^{(2j,2l-1)}(x_0,b)$$

$$+\cos(\alpha_m x_0)\cos(\beta_n y_0)u^{(2j,2l-1)}(x_0,y_0)]$$

$$+\sum_{j=0}^{k_1}(-1)^{k_1+k_2-j+1}\alpha_m^{2k_1-2j}\beta_n^{2k_2+1}\int_{y_0}^{b}[(-1)^m u^{(2j,0)}(a,y)$$

$$-\cos(\alpha_m x_0)u^{(2j,0)}(x_0,y)]\cos(\beta_n y)dy$$

$$+\sum_{l=1}^{k_2}(-1)^{k_1+k_2-l+1}\alpha_m^{2k_1+1}\beta_n^{2k_2-2l+1}\int_{x_0}^{a}[(-1)^n u^{(0,2l-1)}(x,b)$$

$$-\cos(\beta_n y_0)u^{(0,2l-1)}(x,y_0)]\sin(\alpha_m x)dx$$

$$+(-1)^{k_1+k_2+1}\alpha_m^{2k_1+1}\beta_n^{2k_2+1}I_{SC}(0,0,x_0,y_0), \qquad (A.14)$$

and

$$I_{SS}(2k_1+1,2k_2+1,x_0,y_0) = \sum_{j=1}^{k_1}\sum_{l=1}^{k_2}(-1)^{k_1+k_2-j-l+2}\alpha_m^{2k_1-2j+1}\beta_n^{2k_2-2l+1}$$

$$\cdot[(-1)^{m+n}u^{(2j-1,2l-1)}(a,b)-(-1)^m\cos(\beta_n y_0)u^{(2j-1,2l-1)}(a,y_0)$$

$$-(-1)^n\cos(\alpha_m x_0)u^{(2j-1,2l-1)}(x_0,b)$$

$$+\cos(\alpha_m x_0)\cos(\beta_n y_0)u^{(2j-1,2l-1)}(x_0,y_0)]$$

$$+\sum_{j=1}^{k_1}(-1)^{k_1+k_2-j+2}\alpha_m^{2k_1-2j+1}\beta_n^{2k_2+1}\int_{y_0}^{b}[(-1)^m u^{(2j-1,0)}(a,y)$$

$$-\cos(\alpha_m x_0)u^{(2j-1,0)}(x_0,y)]\cos(\beta_n y)dy$$

$$+\sum_{l=1}^{k_2}(-1)^{k_1+k_2-l+2}\alpha_m^{2k_1+1}\beta_n^{2k_2-2l+1}\int_{x_0}^{a}[(-1)^n u^{(0,2l-1)}(x,b)$$

$$-\cos(\beta_n y_0)u^{(0,2l-1)}(x,y_0)]\cos(\alpha_m x)dx$$

$$+(-1)^{k_1+k_2+2}\alpha_m^{2k_1+1}\beta_n^{2k_2+1}I_{CC}(0,0,x_0,y_0). \qquad (A.15)$$

7 The Generalized Fourier Series of Functions

Thus far, it should have become clear that for a sufficiently smooth function defined on a compact domain, the convergence characteristics of the corresponding Fourier series expansions are dictated by the smoothness of its periodically-extended versions. In other words, preserving the smoothness of the original function as much as possible (or needed) during its periodic-extension is critical to ensuring a desired convergence behavior of the resulting Fourier series. Given a sufficiently smooth function, however, the process of periodic extension itself can do little toward realizing this goal. Instead, we have to resort to other means, such as, the use of appropriate supplementary terms, to achieve the desired degree of smoothness of the residual function which is subsequently to be expanded into a Fourier series, as described below.

7.1 STRUCTURAL DECOMPOSITIONS OF ONE-DIMENSIONAL FUNCTIONS

We denote by $C^{2r-1}([-a,a])$ (or $C^{2r-1}([0,a])$) the set of functions which have $2r-1$ continuous derivatives over the interval $[-a,a]$(or $[0,a]$). Based on the sufficient conditions for the $2r$ times term-by-term differentiations of a Fourier series, the one-dimensional function $u(x)$ is herein decomposed into two parts

$$u(x) = \varphi_0(x) + \varphi_1(x), \qquad (7.1)$$

where $\varphi_0(x)$ is referred to as the internal function and $\varphi_1(x)$ as the boundary function.

In what follows, this decomposition scheme will be implemented in three different ways.

First, we assume $u(x) \in C^{2r-1}([-a,a])$ and its $2r$-th order derivative is absolutely integrable over the interval $[-a,a]$. The boundary function $\varphi_1(x)$ is then such constructed that the internal function $\varphi_0(x)$ is forced to satisfy the sufficient conditions to ensure its (full-range) Fourier series to be $2r$ times termwise differentiable.

More explicitly, the sufficient conditions for the $2r$ times term-by-term differentiations can be expressed as

$$\varphi_0^{(k)}(a) - \varphi_0^{(k)}(-a) = 0, \, k = 0, 1, \cdots, 2r-1, \qquad (7.2)$$

or

$$\varphi_1^{(k)}(a) - \varphi_1^{(k)}(-a) = u^{(k)}(a) - u^{(k)}(-a), \, k = 0, 1, \cdots, 2r-1. \qquad (7.3)$$

DOI: 10.1201/9781003194859-7

In the second way, it is assumed that the function $u(x) \in C^{2r-1}([0,a])$ and its $2r$-th order derivative is absolutely integrable over the interval $[0,a]$. The boundary function $\varphi_1(x)$ is then such constructed that the internal function $\varphi_0(x)$ is forced to satisfy the sufficient conditions for its half-range Fourier cosine series to be $2r$ times termwise differentiable.

The sufficient conditions for the $2r$ times term-by-term differentiations can be accordingly written as

$$\varphi_0^{(2k+1)}(a) = 0, \varphi_0^{(2k+1)}(0) = 0, k = 0, 1, \cdots, r-1, \tag{7.4}$$

or

$$\varphi_1^{(2k+1)}(a) = u^{(2k+1)}(a), \varphi_1^{(2k+1)}(0) = 0, k = 0, 1, \cdots, r-1, \tag{7.5}$$

Last, assume $u(x) \in C^{2r-1}([0,a])$ and its $2r$-th order derivative is absolutely integrable over the interval $[0,a]$, the boundary function $\varphi_1(x)$ is then constructed in such a way that the internal function $\varphi_0(x)$ is forced to satisfy the sufficient conditions for its half-range Fourier sine series to be $2r$ times termwise differentiable.

Accordingly, the sufficient conditions for the $2r$ times term-by-term differentiations are given as

$$\varphi_0^{(2k)}(a) = 0, \varphi_0^{(2k)}(0) = 0, k = 0, 1, \cdots, r-1, \tag{7.6}$$

or

$$\varphi_1^{(2k)}(a) = u^{(2k)}(a), \varphi_1^{(2k)}(0) = u^{(2k)}(0), k = 0, 1, \cdots, r-1. \tag{7.7}$$

7.2 GENERALIZED FOURIER SERIES OF ONE-DIMENSIONAL FUNCTIONS

On the basis of structural decompositions of a one-dimensional function described in the previous section, we now discuss how to select the boundary and internal functions for the simultaneous series expansions of the function and its derivatives.

7.2.1 THE GENERALIZED FULL-RANGE FOURIER SERIES OF ONE-DIMENSIONAL FUNCTIONS

Let's first consider the generalized Fourier series (GFS) for the one-dimensional function $u(x) \in C^{2r-1}([-a,a])$ and its derivatives.

In constructing the boundary function $\varphi_1(x)$, the vector of supplementary functions is defined as:

$$\mathbf{p}_1^T(x) = [p_{1,1}(x) \quad p_{1,2}(x) \quad \cdots \quad p_{1,2r}(x)], \tag{7.8}$$

where $p_{1,j}(x)$, $j = 1, 2, \cdots, 2r$, represent a set of closed-form functions which are linearly independent and sufficiently smooth on the interval $[-a, a]$. The term "sufficiently smooth" here implies that each of these functions is, at least, $2r$ times continuously differentiable on the interval. Such a requirement can be readily satisfied by many forms of functions such as polynomials; theoretically, there exists a large (or even an infinite) number of possible choices for such supplementary functions.

The differential forms of (7.8) are simply given as

$$\mathbf{p}_1^{(k)\mathrm{T}}(x) = [p_{1,1}^{(k)}(x) \quad p_{1,2}^{(k)}(x) \quad \cdots \quad p_{1,2r}^{(k)}(x)], \quad \text{for } k = 0, 1, \cdots, 2r. \tag{7.9}$$

Let $\mathbf{a}_1^{\mathrm{T}} = [G_{1,1} \quad G_{1,2} \quad \cdots \quad G_{1,2r}]$ be a vector of the undetermined constants. Then the boundary function $\varphi_1(x)$ can be generically expressed as

$$\varphi_1(x) = \mathbf{p}_1^{\mathrm{T}}(x) \cdot \mathbf{a}_1. \tag{7.10}$$

Inserting (7.10) into the sufficient conditions (7.3) for the boundary function $\varphi_1(x)$ results in

$$\mathbf{R}_1 \mathbf{a}_1 = \mathbf{q}_1, \tag{7.11}$$

where

$$\mathbf{R}_1 = \begin{bmatrix} \mathbf{p}_1^{(0)\mathrm{T}}(a) - \mathbf{p}_1^{(0)\mathrm{T}}(-a) \\ \mathbf{p}_1^{(1)\mathrm{T}}(a) - \mathbf{p}_1^{(1)\mathrm{T}}(-a) \\ \vdots \\ \mathbf{p}_1^{(2r-1)\mathrm{T}}(a) - \mathbf{p}_1^{(2r-1)\mathrm{T}}(-a) \end{bmatrix} \tag{7.12}$$

and

$$\mathbf{q}_1^{\mathrm{T}} = [u^{(0)}(a) - u^{(0)}(-a) \quad u^{(1)}(a) - u^{(1)}(-a) \quad \cdots \quad u^{(2r-1)}(a) - u^{(2r-1)}(-a)]. \tag{7.13}$$

From (7.11) we have

$$\mathbf{a}_1 = \mathbf{R}_1^{-1} \mathbf{q}_1. \tag{7.14}$$

Define the vector of the basis functions as

$$\mathbf{\Phi}_1^{\mathrm{T}}(x) = \mathbf{p}_1^{\mathrm{T}}(x) \cdot \mathbf{R}_1^{-1}, \tag{7.15}$$

the boundary function $\varphi_1(x)$ can be expressed as

$$\varphi_1(x) = \mathbf{\Phi}_1^{\mathrm{T}}(x) \cdot \mathbf{q}_1. \tag{7.16}$$

In light of (7.2), the internal function $\varphi_0(x)$ can be expanded into a full-range Fourier series as

$$\varphi_0(x) = \sum_{m=0}^{\infty} \mu_m [V_{1m} \cos(\alpha_m x) + V_{2m} \sin(\alpha_m x)], \qquad (7.17)$$

where $\alpha_m = m\pi/a$, $\mu_m = \begin{cases} 1/2 & m = 0 \\ 1 & m > 0 \end{cases}$, and V_{1m} and V_{2m} are the Fourier coefficients.

Denote

$$\varphi_{01m}(x) = \mu_m \cos(\alpha_m x), m = 0, 1, 2, \cdots \qquad (7.18)$$

and

$$\varphi_{02m}(x) = \mu_m \sin(\alpha_m x), m = 1, 2, \cdots, \qquad (7.19)$$

then we have:

for a nonnegative even integer k,

$$\varphi_{01m}^{(k)}(x) = \mu_m (-1)^{k/2} \alpha_m^k \cos(\alpha_m x) \qquad (7.20)$$

and

$$\varphi_{02m}^{(k)}(x) = \mu_m (-1)^{k/2} \alpha_m^k \sin(\alpha_m x); \qquad (7.21)$$

for a nonnegative odd integer k,

$$\varphi_{01m}^{(k)}(x) = \mu_m (-1)^{(k+1)/2} \alpha_m^k \sin(\alpha_m x) \qquad (7.22)$$

and

$$\varphi_{02m}^{(k)}(x) = \mu_m (-1)^{(k-1)/2} \alpha_m^k \cos(\alpha_m x). \qquad (7.23)$$

Further, define the vector of the basis functions

$$\Phi_0^T(x) = [\Phi_{01}^T(x) \quad \Phi_{02}^T(x)], \qquad (7.24)$$

where

$$\Phi_{01}^T(x) = [\varphi_{010}(x) \quad \varphi_{011}(x) \quad \cdots \quad \varphi_{01m}(x) \quad \cdots] \qquad (7.25)$$

and

$$\Phi_{02}^T(x) = [\varphi_{021}(x) \quad \varphi_{022}(x) \quad \cdots \quad \varphi_{02m}(x) \quad \cdots]. \qquad (7.26)$$

Then the internal function $\varphi_0(x)$ can be expanded into a Fourier series as

$$\varphi_0(x) = \mathbf{\Phi}_0^{\mathrm{T}}(x) \cdot \mathbf{q}_0, \tag{7.27}$$

where the vector of Fourier coefficients

$$\mathbf{q}_0^{\mathrm{T}} = [\mathbf{q}_{01}^{\mathrm{T}} \quad \mathbf{q}_{02}^{\mathrm{T}}] \tag{7.28}$$

with

$$\mathbf{q}_{01}^{\mathrm{T}} = [V_{10} \quad V_{11} \quad \cdots \quad V_{1m} \quad \cdots] \tag{7.29}$$

and

$$\mathbf{q}_{02}^{\mathrm{T}} = [V_{21} \quad V_{22} \quad \cdots \quad V_{2m} \quad \cdots]. \tag{7.30}$$

It should be pointed out that the Fourier series (7.27) can be differentiated, term-by-term, up to $2r$ times, and such differential operations can be directly applied to the basis functions, that is,

$$\varphi_0^{(k)}(x) = \mathbf{\Phi}_0^{(k)\mathrm{T}}(x) \cdot \mathbf{q}_0, \tag{7.31}$$

where

$$\mathbf{\Phi}_0^{(k)\mathrm{T}}(x) = [\mathbf{\Phi}_{01}^{(k)\mathrm{T}}(x) \quad \mathbf{\Phi}_{02}^{(k)\mathrm{T}}(x)] \tag{7.32}$$

with

$$\mathbf{\Phi}_{01}^{(k)\mathrm{T}}(x) = [\varphi_{010}^{(k)}(x) \quad \varphi_{011}^{(k)}(x) \quad \cdots \quad \varphi_{01m}^{(k)}(x) \quad \cdots] \tag{7.33}$$

and

$$\mathbf{\Phi}_{02}^{(k)\mathrm{T}}(x) = [\varphi_{021}^{(k)}(x) \quad \varphi_{022}^{(k)}(x) \quad \cdots \quad \varphi_{02m}^{(k)}(x) \quad \cdots]. \tag{7.34}$$

In this way, expression (7.1) actually represents a generalized form of Fourier series expansion of the function $u(x)$, which is a superposition of the internal function $\varphi_0(x)$ in the form of a conventional Fourier series (CFS) and the boundary function $\varphi_1(x)$ made up of the selected supplementary functions. This GFS can be termwise differentiated up to $2r$ times to obtain its counterparts for the corresponding derivatives of the function $u(x)$, namely,

$$
\begin{aligned}
u^{(k)}(x) &= \varphi_0^{(k)}(x) + \varphi_1^{(k)}(x) \\
&= \mathbf{\Phi}_0^{(k)\mathrm{T}}(x) \cdot \mathbf{q}_0 + \mathbf{\Phi}_1^{(k)\mathrm{T}}(x) \cdot \mathbf{q}_1 \quad (k = 0, 1, \cdots, 2r).
\end{aligned} \tag{7.35}
$$

7.2.2 THE GENERALIZED HALF-RANGE FOURIER COSINE SERIES OF ONE-DIMENSIONAL FUNCTIONS

Next, we consider the generalized half-range Fourier cosine series expansion for the function $u(x) \in C^{2r-1}([0,a])$.

In constructing the boundary function $\varphi_1(x)$, we define the vector of supplementary functions as

$$\mathbf{p}_1^T(x) = [p_{1,1}(x) \quad p_{1,2}(x) \quad \cdots \quad p_{1,2r}(x)], \tag{7.36}$$

where $p_{1,j}(x)$, $j = 1, 2, \cdots, 2r$, represent a set of closed-form sufficiently smooth functions defined over the interval $[0,a]$.

The corresponding differential forms are

$$\mathbf{p}_1^{(k)T}(x) = [p_{1,1}^{(k)}(x) \quad p_{1,2}^{(k)}(x) \quad \cdots \quad p_{1,2r}^{(k)}(x)] \quad (k = 0, 1, \cdots, 2r). \tag{7.37}$$

Similar to the procedures described above, the vector of the basis functions can be expressed as

$$\Phi_1^T(x) = \mathbf{p}_1^T(x) \cdot \mathbf{R}_1^{-1}, \tag{7.38}$$

where

$$\mathbf{R}_1 = \begin{bmatrix} \mathbf{R}_{11} \\ \mathbf{R}_{12} \end{bmatrix} \tag{7.39}$$

with

$$\mathbf{R}_{11} = \begin{bmatrix} \mathbf{p}_1^{(1)T}(a) \\ \mathbf{p}_1^{(3)T}(a) \\ \vdots \\ \mathbf{p}_1^{(2r-1)T}(a) \end{bmatrix} \tag{7.40}$$

and

$$\mathbf{R}_{12} = \begin{bmatrix} \mathbf{p}_1^{(1)T}(0) \\ \mathbf{p}_1^{(3)T}(0) \\ \vdots \\ \mathbf{p}_1^{(2r-1)T}(0) \end{bmatrix}. \tag{7.41}$$

Thus, the boundary function $\varphi_1(x)$ can be obtained as

$$\varphi_1(x) = \Phi_1^T(x) \cdot \mathbf{q}_1, \tag{7.42}$$

where

$$\mathbf{q}_1^T = [\mathbf{q}_{11}^T \quad \mathbf{q}_{12}^T], \tag{7.43}$$

$$\mathbf{q}_{11}^T = [u^{(1)}(a) \quad u^{(3)}(a) \quad \cdots \quad u^{(2r-1)}(a)], \tag{7.44}$$

and

$$\mathbf{q}_{12}^T = [u^{(1)}(0) \quad u^{(3)}(0) \quad \cdots \quad u^{(2r-1)}(0)]. \tag{7.45}$$

The internal function $\varphi_0(x)$ will be sought in the form of half-range Fourier cosine series over the interval $[0, a]$:

$$\varphi_0(x) = \sum_{m=0}^{\infty} \mu_m V_{1m} \cos(\alpha_m x), \tag{7.46}$$

where $\alpha_m = m\pi/a$, $\mu_m = \begin{cases} 1/2 & m = 0 \\ 1 & m > 0 \end{cases}$, and V_{1m} are the Fourier coefficients.

Denote

$$\varphi_{01m}(x) = \mu_m \cos(\alpha_m x), m = 0, 1, 2, \cdots, \tag{7.47}$$

then we have:
for a nonnegative even integer k,

$$\varphi_{01m}^{(k)}(x) = \mu_m (-1)^{k/2} \alpha_m^k \cos(\alpha_m x); \tag{7.48}$$

for a nonnegative odd integer k,

$$\varphi_{01m}^{(k)}(x) = \mu_m (-1)^{(k+1)/2} \alpha_m^k \sin(\alpha_m x). \tag{7.49}$$

The internal function $\varphi_0(x)$ can finally be expressed as

$$\varphi_0(x) = \Phi_{01}^T(x) \cdot \mathbf{q}_{01}, \tag{7.50}$$

where the vector of the basis functions is

$$\Phi_{01}^{T}(x) = [\varphi_{010}(x) \quad \varphi_{011}(x) \quad \cdots \quad \varphi_{01m}(x) \quad \cdots] \tag{7.51}$$

and the vector of Fourier coefficients is

$$\mathbf{q}_{01}^{T} = [V_{10} \quad V_{11} \quad \cdots \quad V_{1m} \quad \cdots]. \tag{7.52}$$

The Fourier series (7.50) can be termwise differentiated up to $2r$ times, namely,

$$\varphi_{0}^{(k)}(x) = \Phi_{01}^{(k)T}(x) \cdot \mathbf{q}_{01}, \tag{7.53}$$

where

$$\Phi_{01}^{(k)T}(x) = [\varphi_{010}^{(k)}(x) \quad \varphi_{011}^{(k)}(x) \quad \cdots \quad \varphi_{01m}^{(k)}(x) \quad \cdots]. \tag{7.54}$$

Evidently, the term-by-term differentiations are accordingly applicable to the generalized Fourier cosine/sine series of the function $u(x)$ and its derivatives:

$$\begin{aligned}
u^{(k)}(x) &= \varphi_{0}^{(k)}(x) + \varphi_{1}^{(k)}(x) \\
&= \Phi_{01}^{(k)T}(x) \cdot \mathbf{q}_{01} + \Phi_{1}^{(k)T}(x) \cdot \mathbf{q}_{1} \quad (k = 0, 1, \cdots, 2r).
\end{aligned} \tag{7.55}$$

7.2.3 The Generalized Half-Range Fourier Sine Series of One-Dimensional Functions

Finally, suppose $u(x) \in C^{2r-1}([0, a])$, and is expanded into the generalized half-range Fourier sine series.

The boundary function is constructed in the same way

$$\varphi_{1}(x) = \Phi_{1}^{T}(x) \cdot \mathbf{q}_{1}. \tag{7.56}$$

However, the vector of the basis functions and the vector of boundary values in (7.56) should be correspondingly modified to

$$\Phi_{1}^{T}(x) = \mathbf{p}_{1}^{T}(x) \cdot \mathbf{R}_{1}^{-1}, \tag{7.57}$$

where $\mathbf{p}_{1}^{T}(x)$ is the same as given in (7.36),

$$\mathbf{R}_{1} = \begin{bmatrix} \mathbf{R}_{11} \\ \mathbf{R}_{12} \end{bmatrix} \tag{7.58}$$

and

$$\mathbf{q}_{1}^{T} = [\mathbf{q}_{11}^{T} \quad \mathbf{q}_{12}^{T}] \tag{7.59}$$

with

$$
\mathbf{R}_{11} = \begin{bmatrix} \mathbf{p}_1^{(0)\mathrm{T}}(a) \\ \mathbf{p}_1^{(2)\mathrm{T}}(a) \\ \vdots \\ \mathbf{p}_1^{(2r-2)\mathrm{T}}(a) \end{bmatrix}, \tag{7.60}
$$

$$
\mathbf{R}_{12} = \begin{bmatrix} \mathbf{p}_1^{(0)\mathrm{T}}(0) \\ \mathbf{p}_1^{(2)\mathrm{T}}(0) \\ \vdots \\ \mathbf{p}_1^{(2r-2)\mathrm{T}}(0) \end{bmatrix}, \tag{7.61}
$$

$$
\mathbf{q}_{11}^{\mathrm{T}} = [u^{(0)}(a) \quad u^{(2)}(a) \quad \cdots \quad u^{(2r-2)}(a)], \tag{7.62}
$$

and

$$
\mathbf{q}_{12}^{\mathrm{T}} = [u^{(0)}(0) \quad u^{(2)}(0) \quad \cdots \quad u^{(2r-2)}(0)]. \tag{7.63}
$$

The internal function $\varphi_0(x)$ is expanded into a half-range sine series on the interval $[0, a]$

$$
\varphi_0(x) = \sum_{m=1}^{\infty} V_{2m} \sin(\alpha_m x), \tag{7.64}
$$

where $\alpha_m = m\pi/a$ and V_{2m} are the Fourier coefficients.

Denote

$$
\varphi_{02m}(x) = \sin(\alpha_m x), m = 1, 2, \cdots, \tag{7.65}
$$

then we have:

for a nonnegative even integer k,

$$
\varphi_{02m}^{(k)}(x) = (-1)^{k/2} \alpha_m^k \sin(\alpha_m x); \tag{7.66}
$$

for a nonnegative odd integer k,

$$
\varphi_{02m}^{(k)}(x) = (-1)^{(k-1)/2} \alpha_m^k \cos(\alpha_m x). \tag{7.67}
$$

The half-range sine series of the internal function $\varphi_0(x)$ can be written as

$$
\varphi_0(x) = \Phi_{02}^{\mathrm{T}}(x) \cdot \mathbf{q}_{02}, \tag{7.68}
$$

where

$$\boldsymbol{\Phi}_{02}^{\mathrm{T}}(x) = [\varphi_{021}(x) \quad \varphi_{022}(x) \quad \cdots \quad \varphi_{02m}(x) \quad \cdots] \tag{7.69}$$

and

$$\mathbf{q}_{02}^{\mathrm{T}} = [V_{21} \quad V_{22} \quad \cdots \quad V_{2m} \quad \cdots]. \tag{7.70}$$

Accordingly, the generalized half-range Fourier sine/cosine series of the function and its derivatives are generically given as

$$\begin{aligned} u^{(k)}(x) &= \varphi_0^{(k)}(x) + \varphi_1^{(k)}(x) \\ &= \boldsymbol{\Phi}_{02}^{(k)\mathrm{T}}(x) \cdot \mathbf{q}_{02} + \boldsymbol{\Phi}_1^{(k)\mathrm{T}}(x) \cdot \mathbf{q}_1 \quad (k = 0, 1, \cdots, 2r). \end{aligned} \tag{7.71}$$

7.3   THE POLYNOMIAL-BASED GENERALIZED FOURIER SERIES FOR ONE-DIMENSIONAL FUNCTIONS

In Sec. 7.2, a framework has been established for the simultaneous series expansions of a sufficiently smooth function and its derivatives defined on a compact interval. The key concept is to supplement the CFS with a set of closed-form of functions to allow performing term-by-term differentiations as many times as needed. Actually, the supplementary functions, $p_{1,j}(x)$, $j = 1, 2, \cdots, 2r$, are only required to be sufficiently smooth and linearly independent. Thus, there is a large (theoretically, an infinite) number of possible choices, and for each choice, the corresponding GFS may be slightly different from others. However, regardless of what supplementary functions are specifically selected, it is evident from the above discussions that the GFS are essentially unaffected in terms of the execution procedures and final forms. For example, if the Lanczos polynomials or trigonometric functions are chosen as the supplementary functions, the current framework will basically lead to the same series expansions as presented in Sec. 3.1–3.4.

In what follows, the simplest set of polynomials will be specifically used:

$$\mathbf{p}_1^{\mathrm{T}}(x) = [p_{1,1}(x) \quad p_{1,2}(x) \quad \cdots \quad p_{1,2r}(x)], \tag{7.72}$$

where

$$p_{1,j}(x) = (x/a)^j \quad (j = 1, 2, \cdots, 2r), \tag{7.73}$$

for the generalized full-range Fourier series over the interval $[-a, a]$ or for the generalized half-range cosine series over the interval $[0, a]$;

$$p_{1,j}(x) = (x/a)^{j-1} \quad (j = 1, 2, \cdots, 2r), \tag{7.74}$$

for the generalized half-range sine series over the interval $[0, a]$.

Among other potential benefits, the polynomial-based GFS expansions are obviously capable of reproducing the complete polynomials of specific degrees, that is:

1. The generalized Fourier series over $[-a,a]$ (or the generalized cosine series over $[0,a]$) can reproduce a complete polynomial of degree $2r$;
2. The generalized sine series over $[0,a]$ can reproduce a complete polynomial of degree $2r-1$.

7.4 EXAMPLES

By setting $2r = 6$, we will here use the functions listed in Table 7.1 to investigate the convergence characteristics and approximation accuracy of the polynomial-based GFS expansions.

7.4.1 ERROR INDEXES OF THE SERIES APPROXIMATIONS

Error indexes will first be defined to assess the convergence and accuracy of the series expansions and, possibly, their spatial dependences.

As illustrated in Figure 7.1, let N_1^t be a positive integer and $\Upsilon^t = \{x_{n_1}, n_1 = 1, 2, \cdots, N_1^t\}$ be a set of evenly distributed sampling points on the interval. Further, the set Υ^t is divided into two subsets Υ_I^t and Υ_B^t: Υ_I^t is the subset of the sampling

---

TABLE 7.1

Sample Functions

No.	Sample Functions
1	$u(x) = \dfrac{1}{2} - \dfrac{x}{a} - \dfrac{1}{2}\left(\dfrac{x}{a}\right)^2 + \left(\dfrac{x}{a}\right)^3, x \in [-a,a]$
2	$u(x) = \sin(\alpha_0 x), \alpha_0 = \pi/2a, x \in [-a,a]$
3	$u(x) = \dfrac{1}{2} - \dfrac{x}{a} - \dfrac{1}{2}\left(\dfrac{x}{a}\right)^2 + \left(\dfrac{x}{a}\right)^3, x \in [0,a]$
4	$u(x) = \cos(\alpha_0 x), \alpha_0 = \pi/2a, x_1 \in [0,a]$

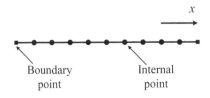

Boundary   Internal
point   point

FIGURE 7.1 Sampling points used for calculating the approximation errors associated with the generalized Fourier series of one-dimensional functions.

points completely inside the interval and Υ'_B is the subset of sampling points on the boundaries. Specifically, for the current one-dimensional case, if the numbers of elements of Υ', Υ'_I, and Υ'_B, are respectively denoted by I'_γ, $I'_{\gamma I}$, and $I'_{\gamma B}$, then $I'_\gamma = N'_1, I'_{\gamma I} = N'_1 - 2$, and $I'_{\gamma B} = 2$.

Let M be the truncation number for (the trigonometric series in) the GFS of the function $u(x)$ and $u_M(x)$ be the corresponding partial sum.

We then define an index (or norm) of approximation error as

$$e^{(k)}(u_M) = \frac{1}{I'_\gamma \cdot u^{(k)}_{max}} \sum_{x \in \Upsilon'} \left| u^{(k)}_M(x) - u^{(k)}(x) \right|, \qquad (7.75)$$

where $u^{(k)}_{max}$ is the maximum value of the derivative $u^{(k)}(x)$ on the interval.

Obviously, the index (7.75) measures the approximation error for a particular partial sum. We can also define an overall error index associated with the simultaneous approximations of a function and its derivatives of orders up to p, that is,

$$\|e\|^p(u_M) = \frac{1}{p+1} \sum_{0 \le k \le p} e^{(k)}(u_M)$$

$$= \frac{1}{p+1} \sum_{0 \le k \le p} \frac{1}{I'_\gamma \cdot u^{(k)}_{max}} \sum_{x \in \Upsilon'} \left| u^{(k)}_M(x) - u^{(k)}(x) \right|. \qquad (7.76)$$

Additionally, the error indexes defined in (7.75) and (7.76) can be spatially refined by replacing the set of sampling points Υ' with Υ'_I or Υ'_B, resulting in the so-called internal and boundary error indexes $e^{(k)}_I(u_M)$ and $e^{(k)}_B(u_M)$, respectively.

7.4.2 CONVERGENCE CHARACTERISTICS

For the sample functions 2 and 4 given in Table 7.1, their GFS expansions are successively truncated to the first 2, 3, 5, 10, 20, 30, and 40 terms. By setting $N'_1 = 101$, the overall, internal and boundary error indexes are respectively plotted in Figures 7.2 and 7.3 for these two sample functions and some of their derivatives of orders up to 6.

It is seen that in both cases the overall and internal approximation errors for the sample functions and their up to third order derivatives decrease rapidly as the truncation number increases. For instance, for the truncation number $M = 10$, the corresponding error indexes are all below 1.0E−4 and the trend of rapid and sustained descending is continued; their values at $M = 40$ are further reduced by more than 2 orders of magnitude. In comparison, the boundary approximations are not as good as their internal counterparts in terms of convergence rates indicated by the steepness of the corresponding error index curves.

As established by Theorem 2.3 in Chapter 2, the convergence rate is determined by the smoothness of a periodic function. Throughout the discussions thus far, it should have become clear that the supplementary terms in the GFS are

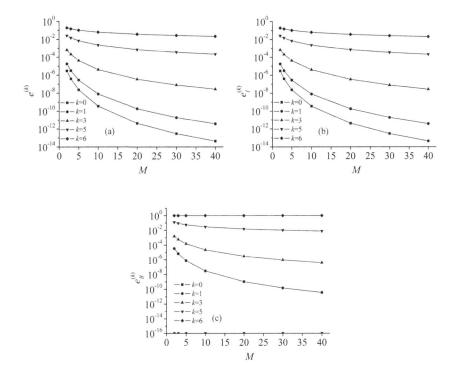

FIGURE 7.2   Convergence of the generalized Fourier series of sample function 2 and its derivatives: (a) $e^{(k)}(u_M)$-M curves; (b) $e_I^{(k)}(u_M)$-M curves; and (c) $e_B^{(k)}(u_M)$-M curves.

literally used to get rid of the possible discontinuities at the boundaries of the interval which potentially result from the periodic extensions of the original (sufficiently smooth) function. Specifically, by setting $2r = 6$ as in the current examples, the internal functions, when periodically extended, are at least ensured to have C^5 continuity. Thus, the Fourier series of the internal functions, and hence the GFS, will converge in accordance to $\mathcal{O}(m^{-7})$; each time the function is differentiated, the convergence will be slowed down by a factor of M. This should have reasonably explained the noticeable deteriorations of the convergence rates with the derivative orders, as shown in Figures 7.2 and 7.3.

Strictly speaking, Theorem 2.3 or the convergence estimate given by (2.46) is specifically about the Fourier coefficients, rather than the above defined error indexes. However, it is not difficult to understand that the direct quote of this theorem, if not completely appropriate, shall not be a serious offense either.

Finally, it needs to be mentioned that even though the error indexes for the sixth order derivatives are included in Figures 7.2 and 7.3, the corresponding GFS, by setting $2r = 6$, will essentially degenerate into the CFS in regard to convergence rates.

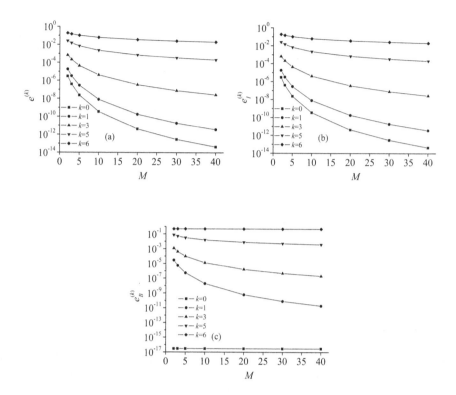

FIGURE 7.3   Convergence of the generalized Fourier series of sample function 4 and its derivatives: (a) $e^{(k)}(u_M)$-M curves; (b) $e_I^{(k)}(u_M)$-M curves; and (c) $e_B^{(k)}(u_M)$-M curves.

7.4.3   REPRODUCING PROPERTY OF POLYNOMIALS

For the polynomial functions 1 and 3 in Table 7.1, the expansion coefficients/constants for their GFS are listed in Table 7.2. For the non-polynomial (or trigonometric) terms, the corresponding coefficients are all essentially zero (numerically, in the order of 1.0E–16 or smaller). For the basis functions in the form of polynomials, the corresponding coefficients can be combined to precisely reconstruct the original sample polynomial functions. This verifies the reproducing property of polynomials for the polynomial-based GFS expansions.

7.4.4   APPROXIMATION ACCURACY

In view of the excellent convergence behavior and numerical stability of the polynomial-based GFS method, the setting $M = 40$ will be used in the subsequent calculations.

Since the polynomial-based GFS has already been shown in Table 7.2 to be capable of precisely approximating the polynomial functions 1 and 3, only the trigonometric functions (No. 2 and No. 4 in Table 7.1) will be here used to

TABLE 7.2
Vectors of Undetermined Constants or Coefficients in the Polynomial-Based Generalized Fourier Series Expansions

Sample Functions	Vectors of Undetermined Constants/Fourier Coefficients
1	a. q_0^T $q_0^T = [q_{01}^T \quad q_{02}^T]$, where $q_{01}^T = [1.0 \quad 1.001396E\text{-}16 \quad 6.468621E\text{-}17 \quad \cdots \quad 7.378673E\text{-}17 \quad 1.233958E\text{-}17]$, $q_{02}^T = [8.012204E\text{-}18 \quad 1.184774E\text{-}16 \quad \cdots \quad -3.757968E\text{-}18 \quad 1.034444E\text{-}16]$ b. q_1^T $q_1^T = [0.0 \quad -2.0 \quad 12.0 \quad 0.0 \quad 0.0 \quad 0.0]$
3	a. q_{02}^T $q_{02}^T = [3.987988E\text{-}17 \quad 1.296269E\text{-}17 \quad \cdots \quad 2.783921E\text{-}19 \quad 4.874507E\text{-}19]$ b. q_1^T $q_1^T = [0.0 \quad 5.0 \quad 0.0 \quad 0.5 \quad -1.0 \quad 0.0]$

address the accuracy issue. Shown in Table 7.3 are the approximation errors associated with the GFS expansions of these two functions and their derivatives, together with those for the CFS for the purpose of comparison. It is observed that for the functions and their up to fourth order derivatives, the overall and internal accuracies of the corresponding GFS are far more better (by 4 orders of magnitude) than their counterparts. Surprisingly, in terms of the boundary

TABLE 7.3
Approximation Errors with the Generalized Fourier Series (GFS) and Its Conventional Counterparts (CFS)

Approximation Errors	Sample Function 2		Sample Function 4	
	GFS	CFS	GFS	CFS
$\|e^4\|(u_{40})$	4.72516E–07	1.24047E–02	4.72516E–07	1.23744E–02
$\|e_I\|^4 (u_{40})$	4.72595E–07	1.22866E–02	4.72603E–07	1.23165E–02
$\|e_B\|^4 (u_{40})$	7.98557E–08	6.03144E–01	4.14033E–08	3.01591E–01
$e^{(5)}(u_{40})$	2.06660E–04	2.06662E–04	2.06266E–04	2.06188E–04
$e_I^{(5)}(u_{40})$	2.05129E–04	2.05131E–04	2.05511E–04	2.05434E–04
$e_B^{(5)}(u_{40})$	7.85952E–03	7.85949E–03	3.97827E–03	3.97717E–03
$e^{(6)}(u_{40})$	2.05366E–02	2.05368E–02	2.04864E–02	2.04799E–02
$e_I^{(6)}(u_{40})$	2.03407E–02	2.03409E–02	2.03904E–02	2.03840E–02
$e_B^{(6)}(u_{40})$	1.00000E+00	1.00000E+00	5.00000E–01	5.00000E–01

errors, the polynomial-based GFS are even better, outperforming the CFS by 7 orders of magnitude.

For the fifth order derivatives, these two different forms of Fourier series expansions start to behave almost the same with respect to the overall, internal and boundary errors. This is easily understandable: the error indexes, $e^{(5)}(u_{40})$, $e_I^{(5)}(u_{40})$, and $e_B^{(5)}(u_{40})$, are dominated by the series expansion of the fifth derivative of the sample function 2 which already has an equal value at $x = a$ and $x = -a$, no further action is needed to ensure the continuities of the internal function at these points. Similarly, the continuities, at $x = 0$ and $x = a$, of the fifth derivative of the sample function 4 are automatically kept by the even-periodic extension, as implied by its Fourier cosine series expansion. As for the sixth order derivative, the setting $2r = 6$ indicates that the sixth (and higher) order derivative is out of bound, and the GFS and the CFS are essentially the same in terms of convergence.

8 The Generalized Fourier Series of Two-Dimensional Functions

A scheme was described in the previous chapter to accelerate the convergence of the Fourier series of a one-dimensional function which is defined sufficiently smooth on a compact interval. The key concept behind it is that a supplementary function (or the boundary function) is used to force the residual part (or the internal function) of the original function to have the desired degree of smoothness when it is periodically extended onto the whole x-axis as described in Sec. 2.1. In this chapter, this convergence acceleration scheme will be extended to the generalized Fourier series expansions of two-dimensional functions and their partial derivatives. The excellent convergence and accuracy of the resulting series expansions are subsequently demonstrated using numerical examples.

8.1 STRUCTURAL DECOMPOSITIONS OF TWO-DIMENSIONAL FUNCTIONS

It should have been well understood by now that the convergence rate and the term-by-term differentiations of the Fourier series are closely related. Since this statement about one-dimensional functions is easily extendable to two-dimensional functions, we will directly focus on the sufficient conditions for the term-by-term differentiations as established in Chapter 6.

According to Theorems 6.2–6.6, the sufficient conditions for term-by-term differentiations actually only regulate the behaviors of the functions on the boundaries of the domain, specifically including four corners and two pairs of parallel boundary lines. Thus, we will correspondingly decompose a given function $u(x, y)$ into four parts:

$$u(x,y) = \varphi_0(x,y) + \varphi_1(x,y) + \varphi_2(x,y) + \varphi_3(x,y), \tag{8.1}$$

where $\varphi_3(x, y)$ is the corner function dealing with the potential jumps of $u(x, y)$ at the corners, $\varphi_1(x, y)$ and $\varphi_2(x, y)$ are boundary functions coping with the potential jumps along the boundary lines, and $\varphi_0(x, y)$ is the internal function which is conditioned to have the desired smoothness after a periodic extension. The word "jumps" here shall be broadly understood as the deviations, if any, from the complete satisfactions of, such as, (6.93)–(6.95), which is then deemed as a violation

DOI: 10.1201/9781003194859-8

of the sufficient conditions for legally performing the term-by-term differentiations of the involved series expansions. Thus, the implications of jumps will vary with the specific form of the Fourier series.

The structural decompositions of a two-dimensional function $u(x,y)$ can manifest themselves in many forms, such as:

1. If $u(x,y)$ belongs to $C^{2r-1}([-a,a]\times[-b,b])$ and all its $2r$-th order partial derivatives are absolutely integrable over the domain $[-a,a]\times[-b,b]$, then the internal function is forced to satisfy the sufficient conditions (refer to Theorem 6.2) so that its full-range Fourier series can be termwise differentiated up to the order of $2r$;
2. If $u(x,y)$ belongs to $C^{2r-1}([0,a]\times[0,b])$ and all its $2r$-th order partial derivatives are absolutely integrable over the domain $[0,a]\times[0,b]$, then the internal function is forced to satisfy the sufficient conditions (refer to Theorem 6.4) so that its half-range Fourier cosine-cosine series can be termwise differentiated up to the order of $2r$;
3. If $u(x,y)$ belongs to $C^{2r-1}([0,a]\times[0,b])$ and all its $2r$-th order partial derivatives are absolutely integrable over the domain $[0,a]\times[0,b]$, then the internal function is forced to satisfy the sufficient conditions (refer to Theorem 6.5) so that its half-range sine-cosine series can be termwise differentiated up to the order of $2r$;
4. If $u(x,y)$ belongs to $C^{2r-1}([0,a]\times[0,b])$ and all its $2r$-th order partial derivatives are absolutely integrable over the domain $[0,a]\times[0,b]$, then the internal function is forced to satisfy the sufficient conditions (refer to Theorem 6.6) so that its half-range cosine-sine series can be termwise differentiated up to the order of $2r$;
5. If $u(x,y)$ belongs to $C^{2r-1}([0,a]\times[0,b])$ and all its $2r$-th order partial derivatives are absolutely integrable over the domain $[0,a]\times[0,b]$, then the internal function is forced to satisfy the sufficient conditions (refer to Theorem 6.3) so that its half-range sine-sine series can be termwise differentiated up to the order of $2r$.

The structural decomposition in the first form is specifically discussed below. First, the corner function $\varphi_3(x,y)$ is selected to satisfy

$$\varphi_3^{(k_1,k_2)}(a,b) - \varphi_3^{(k_1,k_2)}(a,-b) - \varphi_3^{(k_1,k_2)}(-a,b) + \varphi_3^{(k_1,k_2)}(-a,-b) =$$
$$u^{(k_1,k_2)}(a,b) - u^{(k_1,k_2)}(a,-b) - u^{(k_1,k_2)}(-a,b) + u^{(k_1,k_2)}(-a,-b), \qquad (8.2)$$
$$k_1,k_2 = 0,1,2,\cdots,k_1+k_2 \le 2r-2.$$

Denote $\varphi_{012}(x,y) = u(x,y) - \varphi_3(x,y)$, then

$$\varphi_{012}^{(k_1,k_2)}(a,b) - \varphi_{012}^{(k_1,k_2)}(a,-b) - \varphi_{012}^{(k_1,k_2)}(-a,b) + \varphi_{012}^{(k_1,k_2)}(-a,-b) = 0,$$
$$k_1,k_2 = 0,1,2,\cdots,k_1+k_2 \le 2r-2. \qquad (8.3)$$

In other words, if the function $\varphi_{012}(x,y)$ is expanded into full-range Fourier series, then all the corner terms will vanish from the relevant formulas, such as (6.18), (6.21), (6.24), (6.25), and so on, in Sec. 6.2.1 when they are used to calculate the Fourier coefficients for the partial derivatives.

Second, select the boundary function $\varphi_1(x,y)$ such that

$$\varphi_1^{(k_1,0)}(a,y) - \varphi_1^{(k_1,0)}(-a,y) = \varphi_{012}^{(k_1,0)}(a,y) - \varphi_{012}^{(k_1,0)}(-a,y),\ y \in (-b,b), \tag{8.4}$$
$$k_1 = 0, 1, \cdots, 2r-1;$$

$$\varphi_1^{(0,k_2)}(x,b) - \varphi_1^{(0,k_2)}(x,-b) = 0,\ x \in (-a,a),\ k_2 = 0, 1, \cdots, 2r-1; \tag{8.5}$$

$$\varphi_1^{(k_1,k_2)}(a,b) - \varphi_1^{(k_1,k_2)}(a,-b) - \varphi_1^{(k_1,k_2)}(-a,b) + \varphi_1^{(k_1,k_2)}(-a,-b) = 0, \tag{8.6}$$
$$k_1, k_2 = 0, 1, 2, \cdots, k_1 + k_2 \le 2r-2.$$

Similarly, the other boundary function $\varphi_2(x,y)$ is selected to satisfy

$$\varphi_2^{(k_1,0)}(a,y) - \varphi_2^{(k_1,0)}(-a,y) = 0,\ y \in (-b,b),\ k_1 = 0, 1, \cdots, 2r-1; \tag{8.7}$$

$$\varphi_2^{(0,k_2)}(x,b) - \varphi_2^{(0,k_2)}(x,-b) = \varphi_{012}^{(0,k_2)}(x,b) - \varphi_{012}^{(0,k_2)}(x,-b),\ x \in (-a,a), \tag{8.8}$$
$$k_2 = 0, 1, \cdots, 2r-1;$$

$$\varphi_2^{(k_1,k_2)}(a,b) - \varphi_2^{(k_1,k_2)}(a,-b) - \varphi_2^{(k_1,k_2)}(-a,b) + \varphi_2^{(k_1,k_2)}(-a,-b) = 0, \tag{8.9}$$
$$k_1, k_2 = 0, 1, 2, \cdots, k_1 + k_2 \le 2r-2.$$

It is seen that the boundary functions are primarily used to get rid of the possible boundary jumps with the function $\varphi_{012}^{(k_1,k_2)}(x,y)$, and their "properties" are clearly marked to avoid any mutual interferences.

Finally, we have an internal function

$$\varphi_0(x,y) = \varphi_{012}(x,y) - \varphi_1(x,y) - \varphi_2(x,y)$$
$$= u(x,y) - \varphi_1(x,y) - \varphi_2(x,y) - \varphi_3(x,y), \tag{8.10}$$

which satisfies

$$\varphi_0^{(k_1,0)}(a,y) - \varphi_0^{(k_1,0)}(-a,y) = 0,\ y \in (-b,b),\ k_1 = 0, 1, \cdots, 2r-1; \tag{8.11}$$

$$\varphi_0^{(0,k_2)}(x,b) - \varphi_0^{(0,k_2)}(x,-b) = 0,\ x \in (-a,a),\ k_2 = 0, 1, \cdots, 2r-1; \tag{8.12}$$

$$\varphi_0^{(k_1,k_2)}(a,b) - \varphi_0^{(k_1,k_2)}(a,-b) - \varphi_0^{(k_1,k_2)}(-a,b) + \varphi_0^{(k_1,k_2)}(-a,-b) = 0, \tag{8.13}$$
$$k_1, k_2 = 0, 1, 2, \cdots, k_1 + k_2 \le 2r-2.$$

More explicitly, this internal function $\varphi_0(x,y)$ satisfies the sufficient condition laid out in Theorem 6.2 so that its Fourier series can be termwise differentiated up to $2r$ times.

The other four forms involve various half-range Fourier series expansions. As an example, we will use the fifth one to demonstrate the corresponding structural decomposition process.

We first define a corner function $\varphi_3(x,y)$ such that

$$\varphi_3^{(2k_1,2k_2)}(a,b) = u^{(2k_1,2k_2)}(a,b), \varphi_3^{(2k_1,2k_2)}(a,0) = u^{(2k_1,2k_2)}(a,0),$$
$$\varphi_3^{(2k_1,2k_2)}(0,b) = u^{(2k_1,2k_2)}(0,b), \varphi_3^{(2k_1,2k_2)}(0,0) = u^{(2k_1,2k_2)}(0,0), \tag{8.14}$$
$$k_1, k_2 = 0, 1, 2, \cdots, k_1 + k_2 \le r - 1.$$

Denote $\varphi_{012}(x,y) = u(x,y) - \varphi_3(x,y)$, it then represents a function without the corner jumps, that is,

$$\varphi_{012}^{(2k_1,2k_2)}(a,b) = 0, \varphi_{012}^{(2k_1,2k_2)}(a,0) = 0,$$
$$\varphi_{012}^{(2k_1,2k_2)}(0,b) = 0, \varphi_{012}^{(2k_1,2k_2)}(0,0) = 0, \tag{8.15}$$
$$k_1, k_2 = 0, 1, 2, \cdots, k_1 + k_2 \le r - 1.$$

Second, select a boundary function $\varphi_1(x,y)$ in such a way that

$$\varphi_1^{(2k_1,0)}(a,y) = \varphi_{012}^{(2k_1,0)}(a,y), \varphi_1^{(2k_1,0)}(0,y) = \varphi_{012}^{(2k_1,0)}(0,y), y \in (0,b), \tag{8.16}$$
$$k_1 = 0, 1, \cdots, r - 1;$$

$$\varphi_1^{(0,2k_2)}(x,b) = 0, \varphi_1^{(0,2k_2)}(x,0) = 0, x \in (0,a), k_2 = 0, 1, \cdots, r - 1; \tag{8.17}$$

$$\varphi_1^{(2k_1,2k_2)}(a,b) = 0, \varphi_1^{(2k_1,2k_2)}(a,0) = 0,$$
$$\varphi_1^{(2k_1,2k_2)}(0,b) = 0, \varphi_1^{(2k_1,2k_2)}(0,0) = 0, \tag{8.18}$$
$$k_1, k_2 = 0, 1, 2, \cdots, k_1 + k_2 \le r - 1.$$

The other boundary function $\varphi_2(x,y)$ is similarly defined as

$$\varphi_2^{(2k_1,0)}(a,y) = 0, \varphi_2^{(2k_1,0)}(0,y) = 0, y \in (0,b), k_1 = 0, 1, \cdots, r - 1; \tag{8.19}$$

$$\varphi_2^{(0,2k_2)}(x,b) = \varphi_{012}^{(0,2k_2)}(x,b), \varphi_2^{(0,2k_2)}(x,0) = \varphi_{012}^{(0,2k_2)}(x,0), x \in (0,a), \tag{8.20}$$
$$k_2 = 0, 1, \cdots, r - 1;$$

$$\varphi_2^{(2k_1,2k_2)}(a,b) = 0, \varphi_2^{(2k_1,2k_2)}(a,0) = 0,$$
$$\varphi_2^{(2k_1,2k_2)}(0,b) = 0, \varphi_2^{(2k_1,2k_2)}(0,0) = 0, \tag{8.21}$$
$$k_1, k_2 = 0, 1, 2, \cdots, k_1 + k_2 \le r - 1.$$

Finally, an internal function can be constructed as

$$\varphi_0(x,y) = \varphi_{012}(x,y) - \varphi_1(x,y) - \varphi_2(x,y)$$
$$= u(x,y) - \varphi_1(x,y) - \varphi_2(x,y) - \varphi_3(x,y), \tag{8.22}$$

which satisfies

$$\varphi_0^{(2k_1,0)}(a,y) = 0, \ \varphi_0^{(2k_1,0)}(0,y) = 0, \ y \in (0,b), \ k_1 = 0,1,\cdots,r-1; \tag{8.23}$$

$$\varphi_0^{(0,2k_2)}(x,b) = 0, \ \varphi_0^{(0,2k_2)}(x,0) = 0, \ x \in (0,a), \ k_2 = 0,1,\cdots,r-1; \tag{8.24}$$

$$\varphi_0^{(2k_1,2k_2)}(a,b) = 0, \ \varphi_0^{(2k_1,2k_2)}(a,0) = 0,$$
$$\varphi_0^{(2k_1,2k_2)}(0,b) = 0, \ \varphi_0^{(2k_1,2k_2)}(0,0) = 0, \tag{8.25}$$
$$k_1,k_2 = 0,1,2,\cdots, k_1+k_2 \le r-1.$$

The internal function $\varphi_0(x,y)$ obviously satisfies the sufficient condition laid out in Theorem 6.3 and its half-range sine-sine series is termwise differentiable up to $2r$ times.

The structural decompositions for the other half-range series forms are essentially the same and will not be elaborated any further. It suffices to point out that although the sufficient conditions there are slightly different in forms, the process of decomposing a two-dimensional function into four constitutive parts is equally applicable.

8.2 THE GENERALIZED FOURIER SERIES EXPANSIONS FOR TWO-DIMENSIONAL FUNCTIONS

Based on the decomposition schemes described in Sec. 8.2, we will now specifically determine the corner functions, the boundary functions, the internal functions and the corresponding forms of the generalized Fourier series expansions for two-dimensional functions and their partial derivatives.

8.2.1 THE GENERALIZED FULL-RANGE FOURIER SERIES

By assuming $u(x,y) \in C^{2r-1}([-a,a] \times [-b,b])$, we set out to derive the generalized Fourier series for such a function and its partial derivatives.

Let's first consider the corner function $\varphi_3(x,y)$.

Denote

$$\Omega_{31} = \left\{ (2j+2,2l+2) \big| j,l = 0,1,2,\cdots, j+l \le r-2 \right\}, \tag{8.26}$$

$$\Omega_{32} = \left\{ (2j+1,2l+2) \big| j,l = 0,1,2,\cdots, j+l \le r-2 \right\}, \tag{8.27}$$

$$\Omega_{33} = \left\{ (2j+2, 2l+1) \,\big|\, j,l = 0, 1, 2, \cdots,\ j+l \le r-2 \right\}, \tag{8.28}$$

$$\Omega_{34} = \left\{ (2j+1, 2l+1) \,\big|\, j,l = 0, 1, 2, \cdots,\ j+l \le r-1 \right\}, \tag{8.29}$$

$$\Omega_3 = \Omega_{31} \cup \Omega_{32} \cup \Omega_{33} \cup \Omega_{34}, \tag{8.30}$$

and

$$\bar{\Omega}_3 = \left\{ (k_1, k_2) | k_1, k_2 = 0, 1, 2, \cdots, k_1 + k_2 \le 2r-2 \right\}. \tag{8.31}$$

Select a set of basis functions

$$\mathbf{p}_3^{\mathrm{T}}(x,y) = [\quad \cdots \quad p_{3,jl}(x,y) \quad \cdots \quad]_{(j,l)\in\Omega_3}, \tag{8.32}$$

where $p_{3,jl}(x,y) \in C^{2r-1}([-a,a]\times[-b,b])$ for $(j,l) \in \Omega_3$.

For $k_1, k_2 = 0, 1, 2, \cdots$ and $k_1 + k_2 \le 2r$, define the corresponding vectors of partial derivatives of the basis functions as

$$\mathbf{p}_3^{(k_1,k_2)\mathrm{T}}(x,y) = [\quad \cdots \quad p_{3,jl}^{(k_1,k_2)}(x,y) \quad \cdots \quad]_{(j,l)\in\Omega_3}. \tag{8.33}$$

The corner function is then sought in the form of

$$\varphi_3(x,y) = \mathbf{p}_3^{\mathrm{T}}(x,y) \cdot \mathbf{a}_3, \tag{8.34}$$

where

$$\mathbf{a}_3^{\mathrm{T}} = [\quad \cdots \quad G_{3,jl} \quad \cdots \quad]_{(j,l)\in\Omega_3} \tag{8.35}$$

represents a set of unknown constants to be determined.

Substitution of (8.34) into the sufficient conditions (8.2) for the corner function $\varphi_3(x,y)$ results in

$$\mathbf{R}_3 \mathbf{a}_3 = \mathbf{q}_3, \tag{8.36}$$

where

$$\mathbf{R}_3 = \begin{bmatrix} \mathbf{p}_3^{(0,0)\mathrm{T}}(a,b) - \mathbf{p}_3^{(0,0)\mathrm{T}}(a,-b) - \mathbf{p}_3^{(0,0)\mathrm{T}}(-a,b) + \mathbf{p}_3^{(0,0)\mathrm{T}}(-a,-b) \\ \vdots \\ \mathbf{p}_3^{(k_1,k_2)\mathrm{T}}(a,b) - \mathbf{p}_3^{(k_1,k_2)\mathrm{T}}(a,-b) - \mathbf{p}_3^{(k_1,k_2)\mathrm{T}}(-a,b) + \mathbf{p}_3^{(k_1,k_2)\mathrm{T}}(-a,-b) \\ \vdots \\ \mathbf{p}_3^{(0,2r-2)\mathrm{T}}(a,b) - \mathbf{p}_3^{(0,2r-2)\mathrm{T}}(a,-b) - \mathbf{p}_3^{(0,2r-2)\mathrm{T}}(-a,b) + \mathbf{p}_3^{(0,2r-2)\mathrm{T}}(-a,-b) \end{bmatrix}_{(k_1,k_2)\in\bar{\Omega}_3}$$

$$\tag{8.37}$$

and the vector on the right-hand-side contains values of the function $u(x,y)$ and its partial derivatives at the corners, that is:

$$\mathbf{q}_3^{\mathrm{T}} = \left[u^{(0,0)}(a,b) - u^{(0,0)}(a,-b) - u^{(0,0)}(-a,b) + u^{(0,0)}(-a,-b) \quad \cdots \right.$$

$$u^{(k_1,k_2)}(a,b) - u^{(k_1,k_2)}(a,-b) - u^{(k_1,k_2)}(-a,b) + u^{(k_1,k_2)}(-a,-b) \quad \cdots$$

$$\left. u^{(0,2r-2)}(a,b) - u^{(0,2r-2)}(a,-b) - u^{(0,2r-2)}(-a,b) + u^{(0,2r-2)}(-a,-b) \right]_{(k_1,k_2)\in\bar{\Omega}_3}.$$

(8.38)

Then we have

$$\mathbf{a}_3 = \mathbf{R}_3^{-1}\mathbf{q}_3; \tag{8.39}$$

or if desired, the vector of the basis functions can be redefined as

$$\Phi_3^{\mathrm{T}}(x,y) = \mathbf{p}_3^{\mathrm{T}}(x,y) \cdot \mathbf{R}_3^{-1} \tag{8.40}$$

and the corner function $\varphi_3(x,y)$ can be written in matrix form as

$$\varphi_3(x,y) = \Phi_3^{\mathrm{T}}(x,y) \cdot \mathbf{q}_3. \tag{8.41}$$

Next, we will turn to the construction of the boundary functions $\varphi_1(x,y)$ and $\varphi_2(x,y)$.

Expand $\varphi_1(x,y)$ in a one-dimensional Fourier series over the interval $[-b,b]$ along the y-direction

$$\varphi_1(x,y) = \sum_{n=0}^{\infty} \mu_n[\xi_{1n}(x)\cos(\beta_n y) + \xi_{2n}(x)\sin(\beta_n y)], \tag{8.42}$$

where $\beta_n = n\pi/b$, $\mu_n = \begin{cases} 1/2 & n = 0 \\ 1 & n > 0 \end{cases}$, and $\xi_{1n}(x)$ and $\xi_{2n}(x)$ are the corresponding Fourier coefficients depending on x.

This boundary function $\varphi_1(x,y)$ is then required to satisfy the sufficient conditions (8.4)–(8.6). While the conditions, (8.5) and (8.6), are automatically met, substituting (8.42) into (8.4) results in

$$\xi_{1n}^{(k_1)}(a) - \xi_{1n}^{(k_1)}(-a) = c_{1n}^{(k_1)} - d_{1n}^{(k_1)}, n = 0, 1, 2, \cdots, k_1 = 0, 1, \cdots, 2r-1; \tag{8.43}$$

$$\xi_{2n}^{(k_1)}(a) - \xi_{2n}^{(k_1)}(-a) = c_{2n}^{(k_1)} - d_{2n}^{(k_1)}, n = 1, 2, \cdots, k_1 = 0, 1, \cdots, 2r-1; \tag{8.44}$$

where $c_{1n}^{(k_1)}$ and $c_{2n}^{(k_1)}$ are the Fourier coefficients of $\varphi_{012}^{(k_1,0)}(a,y) = u^{(k_1,0)}(a,y) - \varphi_3^{(k_1,0)}(a,y)$, and $d_{1n}^{(k_1)}$ and $d_{2n}^{(k_1)}$ are the Fourier coefficients of $\varphi_{012}^{(k_1,0)}(-a,y) = u^{(k_1,0)}(-a,y) - \varphi_3^{(k_1,0)}(-a,y)$.

Here, the functions $\xi_{1n}(x)$ and $\xi_{2n}(x)$ will be sought in the forms of

$$\xi_{1n}(x) = \mathbf{p}_{1n}^{\mathrm{T}}(x) \cdot \mathbf{a}_{1n,1}, n = 0, 1, 2, \cdots; \tag{8.45}$$

$$\xi_{2n}(x) = \mathbf{p}_{1n}^{\mathrm{T}}(x) \cdot \mathbf{a}_{1n,2}, n = 1, 2, \cdots; \tag{8.46}$$

where the vectors of unknown constants

$$\mathbf{a}_{1n,s}^{\mathrm{T}} = [a_{1n,s}^1 \quad a_{1n,s}^2 \quad \cdots \quad a_{1n,s}^{2r}] \quad (s = 1, 2), \tag{8.47}$$

are yet to be determined, and

$$\mathbf{p}_{1n}^{\mathrm{T}}(x) = [p_{1n,1}(x) \quad p_{1n,2}(x) \quad \cdots \quad p_{1n,2r}(x)] \tag{8.48}$$

is a set of functions which are linearly independent and, at least, $2r$ times continuously differentiable over the interval $[-a, a]$, that is,

$$\mathbf{p}_{1n}^{(k_1)\mathrm{T}}(x) = [p_{1n,1}^{(k_1)}(x) \quad p_{1n,2}^{(k_1)}(x) \quad \cdots \quad p_{1n,2r}^{(k_1)}(x)], \quad k_1 = 0, 1, \cdots, 2r. \tag{8.49}$$

Substituting (8.45) and (8.46) respectively into (8.43) and (8.44), we have

$$\mathbf{R}_{1n}\mathbf{a}_{1n,1} = \mathbf{q}_{1n,1}, n = 0, 1, 2, \cdots; \tag{8.50}$$

$$\mathbf{R}_{1n}\mathbf{a}_{1n,2} = \mathbf{q}_{1n,2}, n = 1, 2, \cdots; \tag{8.51}$$

where

$$\mathbf{R}_{1n} = \begin{bmatrix} \mathbf{p}_{1n}^{(0)\mathrm{T}}(a) - \mathbf{p}_{1n}^{(0)\mathrm{T}}(-a) \\ \mathbf{p}_{1n}^{(1)\mathrm{T}}(a) - \mathbf{p}_{1n}^{(1)\mathrm{T}}(-a) \\ \vdots \\ \mathbf{p}_{1n}^{(2r-1)\mathrm{T}}(a) - \mathbf{p}_{1n}^{(2r-1)\mathrm{T}}(-a) \end{bmatrix} \tag{8.52}$$

and

$$\mathbf{q}_{1n,s}^{\mathrm{T}} = [c_{sn}^{(0)} - d_{sn}^{(0)} \quad c_{sn}^{(1)} - d_{sn}^{(1)} \quad \cdots \quad c_{sn}^{(2r-1)} - d_{sn}^{(2r-1)}] \quad (s = 1, 2), \tag{8.53}$$

denote the vectors of the boundary Fourier coefficients of the function $u(x, y) - \varphi_3(x, y)$ along the two opposite edges $x = a$ and $x = -a$, respectively.

Thus, the vectors of unknown constants can be obtained as

$$\mathbf{a}_{1n,1} = \mathbf{R}_{1n}^{-1}\mathbf{q}_{1n,1} \tag{8.54}$$

and

$$\mathbf{a}_{1n,2} = \mathbf{R}_{1n}^{-1}\mathbf{q}_{1n,2}. \tag{8.55}$$

If preferred, we can define the new sets of basis functions as

$$\Phi_{1n,1}^{\mathrm{T}}(x,y) = \mu_n \cos(\beta_n y) \cdot \mathbf{p}_{1n}^{\mathrm{T}}(x) \cdot \mathbf{R}_{1n}^{-1}, n = 0, 1, 2, \cdots; \tag{8.56}$$

$$\Phi_{1n,2}^{\mathrm{T}}(x,y) = \mu_n \sin(\beta_n y) \cdot \mathbf{p}_{1n}^{\mathrm{T}}(x) \cdot \mathbf{R}_{1n}^{-1}, n = 1, 2, \cdots; \tag{8.57}$$

and accordingly
   for $k_2 = 0, 2, \cdots, 2r$,

$$\Phi_{1n,1}^{(k_1,k_2)\mathrm{T}}(x,y) = \mu_n (-1)^{k_2/2} \beta_n^{k_2} \cos(\beta_n y) \cdot \mathbf{p}_{1n}^{(k_1)\mathrm{T}}(x) \cdot \mathbf{R}_{1n}^{-1}, \tag{8.58}$$

$$\Phi_{1n,2}^{(k_1,k_2)\mathrm{T}}(x,y) = \mu_n (-1)^{k_2/2} \beta_n^{k_2} \sin(\beta_n y) \cdot \mathbf{p}_{1n}^{(k_1)\mathrm{T}}(x) \cdot \mathbf{R}_{1n}^{-1}; \tag{8.59}$$

for $k_2 = 1, 3, \cdots, 2r - 1$,

$$\Phi_{1n,1}^{(k_1,k_2)\mathrm{T}}(x,y) = \mu_n (-1)^{(k_2+1)/2} \beta_n^{k_2} \sin(\beta_n y) \cdot \mathbf{p}_{1n}^{(k_1)\mathrm{T}}(x) \cdot \mathbf{R}_{1n}^{-1}, \tag{8.60}$$

$$\Phi_{1n,2}^{(k_1,k_2)\mathrm{T}}(x,y) = \mu_n (-1)^{(k_2-1)/2} \beta_n^{k_2} \cos(\beta_n y) \cdot \mathbf{p}_{1n}^{(k_1)\mathrm{T}}(x) \cdot \mathbf{R}_{1n}^{-1}. \tag{8.61}$$

More concisely, we combine the vectors of the basis functions into the extended vectors as:

$$\Phi_{1,1}^{\mathrm{T}}(x,y) = [\Phi_{10,1}^{\mathrm{T}}(x,y) \quad \Phi_{11,1}^{\mathrm{T}}(x,y) \quad \cdots \quad \Phi_{1n,1}^{\mathrm{T}}(x,y) \quad \cdots], \tag{8.62}$$

$$\Phi_{1,2}^{\mathrm{T}}(x,y) = [\Phi_{11,2}^{\mathrm{T}}(x,y) \quad \Phi_{12,2}^{\mathrm{T}}(x,y) \quad \cdots \quad \Phi_{1n,2}^{\mathrm{T}}(x,y) \quad \cdots], \tag{8.63}$$

and

$$\Phi_1^{\mathrm{T}}(x,y) = [\Phi_{1,1}^{\mathrm{T}}(x,y) \quad \Phi_{1,2}^{\mathrm{T}}(x,y)]. \tag{8.64}$$

The corresponding (extended) vectors of the partial derivatives of the basis functions are

$$\Phi_{1,1}^{(k_1,k_2)\mathrm{T}}(x,y) = [\Phi_{10,1}^{(k_1,k_2)\mathrm{T}}(x,y) \quad \Phi_{11,1}^{(k_1,k_2)\mathrm{T}}(x,y) \quad \cdots \quad \Phi_{1n,1}^{(k_1,k_2)\mathrm{T}}(x,y) \quad \cdots], \tag{8.65}$$

$$\Phi_{1,2}^{(k_1,k_2)\mathrm{T}}(x,y) = [\Phi_{11,2}^{(k_1,k_2)\mathrm{T}}(x,y) \quad \Phi_{12,2}^{(k_1,k_2)\mathrm{T}}(x,y) \quad \cdots \quad \Phi_{1n,2}^{(k_1,k_2)\mathrm{T}}(x,y) \quad \cdots], \tag{8.66}$$

$$\Phi_1^{(k_1,k_2)\mathrm{T}}(x,y) = [\Phi_{1,1}^{(k_1,k_2)\mathrm{T}}(x,y) \quad \Phi_{1,2}^{(k_1,k_2)\mathrm{T}}(x,y)]. \tag{8.67}$$

The vectors of boundary Fourier coefficients of the function $u(x,y) - \varphi_3(x,y)$ can be rearranged accordingly:

$$\mathbf{q}_{1,1}^{\mathrm{T}} = [\mathbf{q}_{10,1}^{\mathrm{T}} \quad \mathbf{q}_{11,1}^{\mathrm{T}} \quad \cdots \quad \mathbf{q}_{1n,1}^{\mathrm{T}} \quad \cdots], \tag{8.68}$$

$$\mathbf{q}_{1,2}^{\mathrm{T}} = [\mathbf{q}_{11,2}^{\mathrm{T}} \quad \mathbf{q}_{12,2}^{\mathrm{T}} \quad \cdots \quad \mathbf{q}_{1n,2}^{\mathrm{T}} \quad \cdots], \tag{8.69}$$

and

$$\mathbf{q}_1^{\mathrm{T}} = [\mathbf{q}_{1,1}^{\mathrm{T}} \quad \mathbf{q}_{1,2}^{\mathrm{T}}]. \tag{8.70}$$

Then the boundary function $\varphi_1(x, y)$ can be finally expressed as

$$\varphi_1(x, y) = \Phi_1^{\mathrm{T}}(x, y) \cdot \mathbf{q}_1. \tag{8.71}$$

The boundary function $\varphi_2(x, y)$ can be similarly constructed by first expanding it, along the x-direction, into a one-dimensional Fourier series as

$$\varphi_2(x, y) = \sum_{m=0}^{\infty} \mu_m [\zeta_{1m}(y)\cos(\alpha_m x) + \zeta_{2m}(y)\sin(\alpha_m x)] \tag{8.72}$$

where $\alpha_m = m\pi/a$, $\mu_m = \begin{cases} 1/2 & m = 0 \\ 1 & m > 0 \end{cases}$, and $\zeta_{1m}(y)$ and $\zeta_{2m}(y)$ are the corresponding one-dimensional Fourier coefficients.

If $p_{2m,l}(y), l = 1, 2, \cdots, 2r$, denote the functions which are linearly independent and at least $2r$ times continuously differentiable over the interval $[-b, b]$, then the procedures described earlier for determining the boundary function $\varphi_1(x, y)$ are equally applicable here, and the boundary function $\varphi_2(x, y)$ can be directly written as

$$\varphi_2(x, y) = \Phi_2^{\mathrm{T}}(x, y) \cdot \mathbf{q}_2, \tag{8.73}$$

where $\mathbf{q}_2$ is the vector of boundary Fourier coefficients of the function $u(x, y)$ $-\varphi_3(x, y)$ calculated along the edges $y = b$ and $y = -b$, respectively; $\Phi_2(x, y)$ can be simply obtained from $\Phi_1(x, y)$ by switching x and y, and replacing b and $\beta_n = n\pi/b$ with a and $\alpha_m = m\pi/a$, respectively.

Finally, the internal function $\varphi_0(x, y)$ is given in the form of Fourier series over the domain $[-a, a] \times [-b, b]$:

$$\varphi_0(x, y) = \sum_{m=0}^{\infty} \sum_{n=0}^{\infty} \lambda_{mn} [V_{1mn}\cos(\alpha_m x)\cos(\beta_n y) + V_{2mn}\sin(\alpha_m x)\cos(\beta_n y)$$
$$+ V_{3mn}\cos(\alpha_m x)\sin(\beta_n y) + V_{4mn}\sin(\alpha_m x)\sin(\beta_n y)], \tag{8.74}$$

where $\alpha_m = m\pi/a$, $\beta_n = n\pi/b$, $\lambda_{mn} = \mu_m\mu_n$, and V_{smn} $(s = 1, 2, 3, 4)$ are the Fourier coefficients.

More concisely, the internal function $\varphi_0(x, y)$ can be rewritten in matrix form as

$$\varphi_0(x, y) = \Phi_0^{\mathrm{T}}(x, y) \cdot \mathbf{q}_0, \tag{8.75}$$

where

$$\Phi_0^T(x,y) = [\Phi_{01}^T(x,y) \quad \Phi_{02}^T(x,y) \quad \Phi_{03}^T(x,y) \quad \Phi_{04}^T(x,y)] \tag{8.76}$$

and

$$\mathbf{q}_0^T = [\mathbf{q}_{01}^T \quad \mathbf{q}_{02}^T \quad \mathbf{q}_{03}^T \quad \mathbf{q}_{04}^T] \tag{8.77}$$

with

$$\Phi_{0s}^T(x,y) = [\quad \cdots \quad \varphi_{0smn}(x,y) \quad \cdots \quad] \quad (s = 1, 2, 3, 4), \tag{8.78}$$

$$\mathbf{q}_{0s}^T = [\quad \cdots \quad V_{smn} \quad \cdots \quad] \quad (s = 1, 2, 3, 4), \tag{8.79}$$

$$\varphi_{01mn}(x,y) = \lambda_{mn} \cos(\alpha_m x)\cos(\beta_n y) \quad (m = 0, 1, 2, \cdots, n = 0, 1, 2, \cdots), \tag{8.80}$$

$$\varphi_{02mn}(x,y) = \lambda_{mn} \sin(\alpha_m x)\cos(\beta_n y) \quad (m = 1, 2, \cdots, n = 0, 1, 2, \cdots), \tag{8.81}$$

$$\varphi_{03mn}(x,y) = \lambda_{mn} \cos(\alpha_m x)\sin(\beta_n y) \quad (m = 0, 1, 2, \cdots, n = 1, 2, \cdots), \tag{8.82}$$

and

$$\varphi_{04mn}(x,y) = \lambda_{mn} \sin(\alpha_m x)\sin(\beta_n y) \quad (m = 1, 2, \cdots, n = 1, 2, \cdots). \tag{8.83}$$

Accordingly, the partial derivatives of the internal functions are

$$\varphi_0^{(k_1,k_2)}(x,y) = \Phi_0^{(k_1,k_2)T}(x,y) \cdot \mathbf{q}_0, \tag{8.84}$$

where

$$\Phi_0^{(k_1,k_2)T}(x,y) = [\Phi_{01}^{(k_1,k_2)T}(x,y) \quad \Phi_{02}^{(k_1,k_2)T}(x,y) \quad \Phi_{03}^{(k_1,k_2)T}(x,y) \quad \Phi_{04}^{(k_1,k_2)T}(x,y)] \tag{8.85}$$

with its sub-vectors being defined as

$$\Phi_{0s}^{(k_1,k_2)T}(x,y) = [\cdots \quad \varphi_{0smn}^{(k_1,k_2)}(x,y) \quad \cdots] \quad (s = 1, 2, 3, 4). \tag{8.86}$$

The explicit expressions for the partial derivatives of the basis functions can be directly written out. For instance, if the nonnegative integers k_1 and k_2 are both even numbers, then

$$\varphi_{01mn}^{(k_1,k_2)}(x,y) = \lambda_{mn}(-1)^{(k_1+k_2)/2}\alpha_m^{k_1}\beta_n^{k_2}\cos(\alpha_m x)\cos(\beta_n y), \tag{8.87}$$

$$\varphi_{02mn}^{(k_1,k_2)}(x,y) = \lambda_{mn}(-1)^{(k_1+k_2)/2}\alpha_m^{k_1}\beta_n^{k_2}\sin(\alpha_m x)\cos(\beta_n y), \tag{8.88}$$

$$\varphi_{03mn}^{(k_1,k_2)}(x,y) = \lambda_{mn}(-1)^{(k_1+k_2)/2}\alpha_m^{k_1}\beta_n^{k_2}\cos(\alpha_m x)\sin(\beta_n y), \qquad (8.89)$$

$$\varphi_{04mn}^{(k_1,k_2)}(x,y) = \lambda_{mn}(-1)^{(k_1+k_2)/2}\alpha_m^{k_1}\beta_n^{k_2}\sin(\alpha_m x)\sin(\beta_n y). \qquad (8.90)$$

Similar expressions can be readily obtained for other different (k_1,k_2) parity combinations.

With the four constituents being fully determined, the generalized Fourier series for the function $u(x,y)$ and its partial derivatives can be formally expressed as

$$u^{(k_1,k_2)}(x,y) = \varphi_0^{(k_1,k_2)}(x,y) + \varphi_1^{(k_1,k_2)}(x,y) + \varphi_2^{(k_1,k_2)}(x,y) + \varphi_3^{(k_1,k_2)}(x,y)$$

$$= \Phi_0^{(k_1,k_2)\mathrm{T}}(x,y)\cdot\mathbf{q}_0 + \Phi_1^{(k_1,k_2)\mathrm{T}}(x,y)\cdot\mathbf{q}_1 + \Phi_2^{(k_1,k_2)\mathrm{T}}(x,y)\cdot\mathbf{q}_2 + \Phi_3^{(k_1,k_2)\mathrm{T}}(x,y)\cdot\mathbf{q}_3,$$

$$k_1,k_2 = 0,1,2,\cdots, k_1+k_2 \leq 2r.$$

$$(8.91)$$

8.2.2 The Generalized Half-Range Fourier Sine-Sine Series

By assuming $u(x,y)\in C^{2r-1}([0,a]\times[0,b])$, we now expand it into the generalized half-range sine-sine series.

We start with the corner function $\varphi_3(x,y)$.

If the positive integer r is even, then

$$\Omega_3 = \big\{(j,l)\big| j,l = 0,1,2,\cdots, j+l \leq 2r-1; \; j+l = 2r, j = 1,2,\cdots, r/2$$

$$\text{or} \qquad l = 1,2,\cdots, r/2\big\}; \qquad (8.92)$$

otherwise,

$$\Omega_3 = \big\{(j,l)\big| j,l = 0,1,2,\cdots, j+l \leq 2r-1; \; j+l = 2r, j = 1,2,\cdots, (r-1)/2$$

$$\text{or} \qquad l = 1,2,\cdots, (r-1)/2; \; j=l=r\big\}. \qquad (8.93)$$

It is easily verified that for a positive even number r, the number of elements in Ω_3 is $1/2\cdot 2r\cdot(2r+1)+2\cdot r/2 = 2r^2+2r$, and for a positive odd number r, the number of elements in Ω_3 is $1/2\cdot 2r\cdot(2r+1)+2\cdot(r-1)/2+1 = 2r^2+2r$. Meanwhile, the number of equations provided by (8.14) is equal to $4\cdot[r(r+1)/2] = 2r^2+2r$.

Select a set of basis functions

$$\mathbf{p}_3^\mathrm{T}(x,y) = [p_{3,00}(x,y) \quad \cdots \quad p_{3,jl}(x,y) \quad \cdots \quad]_{(j,l)\in\Omega_3}, \qquad (8.94)$$

where $p_{3,jl}(x,y), (j,l)\in\Omega_3$, are linearly independent and at least $2r$ times continuously differentiable over the domain $[0,a]\times[0,b]$.

The vectors of the corresponding partial derivatives are denoted as

$$\mathbf{p}_3^{(k_1,k_2)\mathrm{T}}(x,y) = [p_{3,00}^{(k_1,k_2)}(x,y) \quad \cdots \quad p_{3,jl}^{(k_1,k_2)}(x,y) \quad \cdots \quad]_{(j,l)\in\Omega_3}$$

$$\text{for } k_1,k_2 = 0,1,2,\cdots \text{ and } k_1+k_2 \leq 2r. \qquad (8.95)$$

Basically following the same procedures as described above for the full-range Fourier series case, we can come up with a set of basis functions

$$\mathbf{\Phi}_3^T(x,y) = \mathbf{p}_3^T(x,y) \cdot \mathbf{R}_3^{-1}, \qquad (8.96)$$

where

$$\mathbf{R}_3^T = [\mathbf{R}_{31}^T \quad \mathbf{R}_{32}^T \quad \mathbf{R}_{33}^T \quad \mathbf{R}_{34}^T] \qquad (8.97)$$

with its sub-matrices being defined as

$$\mathbf{R}_{31}^T = [\mathbf{p}_3^{(0,0)}(a,b) \quad \cdots \quad \mathbf{p}_3^{(2k_1,2k_2)}(a,b) \quad \cdots \quad \mathbf{p}_3^{(0,2r-2)}(a,b)]_{(2k_1,2k_2)\in\bar{\Omega}_3}, \qquad (8.98)$$

$$\mathbf{R}_{32}^T = [\mathbf{p}_3^{(0,0)}(a,0) \quad \cdots \quad \mathbf{p}_3^{(2k_1,2k_2)}(a,0) \quad \cdots \quad \mathbf{p}_3^{(0,2r-2)}(a,0)]_{(2k_1,2k_2)\in\bar{\Omega}_3}, \qquad (8.99)$$

$$\mathbf{R}_{33}^T = [\mathbf{p}_3^{(0,0)}(0,b) \quad \cdots \quad \mathbf{p}_3^{(2k_1,2k_2)}(0,b) \quad \cdots \quad \mathbf{p}_3^{(0,2r-2)}(0,b)]_{(2k_1,2k_2)\in\bar{\Omega}_3}, \qquad (8.100)$$

$$\mathbf{R}_{34}^T = [\mathbf{p}_3^{(0,0)}(0,0) \quad \cdots \quad \mathbf{p}_3^{(2k_1,2k_2)}(0,0) \quad \cdots \quad \mathbf{p}_3^{(0,2r-2)}(0,0)]_{(2k_1,2k_2)\in\bar{\Omega}_3}, \qquad (8.101)$$

and

$$\bar{\Omega}_3 = \left\{(2k_1,2k_2) \mid k_1, k_2 = 0, 1, 2, \cdots, k_1 + k_2 \le r-1\right\}. \qquad (8.102)$$

Thus, the corner function $\varphi_3(x,y)$ can be expressed as

$$\varphi_3(x,y) = \mathbf{\Phi}_3^T(x,y) \cdot \mathbf{q}_3, \qquad (8.103)$$

where

$$\mathbf{q}_3^T = [\mathbf{q}_{31}^T \quad \mathbf{q}_{32}^T \quad \mathbf{q}_{33}^T \quad \mathbf{q}_{34}^T], \qquad (8.104)$$

with its sub-vectors being defined as

$$\mathbf{q}_{31}^T = [u^{(0,0)}(a,b) \quad \cdots \quad u^{(2k_1,2k_2)}(a,b) \quad \cdots \quad u^{(0,2r-2)}(a,b)]_{(2k_1,2k_2)\in\bar{\Omega}_3}, \qquad (8.105)$$

$$\mathbf{q}_{32}^T = [u^{(0,0)}(a,0) \quad \cdots \quad u^{(2k_1,2k_2)}(a,0) \quad \cdots \quad u^{(0,2r-2)}(a,0)]_{(2k_1,2k_2)\in\bar{\Omega}_3}, \qquad (8.106)$$

$$\mathbf{q}_{33}^T = [u^{(0,0)}(0,b) \quad \cdots \quad u^{(2k_1,2k_2)}(0,b) \quad \cdots \quad u^{(0,2r-2)}(0,b)]_{(2k_1,2k_2)\in\bar{\Omega}_3}, \qquad (8.107)$$

and

$$\mathbf{q}_{34}^T = [u^{(0,0)}(0,0) \quad \cdots \quad u^{(2k_1,2k_2)}(0,0) \quad \cdots \quad u^{(0,2r-2)}(0,0)]_{(2k_1,2k_2)\in\bar{\Omega}_3}. \qquad (8.108)$$

The corner function is clearly dependent upon the corner values of the function $u(x,y)$ and its partial derivatives.

Next, we proceed to constructing the boundary functions $\varphi_1(x,y)$ and $\varphi_2(x,y)$.

The boundary function $\varphi_1(x,y)$ is here sought as the half-range sine series in y-direction, that is,

$$\varphi_1(x,y) = \sum_{n=1}^{\infty} \xi_{2n}(x)\sin(\beta_n y), \tag{8.109}$$

where $\beta_n = n\pi/b$ and $\xi_{2n}(x)$ are the corresponding Fourier coefficients.

This boundary function $\varphi_1(x,y)$ needs to satisfy the sufficient conditions as set forth in (8.16)–(8.18). While the conditions (8.17) and (8.18) are satisfied automatically, substituting (8.109) into (8.16) results in

$$\xi_{2n}^{(2k_1)}(a) = c_{2n}^{(2k_1)}, n = 1, 2, \cdots, k_1 = 0, 1, \cdots, r-1; \tag{8.110}$$

$$\xi_{2n}^{(2k_1)}(0) = d_{2n}^{(2k_1)}, n = 1, 2, \cdots, k_1 = 0, 1, \cdots, r-1; \tag{8.111}$$

where $c_{2n}^{(2k_1)}$ and $d_{2n}^{(2k_1)}$ are the Fourier coefficients of functions $u^{(2k_1,0)}(a,y) - \varphi_3^{(2k_1,0)}(a,y)$ and $u^{(2k_1,0)}(0,y) - \varphi_3^{(2k_1,0)}(0,y)$.

As before, the function $\xi_{2n}(x)$ is given by

$$\xi_{2n}(x) = \mathbf{p}_{1n}^{T}(x) \cdot \mathbf{a}_{1n}, n = 1, 2, \cdots, \tag{8.112}$$

where

$$\mathbf{a}_{1n}^{T} = [a_{1n}^1 \quad a_{1n}^2 \quad \cdots \quad a_{1n}^{2r}] \tag{8.113}$$

and

$$\mathbf{p}_{1n}^{T}(x) = [p_{1n,1}(x) \quad p_{1n,2}(x) \quad \cdots \quad p_{1n,2r}(x)] \tag{8.114}$$

with $p_{1n,j}(x)$, $j = 1, 2, \cdots, 2r$, being a set of functions which are linearly independent and at least $2r$ times continuously differentiable, that is,

$$\mathbf{p}_{1n}^{(k_1)T}(x) = [p_{1n,1}^{(k_1)}(x) \quad p_{1n,2}^{(k_1)}(x) \quad \cdots \quad p_{1n,2r}^{(k_1)}(x)]. \tag{8.115}$$

By substituting (8.112) into (8.110) and (8.111), we have

$$\mathbf{R}_{1n}\mathbf{a}_{1n} = \mathbf{q}_{1n} \quad (n = 1, 2, \cdots), \tag{8.116}$$

where

$$\mathbf{R}_{1n} = \begin{bmatrix} \mathbf{R}_{1n,1} \\ \mathbf{R}_{1n,2} \end{bmatrix} \tag{8.117}$$

and

$$\mathbf{q}_{1n}^{\mathrm{T}} = [\mathbf{q}_{1n,1}^{\mathrm{T}} \quad \mathbf{q}_{1n,2}^{\mathrm{T}}] \tag{8.118}$$

with the involved sub-matrices and sub-vectors being defined as

$$\mathbf{R}_{1n,1} = \begin{bmatrix} \mathbf{p}_{1n}^{(0)\mathrm{T}}(a) \\ \mathbf{p}_{1n}^{(2)\mathrm{T}}(a) \\ \vdots \\ \mathbf{p}_{1n}^{(2r-2)\mathrm{T}}(a) \end{bmatrix}, \tag{8.119}$$

$$\mathbf{R}_{1n,2} = \begin{bmatrix} \mathbf{p}_{1n}^{(0)\mathrm{T}}(0) \\ \mathbf{p}_{1n}^{(2)\mathrm{T}}(0) \\ \vdots \\ \mathbf{p}_{1n}^{(2r-2)\mathrm{T}}(0) \end{bmatrix}, \tag{8.120}$$

$$\mathbf{q}_{1n,1}^{\mathrm{T}} = [c_{2n}^{(0)} \quad c_{2n}^{(2)} \quad \cdots \quad c_{2n}^{(2r-2)}], \tag{8.121}$$

and

$$\mathbf{q}_{1n,2}^{\mathrm{T}} = [d_{2n}^{(0)} \quad d_{2n}^{(2)} \quad \cdots \quad d_{2n}^{(2r-2)}]. \tag{8.122}$$

The elements in (8.121) and (8.122) are the boundary Fourier coefficients of the function $u(x,y) - \varphi_3(x,y)$ calculated along the edges $x = a$ and $x = 0$, respectively.

From (8.116), we have

$$\mathbf{a}_{1n} = \mathbf{R}_{1n}^{-1}\mathbf{q}_{1n}. \tag{8.123}$$

If desired, a new set of basis functions can be defined as

$$\mathbf{\Phi}_{1n}^{\mathrm{T}}(x,y) = \sin(\beta_n y) \cdot \mathbf{p}_{1n}^{\mathrm{T}}(x) \cdot \mathbf{R}_{1n}^{-1} \quad (n = 1, 2, \cdots), \tag{8.124}$$

and its differential forms are:

$$\mathbf{\Phi}_{1n}^{(k_1,k_2)\mathrm{T}}(x,y) = (-1)^{k_2/2} \beta_n^{k_2} \sin(\beta_n y) \cdot \mathbf{p}_{1n}^{(k_1)\mathrm{T}}(x) \cdot \mathbf{R}_{1n}^{-1} \tag{8.125}$$

for $k_1 = 0, 2, \cdots, 2r$;

$$\mathbf{\Phi}_n^{(k_1,k_2)\mathrm{T}}((x,y)) = (-1)^{(k_2-1)/2} \beta_n^{k_2} \cos(\beta_n y) \cdot \mathbf{p}_{1n}^{(k_1)\mathrm{T}}(x) \cdot \mathbf{R}_{1n}^{-1} \tag{8.126}$$

for $k_1 = 1, 3, \cdots, 2r - 1$.

Denote

$$\Phi_1^T(x,y) = [\Phi_{11}^T(x,y) \quad \Phi_{12}^T(x,y) \quad \cdots \quad \Phi_{1n}^T(x,y) \quad \cdots] \tag{8.127}$$

and

$$\mathbf{q}_1^T = [\mathbf{q}_{11}^T \quad \mathbf{q}_{12}^T \quad \cdots \quad \mathbf{q}_{1n}^T \quad \cdots], \tag{8.128}$$

the boundary function $\varphi_1(x,y)$ can be finally expressed as

$$\varphi_1(x,y) = \Phi_1^T(x,y) \cdot \mathbf{q}_1. \tag{8.129}$$

Similarly, the boundary function $\varphi_2(x,y)$ is expanded into a one-dimensional half-range sine series along the x-direction as

$$\varphi_2(x,y) = \sum_{m=1}^{\infty} \zeta_{2m}(y)\sin(\alpha_m x), \tag{8.130}$$

where $\alpha_m = m\pi/a$, and $\zeta_{2m}(y)$ are the corresponding Fourier coefficients.

Let $p_{2m,l}(y)$, $l = 1, 2, \cdots, 2r$, be a set of basis functions which are at least $2r$ times continuously differentiable over the interval $[0,b]$. By following the same procedures as described above, we can express the boundary function $\varphi_2(x,y)$ as

$$\varphi_2(x,y) = \Phi_2^T(x,y) \cdot \mathbf{q}_2, \tag{8.131}$$

where $\mathbf{q}_2$ is the vector of boundary Fourier coefficients of the function $u(x,y)$ $-\varphi_3(x,y)$ calculated along the edges $y = b$ and $y = 0$, respectively; $\Phi_2(x,y)$ is the vector of basis functions which can be simply obtained from $\Phi_1(x,y)$ by switching x and y and replacing b and $\beta_n = n\pi/b$ with a and $\alpha_m = m\pi/a$, respectively.

Finally, the internal function $\varphi_0(x,y)$ is expanded into a half-range sine-sine series on the domain $[0,a] \times [0,b]$

$$\varphi_0(x,y) = \sum_{m=1}^{\infty}\sum_{n=1}^{\infty} V_{4mn}\sin(\alpha_m x)\sin(\beta_n y), \tag{8.132}$$

where $\alpha_m = m\pi/a$, $\beta_n = n\pi/b$, and V_{4mn} are the Fourier coefficients of $\varphi_0(x,y)$.

Let k_1 and k_2 be nonnegative integers and

$$\varphi_{04mn}(x,y) = \sin(\alpha_m x)\sin(\beta_n y) \quad (m = 1, 2, \cdots, n = 1, 2, \cdots). \tag{8.133}$$

Then we have:

$$\varphi_{04mn}^{(k_1,k_2)}(x,y) = (-1)^{(k_1+k_2)/2}\alpha_m^{k_1}\beta_n^{k_2}\sin(\alpha_m x)\sin(\beta_n y), \tag{8.134}$$

if k_1 and k_2 are both even;

$$\varphi_{04mn}^{(k_1,k_2)}(x,y) = (-1)^{(k_1+k_2-1)/2} \alpha_m^{k_1} \beta_n^{k_2} \sin(\alpha_m x)\cos(\beta_n y), \qquad (8.135)$$

if k_1 is even and k_2 is odd;

$$\varphi_{04mn}^{(k_1,k_2)}(x,y) = (-1)^{(k_1+k_2-1)/2} \alpha_m^{k_1} \beta_n^{k_2} \cos(\alpha_m x)\sin(\beta_n y), \qquad (8.136)$$

if k_1 is odd and k_2 is even;

$$\varphi_{04mn}^{(k_1,k_2)}(x,y) = (-1)^{(k_1+k_2-2)/2} \alpha_m^{k_1} \beta_n^{k_2} \cos(\alpha_m x)\cos(\beta_n y), \qquad (8.137)$$

if k_1 and k_2 are both odd.
  Define

$$\boldsymbol{\Phi}_{04}^{\mathrm{T}}(x,y) = [\varphi_{0411}(x,y) \quad \cdots \quad \varphi_{04mn}(x,y) \quad \cdots] \qquad (8.138)$$

and

$$\mathbf{q}_{04}^{\mathrm{T}} = [V_{411} \quad \cdots \quad V_{4mn} \quad \cdots]. \qquad (8.139)$$

The internal function $\varphi_0(x,y)$ can be formally expressed as

$$\varphi_0(x,y) = \boldsymbol{\Phi}_{04}^{\mathrm{T}}(x,y) \cdot \mathbf{q}_{04}. \qquad (8.140)$$

With the four constituents in the structural decomposition (8.22) being fully determined, the generalized half-range sine-sine series for the two-dimensional function $u(x,y)$ and its partial derivatives of orders up to $2r$ can be simultaneously obtained as

$$\begin{aligned}
u^{(k_1,k_2)}(x,y) &= \varphi_0^{(k_1,k_2)}(x,y) + \varphi_1^{(k_1,k_2)}(x,y) + \varphi_2^{(k_1,k_2)}(x,y) + \varphi_3^{(k_1,k_2)}(x,y) \\
&= \boldsymbol{\Phi}_{04}^{(k_1,k_2)\mathrm{T}}(x,y) \cdot \mathbf{q}_{04} + \boldsymbol{\Phi}_1^{(k_1,k_2)\mathrm{T}}(x,y) \cdot \mathbf{q}_1 + \boldsymbol{\Phi}_2^{(k_1,k_2)\mathrm{T}}(x,y) \cdot \mathbf{q}_2 \\
&\quad + \boldsymbol{\Phi}_3^{(k_1,k_2)\mathrm{T}}(x,y) \cdot \mathbf{q}_3, \\
k_1,k_2 &= 0,1,2,\cdots, k_1+k_2 \le 2r.
\end{aligned} \qquad (8.141)$$

In the above discussions, the objective for performing the structural decomposition is primarily stated from the perspective of ensuring the involved series expansions to be termwise differentiable up to $2r$ times. As evident from Theorem 2.3, the smoothness of a periodic (or, the periodically extended) function dictate not only the term-by-term differentiability, but also the convergence characteristic of its Fourier series expansion. Thus, we will view the above structural decompositions and the accompanying series expansion schemes as a general approach for improving the convergence and accuracy of the Fourier series approximations of the (two-dimensional) functions and their (partial) derivatives.

8.3   THE POLYNOMIAL-BASED GENERALIZED FOURIER SERIES EXPANSIONS

Thus far, we have completed a general framework for the simultaneous series expansions of a two-dimensional function and its partial derivatives. The primary difference between the generalized and the conventional Fourier series expansions is that three supplementary or conditioning functions are literally used to force the internal (or residual) function, and hence the corresponding series, to behave in certain ways or with the desired characteristics. In the above discussions, the basis functions $p_{1n,j}(x)$, $p_{2m,l}(y)$, and $p_{3,jl}(x,y)$ (or, $\Phi_1(x,y), \Phi_2(x,y)$, and $\Phi_3(x,y)$) are only required to be sufficiently smooth over a compact domain, without regulating them to any particular forms. This implies that there is actually a large (theoretically, an infinite) number of possible choices for such basis functions in the process of implementing the generalized Fourier series expansions.

In what follows, the simple polynomials given in Table 8.1 are specifically employed for the construction of the supplementary functions.

In comparison with the conventional Fourier series method, the polynomial-based generalized Fourier series methods typically involve three kinds of basis

---

TABLE 8.1

The Simple Polynomial Sets Used in the Generalized Fourier Series Expansions

The Generalized Fourier Series Expansions	Basis Functions
The full-range Fourier series over the domain $[-a,a] \times [-b,b]$	(1) Selection of $\mathbf{p}_3^T(x,y)$ $p_{3,jl}(x,y) = (x/a)^j (y/b)^l$, $(j,l) \in \Omega_3$, $\mathbf{p}_3^T(x,y) = [\cdots p_{3,jl}(x,y) \cdots]_{(j,l) \in \Omega_3}$, where the index set Ω_3 is defined as in (8.30). (2) Selection of $\mathbf{p}_{1n}^T(x)$, $n = 0, 1, 2, \cdots$ $p_{1n,j}(x) = (x/a)^j$, $j = 1, 2, \cdots, 2r$, $\mathbf{p}_{1n}^T(x) = [p_{1n,1}(x) \quad p_{1n,2}(x) \cdots p_{1n,2r}(x)]$. (3) Selection of $\mathbf{p}_{2m}^T(y)$, $m = 0, 1, 2, \cdots$ $p_{2m,l}(y) = (y/b)^l$, $l = 1, 2, \cdots, 2r$, $\mathbf{p}_{2m}^T(y) = [p_{2m,1}(y) \quad p_{2m,2}(y) \cdots p_{2m,2r}(y)]$.
The half-range sine-sine series over the domain $[0,a] \times [0,b]$	(1) Selection of $\mathbf{p}_3^T(x,y)$ $p_{3,jl}(x,y) = (x/a)^j (y/b)^l$, $(j,l) \in \Omega_3$, $\mathbf{p}_3^T(x,y) = [p_{3,00}(x,y) \cdots p_{3,jl}(x,y) \cdots]_{(j,l) \in \Omega_3}$, where the index set Ω_3 is defined as in (8.92) or (8.93). (2) Selection of $\mathbf{p}_{1n}^T(x)$, $n = 1, 2, \cdots$ $p_{1n,j}(x) = (x/a)^{j-1}$, $j = 1, 2, \cdots, 2r$, $\mathbf{p}_{1n}^T(x) = [p_{1n,1}(x) \quad p_{1n,2}(x) \cdots p_{1n,2r}(x)]$. (3) Selection of $\mathbf{p}_{2m}^T(y)$, $m = 1, 2, \cdots$ $p_{2m,l}(y) = (y/b)^{l-1}$, $l = 1, 2, \cdots, 2r$, $\mathbf{p}_{2m}^T(y) = [p_{2m,1}(y) \quad p_{2m,2}(y) \cdots p_{2m,2r}(y)]$.

functions: pure trigonometric functions, pure polynomials, and the products of polynomials and trigonometric functions. Thus, it can be expected that the generalized Fourier series not only are able to retain the essences of the conventional Fourier series, but also have the reproducing property of complete polynomials of certain degrees.

Specifically, the reproducing property implies:

1. the polynomial-based generalized full-range Fourier series is able to reproduce any two-dimensional polynomial of degree $2r$ over the domain $[-a,a] \times [-b,b]$;
2. the polynomial-based generalized half-range sine-sine series is able to reproduce any two-dimensional polynomial of degree $2r - 1$ over the domain $[0,a] \times [0,b]$.

While the second case is easily verified, a comprehensive proof of the first case involves collecting together all the polynomials from the corner function, the boundary functions and the internal function in the polynomial-based generalized full-range Fourier series expansion. This process is understandably lengthy and tedious, and will not presented herein for the sake of compactness.

8.4 NUMERICAL CHARACTERISTICS OF THE GENERALIZED FOURIER SERIES

In this section, by setting $2r = 6$ and length-to-width ratio $a/b = 1.0$, we will use the four sample functions in Table 8.2 to investigate the convergence behaviors and approximation accuracy of the polynomial-based generalized Fourier series expansions.

TABLE 8.2
Sample Functions

No.	Sample Functions
1	$u(x,y) = \left[\dfrac{1}{2} - \dfrac{x}{a} - \dfrac{1}{2}\left(\dfrac{x}{a}\right)^2 + \left(\dfrac{x}{a}\right)^3 \right]\left[\dfrac{1}{2} - \dfrac{y}{b} - \dfrac{1}{2}\left(\dfrac{y}{b}\right)^2 + \left(\dfrac{y}{b}\right)^3 \right]$,
	$(x,y) \in [-a,a] \times [-b,b]$
2	$u(x,y) = \sin(\alpha_0 x)\cos(\beta_0 y)$, $\alpha_0 = \pi/2a$, $\beta_0 = \pi/2b$,
	$(x,y) \in [-a,a] \times [-b,b]$
3	$u(x,y) = \left[\dfrac{1}{2} - \dfrac{x}{a} - \dfrac{1}{2}\left(\dfrac{x}{a}\right)^2 + \left(\dfrac{x}{a}\right)^3 \right]\left[\dfrac{1}{2} - \dfrac{y}{b} - \dfrac{1}{2}\left(\dfrac{y}{b}\right)^2 + \left(\dfrac{y}{b}\right)^3 \right]$,
	$(x,y) \in [0,a] \times [0,b]$
4	$u(x,y) = \sin(\alpha_0 x)\cos(\beta_0 y)$, $\alpha_0 = \pi/2a$, $\beta_0 = \pi/2b$,
	$(x,y) \in [0,a] \times [0,b]$

8.4.1   Error Norms of Simultaneous Series Approximations

Some error norms will be first defined to assess the convergence characteristics and numerical accuracy of the generalized Fourier series when used to expand two-dimensional functions and their partial derivatives.

For positive integers N_1^t and N_2^t, let $\Upsilon^t = \{(x_{n_1}, y_{n_2}), n_1 = 1, 2, \cdots, N_1^t, n_2 = 1, 2, \cdots, N_2^t\}$ be the set of uniformly distributed sampling points over a rectangular domain, as shown in Figure 8.1. Further, we divide the set Υ^t into three subsets: Υ_I^t, Υ_B^t, and Υ_C^t, containing the sampling points completely inside the domain, the sampling points on the boundaries (excluding the four corner points) and the sampling points at the corners, respectively. If I_γ^t, $I_{\gamma I}^t$, $I_{\gamma B}^t$, and $I_{\gamma C}^t$ denote the numbers of elements of Υ^t and its three subsets, respectively, then we have: $I_\gamma^t = N_1^t N_2^t$, $I_{\gamma I}^t = (N_1^t - 2)(N_2^t - 2)$, $I_{\gamma B}^t = 2(N_1^t - 2) + 2(N_2^t - 2)$, and $I_{\gamma C}^t = 4$.

For a two-dimensional function $u(x, y)$, let M and N be the truncation numbers for its generalized Fourier series in the x- and y-direction, respectively, and $u_{M,N}(x, y)$ be the corresponding partial sum.

Suppose that k_1 and k_2 are nonnegative integers and $k_1 + k_2 \leq 2r$. The overall error for the partial derivative $u^{(k_1, k_2)}(x, y)$ is here define as

$$e^{(k_1, k_2)}(u_{M,N}) = \frac{1}{I_\gamma^t \cdot u_{\max}^{(k_1, k_2)}} \sum_{(x,y) \in \Upsilon^t} \left| u_{M,N}^{(k_1, k_2)}(x, y) - u^{(k_1, k_2)}(x, y) \right|, \qquad (8.142)$$

where $u_{\max}^{(k_1, k_2)}$ is the maximum value of the partial derivative $u^{(k_1, k_2)}(x, y)$ over the expansion domain.

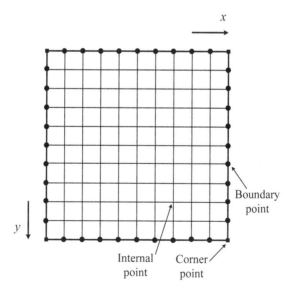

FIGURE 8.1   Sampling points used for calculating the approximation errors associated with the generalized Fourier series of two-dimensional functions.

For $p \leq 2r$, the overall error for the p-th order partial derivatives of the function $u(x, y)$ is accordingly defined by

$$
\begin{aligned}
\left|e\right|^p (u_{M,N}) &= \frac{1}{p+1} \sum_{\substack{k_1 \geq 0, k_2 \geq 0 \\ k_1 + k_2 = p}} e^{(k_1,k_2)}(u_{M,N}) \\
&= \frac{1}{p+1} \sum_{\substack{k_1 \geq 0, k_2 \geq 0 \\ k_1 + k_2 = p}} \frac{1}{I_\gamma^t \cdot u_{\max}^{(k_1,k_2)}} \sum_{(x,y) \in \Upsilon^t} \left| u_{M,N}^{(k_1,k_2)}(x,y) - u^{(k_1,k_2)}(x,y) \right|.
\end{aligned}
\tag{8.143}
$$

The overall error for the partial derivatives of up to the p-th order is given as

$$
\begin{aligned}
\left\|e\right\|^p (u_{M,N}) &= \frac{2}{(p+1)(p+2)} \sum_{\substack{k_1 \geq 0, k_2 \geq 0 \\ k_1 + k_2 \leq p}} e^{(k_1,k_2)}(u_{M,N}) \\
&= \frac{2}{(p+1)(p+2)} \sum_{\substack{k_1 \geq 0, k_2 \geq 0 \\ k_1 + k_2 \leq p}} \frac{1}{I_\gamma^t \cdot u_{\max}^{(k_1,k_2)}} \sum_{(x,y) \in \Upsilon^t} \left| u_{M,N}^{(k_1,k_2)}(x,y) - u^{(k_1,k_2)}(x,y) \right|.
\end{aligned}
\tag{8.144}
$$

By replacing the set (of sampling points) Υ^t in (8.142)–(8.144) with Υ_I^t, Υ_B^t or Υ_C^t, we will have the correspondingly internal, boundary and corner errors to account for their spatial characteristics, as summarized in Table 8.3.

8.4.2 CONVERGENCE CHARACTERISTICS

The above defined error indexes obviously depend upon the truncation numbers and convergence characteristics of the generalized Fourier series expansions. The term convergence is originally associated with the decay rates of the Fourier coefficients as $m \to \infty$ and/or $n \to \infty$. However, it is understandable that

---

TABLE 8.3

Error Indexes of the Generalized Fourier Series Expansions of the Partial Derivatives

	Sets of Sampling Points			
	Υ^t	Υ_I^t	Υ_B^t	Υ_C^t
The partial derivative $u^{(k_1,k_2)}(x,y)$	$e^{(k_1,k_2)}(u_{M,N})$	$e_I^{(k_1,k_2)}(u_{M,N})$	$e_B^{(k_1,k_2)}(u_{M,N})$	$e_C^{(k_1,k_2)}(u_{M,N})$
The p-th order partial derivatives	$\left|e\right|^p (u_{M,N})$	$\left|e_I\right|^p (u_{M,N})$	$\left|e_B\right|^p (u_{M,N})$	$\left|e_C\right|^p (u_{M,N})$
Up to the p-th order partial derivatives	$\left\|e\right\|^p (u_{M,N})$	$\left\|e_I\right\|^p (u_{M,N})$	$\left\|e_B\right\|^p (u_{M,N})$	$\left\|e_C\right\|^p (u_{M,N})$

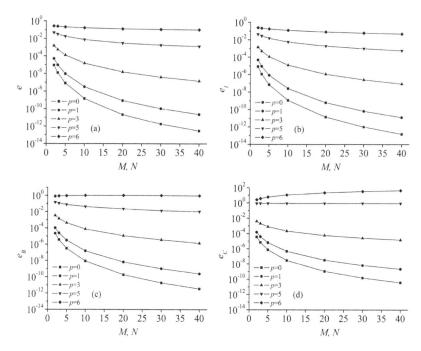

FIGURE 8.2   Convergence of the generalized Fourier series of sample function 2 and its partial derivatives: (a) $|e|^P$ $(u_{M,N})$-M,N curves; (b) $|e_I|^P$ $(u_{M,N})$-M,N curves; (c) $|e_B|^P$ $(u_{M,N})$-M, N curves; and (d) $|e_C|^P$ $(u_{M,N})$-M, N curves.

the convergence characteristic can be assessed by using the error indexes, perhaps even more rigorously from a statistical point of view.

In the following calculations, the generalized Fourier series of the sample functions are successively truncated, in each direction, to the first 2, 3, 5, 10, 20, 30, and 40 terms. By setting $N_1^t = N_2^t = 101$, the error indexes are then determined for different sets of sampling points and partial derivatives of various orders: $p = 0, 1, \cdots, 6$. The results are selectively presented in Figures 8.2 and 8.3.

It is seen from these plots that the overall, internal, boundary, and corner error indexes associated with the sample functions and their up to fifth order derivatives all decrease nicely as the truncation numbers increases. The corner error with the fifth order partial derivatives has failed to converge for the generalized Fourier series (see Figure 8.2(d)); in comparison, the generalized sine-sine series still displays a decent trend of converging (refer to Figure 8.3(d)).

While the internal errors with the sixth order derivatives may be considered acceptable in both cases, their boundary and corner counterparts have clearly failed the convergence tests. It is generally observed that for the given p-th order derivatives, the error indexes for the corresponding partial sums tend to behave better in the interior region, and worst at the corners. While it may well be true that a function and its derivatives are better approximated at interior points, we here have no intention of proving or disproving it even though the adverse impacts of the potential discontinuities with higher order derivatives

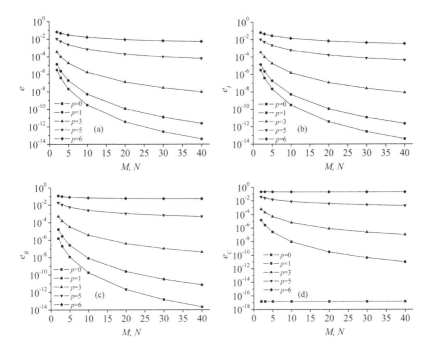

FIGURE 8.3 Convergence of the generalized Fourier series of sample function 4 and its partial derivatives: (a) $|e|^p$ $(u_{M,N})$-M, N curves; (b) $|e_I|^p$ $(u_{M,N})$-M, N curves; (c) $|e_B|^p$ $(u_{M,N})$-M, N curves; and (d) $|e_C|^p$ $(u_{M,N})$-M, N curves.

are probably confined locally to the boundary areas. Besides, since the total error index is primarily dominated by the errors calculated at the interior sampling points, it will typically conform to its interior counterpart.

It is also shown that the convergence of the half-range Fourier series appears to outperform its full-range counterpart in both cases by more than an order of magnitude. This is indeed true for the present setting $2r = 6$. In the generalized half-range sine-sine series expansion of the function $u(x,y)$, the corresponding internal function $\varphi_0(x,y)$ is explicitly made to belong to $C^{2r-2}([-\infty,\infty]\times[-\infty,\infty])$. However, the odd-periodic extension of $\varphi_0(x,y)$ (namely, $\varphi_0(-x,y) = -\varphi_0(x,y)$, $\varphi_0(x,-y) = -\varphi_0(x,y)$, and $\varphi_0(-x,-y) = \varphi_0(x,y)$) will actually elevate its degree of continuity to $\varphi_0(x,y) \in C^{2r-1}((-\infty,\infty)\times(-\infty,\infty))$ if $u(x,y)$ is sufficiently smooth over the domain $[0,a]\times[0,b]$. In comparison, such a continuity elevation does not automatically happen to the internal function in the generalized full-range Fourier series expansion (refer to Theorem 6.2).

8.4.3 The Accuracy of the Generalized Fourier Series

Once the convergence characteristic of the generalized Fourier series is well understood, we can now turn to another important issue: the numerical accuracy of the series approximations. Based on the excellent convergence as demonstrated in Figures 8.2 and 8.3, the setting $M = N = 40$ will be chosen in the subsequent numerical calculations.

Let's first check on the reproducing property of simple polynomials. For this purpose, the sample polynomial functions 1 and 3 in Table 8.2 will be considered here. The calculated expansion constants and Fourier coefficients are tabulated in Tables 8.4 and 8.5 respectively for the generalized Fourier series expansions of

---

TABLE 8.4
The Vectors of the Expansion Constants and Fourier Coefficients for the Generalized Fourier Series Expansions of Sample Function 1

Vectors of Unknown Constants/Fourier Coefficients

(a) q_0^T:

$$q_0^T = [q_{01}^T \quad q_{02}^T \quad q_{03}^T \quad q_{04}^T],$$

where

$q_{01}^T = [1.0 \quad 2.070504E\text{-}14 \quad -5.212722E\text{-}15 \quad \cdots \quad -2.718446E\text{-}19 \quad -2.488159E\text{-}20],$
$q_{02}^T = [3.052553E\text{-}13 \quad -1.604313E\text{-}17 \quad \cdots \quad -8.664684E\text{-}19 \quad 7.630576E\text{-}20],$
$q_{03}^T = [1.528638E\text{-}13 \quad -1.907658E\text{-}14 \quad \cdots \quad -3.189244E\text{-}19 \quad 3.043197E\text{-}19],$
$q_{04}^T = [-3.195236E\text{-}17 \quad -5.780299E\text{-}17 \quad \cdots \quad 1.158276E\text{-}19 \quad 3.719174E\text{-}19].$

(b) q_1^T:

$$q_1^T = [q_{11}^T \quad q_{12}^T],$$
$$q_{11}^T = [q_{10,1}^T \quad q_{11,1}^T \quad \cdots \quad q_{140,1}^T], \, q_{12}^T = [q_{11,2}^T \quad q_{12,2}^T \quad \cdots \quad q_{140,2}^T],$$

where

$q_{10,1}^T = [9.313798E\text{-}19 \quad -2.0 \quad 12.0 \quad 1.672389E\text{-}17 \quad 0.0 \quad 0.0],$
$q_{11,1}^T = [-1.117232E\text{-}18 \quad -1.572635E\text{-}16 \quad 3.306817E\text{-}16 \quad 8.645323E\text{-}18 \quad 0.0 \quad 0.0],$

$$\cdots\cdots$$

$q_{140,1}^T = [-4.520731E\text{-}19 \quad 2.358140E\text{-}18 \quad 5.101171E\text{-}16 \quad 3.161865E\text{-}18 \quad 0.0 \quad 0.0],$
$q_{11,2}^T = [1.171811E\text{-}18 \quad 3.756985E\text{-}16 \quad -7.473177E\text{-}16 \quad 1.534161E\text{-}17 \quad 0.0 \quad 0.0],$

$$\cdots\cdots$$

$q_{140,2}^T = [1.292270E\text{-}18 \quad -2.366920E\text{-}16 \quad 3.668047E\text{-}16 \quad -1.166864E\text{-}18 \quad 0.0 \quad 0.0].$

(c) q_2^T:

$$q_2^T = [q_{21}^T \quad q_{22}^T],$$
$$q_{21}^T = [q_{20,1}^T \quad q_{21,1}^T \quad \cdots \quad q_{240,1}^T], \, q_{22}^T = [q_{21,2}^T \quad q_{22,2}^T \quad \cdots \quad q_{240,2}^T],$$

where

$q_{20,1}^T = [1.423594E\text{-}18 \quad -2.0 \quad 12.0 \quad 1.672389E\text{-}17 \quad 0.0 \quad 0.0],$
$q_{21,1}^T = [1.371922E\text{-}18 \quad -2.072452E\text{-}16 \quad 5.442695E\text{-}17 \quad 8.645323E\text{-}18 \quad 0.0 \quad 0.0],$

$$\cdots\cdots$$

$q_{240,1}^T = [1.674491E\text{-}18 \quad -1.761829E\text{-}18 \quad 4.322714E\text{-}16 \quad 3.161865E\text{-}18 \quad 0.0 \quad 0.0],$
$q_{21,2}^T = [-4.583626E\text{-}19 \quad 3.457093E\text{-}17 \quad -7.460165E\text{-}16 \quad 1.534161E\text{-}17 \quad 0.0 \quad 0.0],$

$$\cdots\cdots$$

$q_{240,2}^T = [-1.736237E\text{-}18 \quad -2.090599E\text{-}16 \quad 4.645624E\text{-}16 \quad -1.166864E\text{-}18 \quad 0.0 \quad 0.0].$

(d) q_3^T:

$$q_3^T = [0.0 \quad 0.0 \quad 0.0 \quad 0.0 \quad 4.0 \quad 0.0 \quad 0.0 \quad -24.0 \quad -24.0 \quad 0.0$$
$$0.0 \quad 0.0 \quad 144.0 \quad 0.0 \quad 0.0].$$

TABLE 8.5
The Vectors of the Expansion Constants and Fourier Coefficients for the Generalized Fourier Series Expansions of Sample Function 3

Vectors of Unknown Constants/Fourier Coefficients

(a) q_{04}^T:

$q_{04}^T = [1.339853\text{E-}17 \quad -5.925981\text{E-}18 \quad \cdots \quad 5.802761\text{E-}20 \quad -1.933270\text{E-}19].$

(b) q_1^T:

$q_1^T = [q_{11}^T \quad q_{12}^T \quad \cdots \quad q_{140}^T],$

where

$q_{11}^T = [-3.873731\text{E-}17 \quad -1.770942\text{E-}16 \quad 0.0 \quad -1.993994\text{E-}17 \quad 3.987988\text{E-}17 \quad 0.0],$
$q_{12}^T = [1.097717\text{E-}17 \quad 3.292473\text{E-}17 \quad 0.0 \quad 6.481345\text{E-}18 \quad -1.296269\text{E-}17 \quad 0.0],$

$\qquad \cdots\cdots$

$q_{140}^T = [6.876119\text{E-}19 \quad 4.300453\text{E-}18 \quad 0.0 \quad 2.437254\text{E-}19 \quad -4.874507\text{E-}19 \quad 0.0].$

(c) q_2^T:

$q_2^T = [q_{21}^T \quad q_{22}^T \quad \cdots \quad q_{240}^T],$

where

$q_{21}^T = [-3.706127\text{E-}17 \quad -1.760560\text{E-}16 \quad 0.0 \quad -1.993994\text{E-}17 \quad 3.987988\text{E-}17 \quad 0.0],$
$q_{22}^T = [9.086650\text{E-}18 \quad 2.494088\text{E-}17 \quad 0.0 \quad 6.481345\text{E-}18 \quad -1.296269\text{E-}17 \quad 0.0],$

$\qquad \cdots\cdots$

$q_{240}^T = [5.016186\text{E-}19 \quad 6.721300\text{E-}18 \quad 0.0 \quad 2.437254\text{E-}19 \quad -4.874507\text{E-}19 \quad 0.0].$

(d) q_3^T:

$q_3^T = [0.0 \quad 0.0 \quad 0.0 \quad 0.0 \quad 25.0 \quad 0.0 \quad 0.0 \quad 2.5 \quad 0.0 \quad 0.0 \quad -5.0 \quad 0.0$
$\quad 0.0 \quad 0.0 \quad 2.5 \quad 0.0 \quad -5.0 \quad 0.0 \quad 0.25 \quad -0.5 \quad -0.5 \quad 0.0 \quad 1.0 \quad 0.0].$

these two sample functions. It is evident that for the non-polynomial forms of basis functions, the corresponding expansion constants or coefficients are all virtually zero (smaller than 1.0E–13). In contrast, for the basis functions in the form of polynomials, the corresponding constants or coefficients enable a complete reconstruction of the original sample functions, thus validating the polynomial reproducing property of the polynomial-based generalized Fourier series expansions.

It's worthy of noticing that for the given setting of $2r = 6$, the basis functions in the half-range sine-sine series expansion include not only the complete set of the polynomials of degree 5, but also a polynomial, $(x/a)^3 (y/b)^3$, of degree 6.

Next, we will use the trigonometric functions (No. 2 and No. 4 in Table 8.2) to examine the approximation errors associated with the generalized Fourier series (GFS) of these two functions and their partial derivatives. Presented in Table 8.6 are some of the results together with those for the conventional Fourier series (CFS) which are determined directly from the explicit expressions for these functions and their derivatives. Even though the errors for the derivatives of lower orders are not shown here, it suffices to point out that for the partial derivatives of up to the fourth order, the overall-, interior-, and boundary-errors with

TABLE 8.6
Approximation Errors Associated with the Generalized Fourier Series Expansions

Approximation Errors	Sample Function 2		Sample Function 4	
	GFS	CFS	GFS	CFS
$\|e\|^4 (u_{40,40})$	5.18432E–06	2.01323E–02	2.15001E–07	1.80706E–02
$\|e_I\|^4 (u_{40,40})$	2.43680E–06	7.86272E–03	2.10033E–07	1.08846E–02
$\|e_B\|^4 (u_{40,40})$	4.67352E–05	3.21987E–01	3.25495E–07	1.94078E–01
$\|e_C\|^4 (u_{40,40})$	2.62376E–03	2.00012E–01	1.45041E–06	2.00799E–01
$\|e\|^5 (u_{40,40})$	1.30588E–03	1.88203E–02	6.18574E–05	1.66987E–02
$\|e_I\|^5 (u_{40,40})$	5.83649E–04	6.49644E–03	4.39762E–05	1.06156E–02
$\|e_B\|^5 (u_{40,40})$	1.04781E–02	3.18975E–01	4.84950E–04	1.64899E–01
$\|e_C\|^5 (u_{40,40})$	8.62915E–01	5.00031E–01	1.98913E–03	2.50000E–01
$\|e\|^6 (u_{40,40})$	9.95679E–02	2.10069E–02	5.36005E–03	1.63157E–02
$\|e_I\|^6 (u_{40,40})$	4.75365E–02	8.77358E–03	3.41443E–03	9.51247E–03
$\|e_B\|^6 (u_{40,40})$	9.41960E–01	3.23994E–01	5.15023E–02	1.83399E–01
$\|e_C\|^6 (u_{40,40})$	4.41929E+01	3.39905E–17	2.04528E–01	1.44568E–01

the corresponding generalized Fourier series are much smaller, by orders of magnitude, than those with their conventional counterparts. Comparatively, however, the degrees of the differences in the corner errors are not equally remarkable, especially in the case of the full-range series expansions. For the fifth order partial derivatives, the generalized Fourier series are still, by at least an order of magnitude, better than the conventional Fourier series in terms of the overall-, internal- and boundary-errors. For the sixth order partial derivatives, all of the error indexes (overall, internal, boundary and corner) are of no significant difference between the generalized and conventional Fourier series. This shall not come as a surprise since the corner and boundary functions in the generalized Fourier series expansion can only ensure the C^{2r-2} continuity of the internal function when it is periodically extended onto the entire x–y plane. In other words, by setting $2r = 6$, we have no intention of improving the smoothness of the derivatives of order 6 or higher (or equivalently, the convergence and accuracy of their generalized Fourier series). This statement itself, however, may not actually constitute a practical limit on the orders of derivatives for which the generalized Fourier series methods can be applied effectively. For instance, if better accuracies are desired for the derivatives of order 6 or higher, it can be achieved by setting $2r \geq 8$ so long as the original function $u(x, y)$ is sufficiently smooth in the domain.

9 Multiscale Fourier Series Methods for Linear Differential Equations

Effectively solving linear differential equations has long been an important topic in mathematics and mathematical physics. Various solution methods, including the Fourier series methods, have been developed for a wide spectrum of scientific and engineering problems which are described by linear differential equations subjected to appropriate boundary conditions. However, the basis functions used to construct solutions are often required to explicitly satisfy either the governing differential equation(s) or the boundary conditions, thus making the resulting solutions be case-dependent. The generalized Fourier series expansions discussed in the previous chapters have potentially opened up a new avenue for the simultaneous approximations of a solution function and its related derivatives defined on a compact domain with the desired smoothness and accuracy. Therefore, if the generalized Fourier series are used to represent the solutions of boundary value problems, the solving processes will simply involve determining a specific set of expansion constants or coefficients so that the governing differential equation(s) and the boundary conditions are simultaneously satisfied either exactly or approximately. In addition, the flexibilities offered in selecting the basis functions allow the developments of various implementation schemes to take fully advantages of the intrinsic characteristics of a particular class of problems.

9.1 THE GENERALIZED FOURIER SERIES SOLUTIONS OF ONE-DIMENSIONAL BOUNDARY VALUE PROBLEMS

Based on the structural decomposition schemes discussed earlier, the solution for a one-dimensional boundary value problem can be expanded into the generalized Fourier series in one of the three different forms as described below.

9.1.1 THE GENERALIZED FULL-RANGE FOURIER SERIES SOLUTIONS

Consider a linear differential equation

$$\mathcal{L}u = f, x \in (-a,a), \tag{9.1}$$

where the differential operator

$$\mathcal{L} = \sum_{k=0}^{2r} a_k \frac{d^k}{dx^k},$$ (9.2)

with a_k, $k = 0, 1, \cdots, 2r$, being constant coefficients.

The corresponding boundary conditions are given as

$$\mathbf{B}u = \mathbf{g},$$ (9.3)

where the vector of the differential operators is

$$\mathbf{B}^{\mathrm{T}} = [\mathcal{B}_1 \quad \mathcal{B}_2 \quad \cdots \quad \mathcal{B}_r]$$ (9.4)

and $\mathbf{g}$ is the vector of the prescribed boundary values.

In particular, the differential operators in (9.4) are here specifically defined as

$$\mathcal{B}_l = \sum_{k=0}^{2r-1} b_k^l \frac{d^k}{dx^k} \quad (l = 1, 2, \cdots, r),$$ (9.5)

where b_k^l are constants.

The solution $u(x)$ of differential equation (9.1) is expressed as

$$u(x) = \varphi_0(x) + \varphi_1(x),$$ (9.6)

where $\varphi_1(x)$ is the boundary function, satisfying

$$\varphi_1^{(k)}(a) - \varphi_1^{(k)}(-a) = u^{(k)}(a) - u^{(k)}(-a), \, k = 0, 1, \cdots, 2r-1,$$ (9.7)

and $\varphi_0(x)$ is the internal function.

The internal function $\varphi_0(x)$ is required to satisfy the sufficient conditions, namely:

$$\varphi_0^{(k)}(a) - \varphi_0^{(k)}(-a) = 0, \, k = 0, 1, \cdots, 2r-1,$$ (9.8)

such that its Fourier series is termwise differentiable continuously up to $2r$ times.

Substitution of (9.6) into (9.1) leads to

$$\mathcal{L}(\varphi_0 + \varphi_1) = f.$$ (9.9)

Further, the boundary function is assigned to take the role of the general solution

$$\mathcal{L}\varphi_1 = 0,$$ (9.10)

and the internal function the role of the particular solution

$$\mathcal{L}\varphi_0 = f. \tag{9.11}$$

In light of the conditions, (9.7) and (9.8), pre-imposed respectively upon the boundary and internal functions, the general and particular solutions are then specifically sought to satisfy:

$$\left.\begin{array}{l} \mathcal{L}\varphi_1 = 0 \\ \varphi_1^{(k)}(a) - \varphi_1^{(k)}(-a) = u^{(k)}(a) - u^{(k)}(-a), k = 0, 1, \cdots, 2r - 1 \end{array}\right\}, \tag{9.12}$$

and

$$\left.\begin{array}{l} \mathcal{L}\varphi_0 = f \\ \varphi_0^{(k)}(a) - \varphi_0^{(k)}(-a) = 0, k = 0, 1, \cdots, 2r - 1 \end{array}\right\}. \tag{9.13}$$

It needs to be pointed out that the general and the particular solutions together will also have to satisfy the prescribed boundary conditions (9.3).

9.1.2   THE GENERALIZED HALF-RANGE FOURIER COSINE SERIES SOLUTIONS

Consider a $2r$-th order linear differential equation over the interval $[0, a]$

$$\mathcal{L}u = f, x \in (0, a), \tag{9.14}$$

where the differential operator

$$\mathcal{L} = \sum_{k=0}^{r} a_{2k} \frac{d^{2k}}{dx^{2k}}, \tag{9.15}$$

and $a_{2k}, k = 0, 1, \cdots, r$, are constant coefficients.

The boundary conditions are prescribed as

$$\mathbf{B}u = \mathbf{g}, \tag{9.16}$$

where the vector of the differential operators is

$$\mathbf{B}^{\mathrm{T}} = [\mathcal{B}_1 \quad \mathcal{B}_2 \quad \cdots \quad \mathcal{B}_r], \tag{9.17}$$

and $\mathbf{g}$ is the vector of the prescribed boundary values.

The differential operators in (9.17) are here given as

$$\mathcal{B}_l = \sum_{k=0}^{r-1} b_{2k}^l \frac{d^{2k}}{dx^{2k}} \quad (l = 1, 2, \cdots, r), \tag{9.18}$$

or

$$\mathcal{B}_l = \sum_{k=0}^{r-1} b_{2k+1}^l \frac{d^{2k+1}}{dx^{2k+1}} \quad (l = 1, 2, \cdots, r), \qquad (9.19)$$

where b_{2k}^l and b_{2k+1}^l are constant coefficients.

Suppose that the solution $u(x)$ is expanded into the generalized half-range cosine series over the interval $[0, a]$, and its derivatives are also expanded into the generalized half-range Fourier series accordingly. Like before, we can divide the solution $u(x)$ into two parts: the general solution and the particular solution. The general solution corresponds to the boundary function $\varphi_1(x)$, satisfying

$$\left. \begin{array}{l} \mathcal{L}\varphi_1 = 0 \\ \varphi_1^{(2k+1)}(a) = u^{(2k+1)}(a), \; \varphi_1^{(2k+1)}(0) = u^{(2k+1)}(0), \, k = 0, 1, \cdots, r-1 \end{array} \right\}. \qquad (9.20)$$

The particular solution corresponds to the internal function $\varphi_0(x)$, satisfying

$$\left. \begin{array}{l} \mathcal{L}\varphi_0 = f \\ \varphi_0^{(2k+1)}(a) = 0, \; \varphi_0^{(2k+1)}(0) = 0, \, k = 0, 1, \cdots, r-1 \end{array} \right\}. \qquad (9.21)$$

Again, the general solution and the particular solution together have to satisfy the prescribed boundary conditions (9.16).

9.1.3 THE GENERALIZED HALF-RANGE FOURIER SINE SERIES SOLUTIONS

Suppose that the solution $u(x)$ of the linear differential equation, (9.14), is now expanded into the generalized half-range sine series on the interval $[0, a]$. It can be similarly decomposed into the general solution $\varphi_1(x)$ and the particular solution $\varphi_0(x)$. The general solution corresponds to the boundary function $\varphi_1(x)$, satisfying

$$\left. \begin{array}{l} \mathcal{L}\varphi_1 = 0 \\ \varphi_1^{(2k)}(a) = u^{(2k)}(a), \; \varphi_1^{(2k)}(0) = u^{(2k)}(0), \, k = 0, 1, \cdots, r-1 \end{array} \right\}, \qquad (9.22)$$

and the particular solution corresponds to the internal function $\varphi_0(x)$, satisfying

$$\left. \begin{array}{l} \mathcal{L}\varphi_0 = f \\ \varphi_0^{(2k)}(a) = 0, \; \varphi_0^{(2k)}(0) = 0, \, k = 0, 1, \cdots, r-1 \end{array} \right\}. \qquad (9.23)$$

Obviously, the general solution and the particular solution combined will have to satisfy the prescribed boundary conditions (9.16).

9.2 THE GENERALIZED FOURIER SERIES SOLUTIONS FOR TWO-DIMENSIONAL BOUNDARY VALUE PROBLEMS

9.2.1 THE GENERALIZED FULL-RANGE FOURIER SERIES SOLUTIONS

Consider a $2r$-th order linear differential equation on the domain $[-a,a]\times[-b,b]$

$$\mathcal{L}u = f, (x,y) \in (-a,a)\times(-b,b), \tag{9.24}$$

where the differential operator

$$\mathcal{L} = \sum_{\substack{k_1\geq 0, k_2\geq 0 \\ k_1+k_2\leq 2r}} a_{k_1,k_2} \frac{\partial^{k_1+k_2}}{\partial x^{k_1}\partial y^{k_2}}, \tag{9.25}$$

and a_{k_1,k_2} $(k_1,k_2 = 0,1,2,\cdots, k_1+k_2 \leq 2r)$ are constant coefficients. The corresponding boundary conditions are given as

$$\mathbf{B}u = \mathbf{g}, \tag{9.26}$$

where the vector of the differential operators is

$$\mathbf{B}^{\mathrm{T}} = [\mathcal{B}_1 \quad \mathcal{B}_2 \quad \cdots \quad \mathcal{B}_r], \tag{9.27}$$

and $\mathbf{g}$ is a vector of functions corresponding to the prescribed boundary conditions. The differential operators in (9.27) are defined as

$$\mathcal{B}_l = \sum_{\substack{k_1\geq 0, k_2\geq 0 \\ k_1+k_2\leq 2r-1}} b^l_{k_1,k_2} \frac{\partial^{k_1+k_2}}{\partial x^{k_1}\partial y^{k_2}} \quad (l=1,2,\cdots,r), \tag{9.28}$$

where $b^l_{k_1,k_2}$ are constant coefficients.

Suppose that the solution $u(x,y)$ is sought in form of the generalized full-range Fourier series over the domain $[-a,a]\times[-b,b]$. It will be then decomposed into four parts:

$$u(x,y) = \varphi_0(x,y) + \varphi_1(x,y) + \varphi_2(x,y) + \varphi_3(x,y). \tag{9.29}$$

The corner function $\varphi_3(x,y)$ in (9.29) satisfies the conditions:

$$\varphi_3^{(k_1,k_2)}(a,b) - \varphi_3^{(k_1,k_2)}(a,-b) - \varphi_3^{(k_1,k_2)}(-a,b) + \varphi_3^{(k_1,k_2)}(-a,-b)$$
$$= u^{(k_1,k_2)}(a,b) - u^{(k_1,k_2)}(a,-b) - u^{(k_1,k_2)}(-a,b) + u^{(k_1,k_2)}(-a,-b), \tag{9.30}$$
$$k_1,k_2 = 0,1,2,\cdots, k_1+k_2 \leq 2r-2.$$

The boundary functions $\varphi_1(x,y)$ and $\varphi_2(x,y)$ respectively satisfy:

$$\varphi_1^{(k_1,0)}(a,y) - \varphi_1^{(k_1,0)}(-a,y) = [u^{(k_1,0)}(a,y) - \varphi_3^{(k_1,0)}(a,y)]$$
$$- [u^{(k_1,0)}(-a,y) - \varphi_3^{(k_1,0)}(-a,y)], \; y \in (-b,b), \quad (9.31)$$
$$k_1 = 0, 1, \cdots, 2r-1,$$

$$\varphi_1^{(0,k_2)}(x,b) - \varphi_1^{(0,k_2)}(x,-b) = 0, \; x \in (-a,a), k_2 = 0, 1, \cdots, 2r-1, \quad (9.32)$$

$$\varphi_1^{(k_1,k_2)}(a,b) - \varphi_1^{(k_1,k_2)}(a,-b) - \varphi_1^{(k_1,k_2)}(-a,b) + \varphi_1^{(k_1,k_2)}(-a,-b) = 0,$$
$$k_1,k_2 = 0, 1, 2, \cdots, k_1+k_2 \le 2r-2, \quad (9.33)$$

and

$$\varphi_2^{(k_1,0)}(a,y) - \varphi_2^{(k_1,0)}(-a,y) = 0, \; y \in (-b,b), k_1 = 0, 1, \cdots, 2r-1, \quad (9.34)$$

$$\varphi_2^{(0,k_2)}(x,b) - \varphi_2^{(0,k_2)}(x,-b) = [u^{(0,k_2)}(x,b) - \varphi_3^{(0,k_2)}(x,b)]$$
$$- [u^{(0,k_2)}(x,-b) - \varphi_3^{(0,k_2)}(x,-b)], \; x \in (-a,a), \quad (9.35)$$
$$k_2 = 0, 1, \cdots, 2r-1,$$

$$\varphi_2^{(k_1,k_2)}(a,b) - \varphi_2^{(k_1,k_2)}(a,-b) - \varphi_2^{(k_1,k_2)}(-a,b) + \varphi_2^{(k_1,k_2)}(-a,-b) = 0,$$
$$k_1,k_2 = 0, 1, 2, \cdots, k_1+k_2 \le 2r-2. \quad (9.36)$$

The internal function $\varphi_0(x,y)$ satisfies the sufficient conditions to ensure its Fourier series to be termwise differentiable continuously up to $2r$ times:

$$\varphi_0^{(k_1,0)}(a,y) - \varphi_0^{(k_1,0)}(-a,y) = 0, \; y \in (-b,b), k_1 = 0, 1, \cdots, 2r-1, \quad (9.37)$$

$$\varphi_0^{(0,k_2)}(x,b) - \varphi_0^{(0,k_2)}(x,-b) = 0, \; x \in (-a,a), k_2 = 0, 1, \cdots, 2r-1, \quad (9.38)$$

$$\varphi_0^{(k_1,k_2)}(a,b) - \varphi_0^{(k_1,k_2)}(a,-b) - \varphi_0^{(k_1,k_2)}(-a,b) + \varphi_0^{(k_1,k_2)}(-a,-b) = 0,$$
$$k_1,k_2 = 0, 1, 2, \cdots, k_1+k_2 \le 2r-2. \quad (9.39)$$

In accordance with (9.24), we further assign these constituents with different functionalities:

$$\mathcal{L}\varphi_1 = 0, \quad (9.40)$$

$$\mathcal{L}\varphi_2 = 0, \quad (9.41)$$

and

$$\mathcal{L}(\varphi_0 + \varphi_3) = f. \quad (9.42)$$

That is, the solution of (9.24) is purposely decomposed into the general solutions and the particular solution. The general solutions correspond to the boundary functions $\varphi_1(x,y)$ and $\varphi_2(x,y)$, satisfying the following equations

$$\mathcal{L}\varphi_1 = 0$$

$$\varphi_1^{(k_1,0)}(a,y) - \varphi_1^{(k_1,0)}(-a,y) = [u^{(k_1,0)}(a,y) - \varphi_3^{(k_1,0)}(a,y)]$$

$$-[u^{(k_1,0)}(-a,y) - \varphi_3^{(k_1,0)}(-a,y)], \; y \in (-b,b), \; k_1 = 0, 1, \cdots, 2r-1$$

$$\varphi_1^{(0,k_2)}(x,b) - \varphi_1^{(0,k_2)}(x,-b) = 0, \; x \in (-a,a), \; k_2 = 0, 1, \cdots, 2r-1$$

$$\varphi_1^{(k_1,k_2)}(a,b) - \varphi_1^{(k_1,k_2)}(a,-b) - \varphi_1^{(k_1,k_2)}(-a,b) + \varphi_1^{(k_1,k_2)}(-a,-b) = 0,$$

$$k_1, k_2 = 0, 1, 2, \cdots, k_1 + k_2 \le 2r-2$$

$$(9.43)$$

and

$$\mathcal{L}\varphi_2 = 0$$

$$\varphi_2^{(k_1,0)}(a,y) - \varphi_2^{(k_1,0)}(-a,y) = 0, \; y \in (-b,b), \; k_1 = 0, 1, \cdots, 2r-1$$

$$\varphi_2^{(0,k_2)}(x,b) - \varphi_2^{(0,k_2)}(x,-b) = [u^{(0,k_2)}(x,b) - \varphi_3^{(0,k_2)}(x,b)]$$

$$-[u^{(0,k_2)}(x,-b) - \varphi_3^{(0,k_2)}(x,-b)], \; x \in (-a,a), \; k_2 = 0, 1, \cdots, 2r-1$$

$$\varphi_2^{(k_1,k_2)}(a,b) - \varphi_2^{(k_1,k_2)}(a,-b) - \varphi_2^{(k_1,k_2)}(-a,b) + \varphi_2^{(k_1,k_2)}(-a,-b) = 0,$$

$$k_1, k_2 = 0, 1, 2, \cdots, k_1 + k_2 \le 2r-2$$

$$(9.44)$$

Again, it is seen that each of these two general solutions is clearly in charge of its own territory, without interfering with each other.

The particular solution consists of the internal function $\varphi_0(x,y)$ and the corner function $\varphi_3(x,y)$, satisfying

$$\mathcal{L}(\varphi_0 + \varphi_3) = f$$

$$\varphi_0^{(k_1,0)}(a,y) - \varphi_0^{(k_1,0)}(-a,y) = 0, \; y \in (-b,b), \; k_1 = 0, 1, \cdots, 2r-1$$

$$\varphi_0^{(0,k_2)}(x,b) - \varphi_0^{(0,k_2)}(x,-b) = 0, \; x \in (-a,a), \; k_2 = 0, 1, \cdots, 2r-1$$

$$\varphi_0^{(k_1,k_2)}(a,b) - \varphi_0^{(k_1,k_2)}(a,-b) - \varphi_0^{(k_1,k_2)}(-a,b) + \varphi_0^{(k_1,k_2)}(-a,-b) = 0,$$

$$k_1, k_2 = 0, 1, 2, \cdots, k_1 + k_2 \le 2r-2$$

$$\varphi_3^{(k_1,k_2)}(a,b) - \varphi_3^{(k_1,k_2)}(a,-b) - \varphi_3^{(k_1,k_2)}(-a,b) + \varphi_3^{(k_1,k_2)}(-a,-b)$$

$$= u^{(k_1,k_2)}(a,b) - u^{(k_1,k_2)}(a,-b) - u^{(k_1,k_2)}(-a,b) + u^{(k_1,k_2)}(-a,-b),$$

$$k_1, k_2 = 0, 1, 2, \cdots, k_1 + k_2 \le 2r-2$$

$$(9.45)$$

Evidently, the general solutions and the particular solution together have to satisfy the prescribed boundary conditions (9.26).

9.2.2 THE GENERALIZED HALF-RANGE FOURIER SINE-SINE SERIES SOLUTIONS

Consider the $2r$-th order linear differential equation

$$\mathcal{L}u = f, (x,y) \in (0,a) \times (0,b), \tag{9.46}$$

where the differential operator

$$\mathcal{L} = \sum_{\substack{k_1 \geq 0, k_2 \geq 0 \\ k_1 + k_2 \leq r}} a_{2k_1,2k_2} \frac{\partial^{2k_1 + 2k_2}}{\partial x^{2k_1} \partial y^{2k_2}}, \tag{9.47}$$

and $a_{2k_1,2k_2}$ $(k_1, k_2 = 0, 1, 2, \cdots, k_1 + k_2 \leq r)$ are constant coefficients. The corresponding boundary conditions are prescribed as

$$\mathbf{B}u = \mathbf{g}, \tag{9.48}$$

where $\mathbf{g}$ is a vector of functions related to the prescribed boundary conditions and the vector of the differential operators is

$$\mathbf{B}^T = [\mathcal{B}_1 \quad \mathcal{B}_2 \quad \cdots \quad \mathcal{B}_r]. \tag{9.49}$$

The differential operators in (9.49) are specifically defined as:
for $l = 1, 2, \cdots, r,$

$$\mathcal{B}_l = \sum_{\substack{k_1 \geq 0, k_2 \geq 0 \\ k_1 + k_2 \leq r-1}} b^l_{2k_1,2k_2} \frac{\partial^{2k_1 + 2k_2}}{\partial x^{2k_1} \partial y^{2k_2}}, \tag{9.50}$$

$$\mathcal{B}_l = \sum_{\substack{k_1 \geq 0, k_2 \geq 0 \\ k_1 + k_2 \leq r-1}} b^l_{2k_1+1,2k_2} \frac{\partial^{2k_1 + 2k_2 + 1}}{\partial x^{2k_1+1} \partial y^{2k_2}}, \tag{9.51}$$

$$\mathcal{B}_l = \sum_{\substack{k_1 \geq 0, k_2 \geq 0 \\ k_1 + k_2 \leq r-1}} b^l_{2k_1,2k_2+1} \frac{\partial^{2k_1 + 2k_2 + 1}}{\partial x^{2k_1} \partial y^{2k_2+1}}, \tag{9.52}$$

or

$$\mathcal{B}_l = \sum_{\substack{k_1 \geq 0, k_2 \geq 0 \\ k_1 + k_2 \leq r-2}} b^l_{2k_1+1,2k_2+1} \frac{\partial^{2k_1 + 2k_2 + 2}}{\partial x^{2k_1+1} \partial y^{2k_2+1}}, \tag{9.53}$$

where $b^l_{2k_1,2k_2}$, $b^l_{2k_1+1,2k_2}$, $b^l_{2k_1,2k_2+1}$, and $b^l_{2k_1+1,2k_2+1}$ are constant coefficients.

Suppose that the solution $u(x,y)$ is expanded into the generalized half-range sine-sine series on the domain $[0,a] \times [0,b]$. Then the solution $u(x,y)$ will also be decomposed into the general solutions and the particular solution. The general solution corresponds to the boundary functions $\varphi_1(x,y)$ and $\varphi_2(x,y)$ which respectively satisfy the following equations

$$
\left.
\begin{aligned}
&\mathcal{L}\varphi_1 = 0 \\
&\varphi_1^{(2k_1,0)}(a,y) = u^{(2k_1,0)}(a,y) - \varphi_3^{(2k_1,0)}(a,y), \\
&\varphi_1^{(2k_1,0)}(0,y) = u^{(2k_1,0)}(0,y) - \varphi_3^{(2k_1,0)}(0,y),\, y \in (0,b),\, k_1 = 0,1,\cdots,r-1 \\
&\varphi_1^{(0,2k_2)}(x,b) = 0,\, \varphi_1^{(0,2k_2)}(x,0) = 0,\, x \in (0,a),\, k_2 = 0,1,\cdots,r-1 \\
&\varphi_1^{(2k_1,2k_2)}(a,b) = 0,\, \varphi_1^{(2k_1,2k_2)}(a,0) = 0,\, \varphi_1^{(2k_1,2k_2)}(0,b) = 0,\, \varphi_1^{(2k_1,2k_2)}(0,0) = 0, \\
&k_1,k_2 = 0,1,2,\cdots,k_1+k_2 \le r-1
\end{aligned}
\right\} \quad (9.54)
$$

and

$$
\left.
\begin{aligned}
&\mathcal{L}\varphi_2 = 0 \\
&\varphi_2^{(2k_1,0)}(a,y) = 0,\, \varphi_2^{(2k_1,0)}(0,y) = 0,\, y \in (0,b),\, k_1 = 0,1,\cdots,r-1 \\
&\varphi_2^{(0,2k_2)}(x,b) = u^{(0,2k_2)}(x,b) - \varphi_3^{(0,2k_2)}(x,b), \\
&\varphi_2^{(0,2k_2)}(x,0) = u^{(0,2k_2)}(x,0) - \varphi_3^{(0,2k_2)}(x,0),\, x \in (0,a),\, k_2 = 0,1,\cdots,r-1 \\
&\varphi_2^{(2k_1,2k_2)}(a,b) = 0,\, \varphi_2^{(2k_1,2k_2)}(a,0) = 0,\, \varphi_2^{(2k_1,2k_2)}(0,b) = 0,\, \varphi_2^{(2k_1,2k_2)}(0,0) = 0, \\
&k_1,k_2 = 0,1,2,\cdots,k_1+k_2 \le r-1
\end{aligned}
\right\} \cdot \quad (9.55)
$$

The particular solution consists of the internal function $\varphi_0(x,y)$ and the corner function $\varphi_3(x,y)$, satisfying

$$
\left.
\begin{aligned}
&\mathcal{L}(\varphi_0 + \varphi_3) = f \\
&\varphi_0^{(2k_1,0)}(a,y) = 0,\, \varphi_0^{(2k_1,0)}(0,y) = 0,\, y \in (0,b),\, k_1 = 0,1,\cdots,r-1 \\
&\varphi_0^{(0,2k_2)}(x,b) = 0,\, \varphi_0^{(0,2k_2)}(x,0) = 0,\, x \in (0,a),\, k_2 = 0,1,\cdots,r-1 \\
&\varphi_0^{(2k_1,2k_2)}(a,b) = 0,\, \varphi_0^{(2k_1,2k_2)}(a,0) = 0,\, \varphi_0^{(2k_1,2k_2)}(0,b) = 0,\, \varphi_0^{(2k_1,2k_2)}(0,0) = 0, \\
&k_1,k_2 = 0,1,2,\cdots,k_1+k_2 \le r-1 \\
&\varphi_3^{(2k_1,2k_2)}(a,b) = u^{(2k_1,2k_2)}(a,b),\, \varphi_3^{(2k_1,2k_2)}(a,0) = u^{(2k_1,2k_2)}(a,0), \\
&\varphi_3^{(2k_1,2k_2)}(0,b) = u^{(2k_1,2k_2)}(0,b),\, \varphi_3^{(2k_1,2k_2)}(0,0) = u^{(2k_1,2k_2)}(0,0), \\
&k_1,k_2 = 0,1,2,\cdots,k_1+k_2 \le r-1
\end{aligned}
\right\} \cdot \quad (9.56)
$$

Again, the general solutions and the particular solution together have to satisfy the prescribed boundary conditions (9.48).

9.3 LIMITATIONS OF THE POLYNOMIAL-BASED GENERALIZED FOURIER SERIES METHODS

In Sec. 3.1, 7.3, and 8.3, polynomials were specifically used as the basis functions in the generalized Fourier series expansions of several one- and two-dimensional sample functions given in the polynomial and trigonometric forms. It has been demonstrated that the generalized Fourier series expansions do not only have the reproducing property of complete polynomials, but also are capable of simultaneously approximating functions and their (partial) derivatives of up to $2r$-th order with excellent accuracy and convergence characteristics. However, the simple polynomials and trigonometric functions may not (always) be able to effectively capture or resolute the rapid spatial fluctuations or local behaviors of a function or solution. We will elucidate this point by considering a simple function below

$$u(x) = \frac{\sinh[\alpha_0(a-x)]}{\sinh(\alpha_0 a)}, \, x \in [0,a], \qquad (9.57)$$

where the parameter $\alpha_0 a$ is used to control the shape of this function. For instance, when $\alpha_0 a$ takes the values of 0.01, 1.00, 10.0, and 100 successively, it is seen from Figure 9.1 that this function can display quite different characteristics: for smaller values 0.01 and 1.00, the function spreads pretty evenly over the interval; as $\alpha_0 a$ further increases, its activities are clearly confined within a narrow region, exhibiting a strong boundary layer type of behaviors. If this function is expanded into the polynomial-based generalized half-range sine series by setting $2r = 6$, the corresponding error indexes are plotted in Figure 9.2 as the functions of the truncation number M.

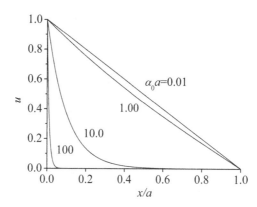

FIGURE 9.1   Sample functions with different values of parameter $\alpha_0 a$.

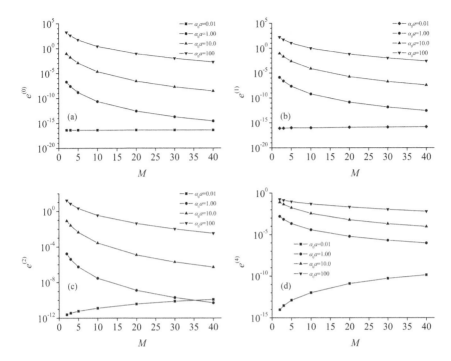

FIGURE 9.2 Approximation errors of the polynomial-based generalized half-range sine series for sample functions with different $\alpha_0 a$ values: (a) $e^{(0)}(u_M)$-M curve; (b) $e^{(1)}(u_M)$-M curve; (c) $e^{(2)}(u_M)$-M curve; and (d) $e^{(4)}(u_M)$-M curve.

It is observed that for the smaller $\alpha_0 a$ values, 0.01 and 1.00, the polynomial-based generalized sine series converges rapidly and approximates the function and its derivatives well with decent accuracies. Even for the case of $\alpha_0 a = 10$ when the sample function descends down to zero very quickly, the polynomial-based generalized sine series still behaves satisfactorily in terms of the convergence and accuracy. For $\alpha_0 a = 100$, the sample function starts to show some strong presence of the boundary layer on the interval $[0, a]$. Consequently, the approximation errors for the generalized half-range Fourier series become substantially higher, even though the trends of converging are still noticeable.

As in the previous cases, this example has again demonstrated the excellent convergence characteristics of the polynomial-based generalized Fourier series. However, it also has revealed that the numerical accuracies of the generalized series expansions can deteriorate meaningfully when the involved functions exhibit some strong local variations or multiscale behaviors. This calls for an alternative implementation scheme to make the generalized Fourier series method better suited to this type of multiscale problems.

As illustrated in this example, the multiscale characteristic of the function (9.57) is primarily dictated by the parameter $\alpha_0 a$. In practical terms, the parameters which affect the behaviors of the solution of a given boundary value problem are often of clear physical or geometrical meanings as defined in the governing differential equation(s) and/or the boundary conditions. Thus, if it is possible to include these parameters, explicitly or implicitly, in the basis functions in accordance with the inherent rules imbedded in the mathematical physics equations, we believe that the multiscale characteristic of the solution can be, at least, partially captured by the generalized Fourier series representation. This sets up a framework for transforming the generalized Fourier series method into the so-called multiscale Fourier series method as described below.

9.4 DETERMINATION OF THE GENERAL SOLUTION

In the above discussions, the solution of a boundary value problem is considered as the combination of the general solution(s) and the particular solution. In a two-dimensional case, the general solutions actually represent two independent components; each of them satisfies the same homogeneous form of the governing differential equation, but with different sufficient conditions to ensure the term-by-term differentiations of the generalized Fourier series expansions. In what follows, we will focus on how to find the general solutions for one- and two-dimensional problems.

9.4.1 THE GENERAL SOLUTIONS OF ONE-DIMENSIONAL BOUNDARY VALUE PROBLEMS

Without loss of generality, the general (or homogeneous) solution involved in the generalized full-range Fourier series solution over the interval $[-a, a]$ is sought in the exponential form as

$$p_H(x) = \exp(\eta x), \tag{9.58}$$

where η is a yet-to-be-determined constant.
Substituting (9.58) into (9.10) results in

$$\sum_{k=0}^{2r} a_k \eta^k = 0. \tag{9.59}$$

Suppose that (9.59) has $2r$ roots denoted as η_l ($l = 1, 2, \cdots, 2r$). We then have a system of $2r$ linearly independent basis functions

$$\mathbf{p}_1^T(x) = [p_{1,H}(x) \quad p_{2,H}(x) \quad \cdots \quad p_{2r,H}(x)]. \tag{9.60}$$

Thus, the general solution can be accordingly expressed as

$$\varphi_1(x) = \mathbf{\Phi}_1^T(x) \cdot \mathbf{q}_1, \tag{9.61}$$

where the vectors of basis function $\Phi_1^T(x)$ and the unknown constants $\mathbf{q}_1^T$ are defined by (7.15) and (7.13), respectively.

9.4.2   THE GENERAL SOLUTIONS OF TWO-DIMENSIONAL BOUNDARY VALUE PROBLEMS

For two-dimensional problems, we will only focus on the procedures for constructing the general solution $\varphi_1(x,y)$ involved in the generalized full-range Fourier series solution over the domain $[-a,a] \times [-b,b]$.

The general solution $\varphi_1(x,y)$ is expanded, along the y-direction, into a one-dimensional Fourier series

$$\varphi_1(x,y) = \sum_{n=0}^{\infty} \mu_n[\xi_{1n}(x)\cos(\beta_n y) + \xi_{2n}(x)\sin(\beta_n y)], \tag{9.62}$$

where $\beta_n = n\pi/b$, $\mu_n = \begin{cases} 1/2 & n=0 \\ 1 & n>0 \end{cases}$, $\xi_{1n}(x)$, and $\xi_{2n}(x)$ are the corresponding Fourier coefficients which are obviously the functions of x.

Substituting (9.62) into (9.40) leads to:

for $n > 0$, the equations for the functions $\xi_{1n}(x)$ and $\xi_{2n}(x)$ are as below:

$$\begin{bmatrix} \mathcal{L}_{1n,1} & \mathcal{L}_{1n,2} \\ \mathcal{L}_{2n,1} & \mathcal{L}_{2n,2} \end{bmatrix} \begin{bmatrix} \xi_{1n} \\ \xi_{2n} \end{bmatrix} = \mathbf{0}, \tag{9.63}$$

where

$$\mathcal{L}_{1n,1} = \mathcal{L}_{2n,2} = \sum_{\substack{k_1 \geq 0, k_2 \geq 0 \\ k_1+k_2 \leq r}} a_{2k_1,2k_2}(-1)^{k_2}\beta_n^{2k_2}\frac{d^{2k_1}}{dx^{2k_1}} + \sum_{\substack{k_1 \geq 0, k_2 \geq 0 \\ k_1+k_2 \leq r-1}} a_{2k_1+1,2k_2}(-1)^{k_2}\beta_n^{2k_2}\frac{d^{2k_1+1}}{dx^{2k_1+1}},$$

$$\tag{9.64}$$

$$\mathcal{L}_{1n,2} = \sum_{\substack{k_1 \geq 0, k_2 \geq 0 \\ k_1+k_2 \leq r-1}} a_{2k_1,2k_2+1}(-1)^{k_2}\beta_n^{2k_2+1}\frac{d^{2k_1}}{dx^{2k_1}} + \sum_{\substack{k_1 \geq 0, k_2 \geq 0 \\ k_1+k_2 \leq r-1}} a_{2k_1+1,2k_2+1}(-1)^{k_2}\beta_n^{2k_2+1}\frac{d^{2k_1+1}}{dx^{2k_1+1}},$$

$$\tag{9.65}$$

and

$$\mathcal{L}_{2n,1} = \sum_{\substack{k_1 \geq 0, k_2 \geq 0 \\ k_1+k_2 \leq r-1}} a_{2k_1,2k_2+1}(-1)^{k_2+1}\beta_n^{2k_2+1}\frac{d^{2k_1}}{dx^{2k_1}} + \sum_{\substack{k_1 \geq 0, k_2 \geq 0 \\ k_1+k_2 \leq r-1}} a_{2k_1+1,2k_2+1}(-1)^{k_2+1}\beta_n^{2k_2+1}\frac{d^{2k_1+1}}{dx^{2k_1+1}};$$

$$\tag{9.66}$$

for $n = 0$, the equation for the function $\xi_{10}(x)$ reduces to

$$\mathcal{L}_{10}\xi_{10} = 0, \tag{9.67}$$

where

$$\mathcal{L}_{10} = \sum_{0 \le k_1 \le 2r} a_{k_1,0} \frac{d^{k_1}}{dx^{k_1}}. \tag{9.68}$$

The solutions of (9.63) are sought in the form of exponentials

$$\left. \begin{aligned} \xi_{1n}(x) &= G_{1n,1} p_{1n}(x) \\ \xi_{2n}(x) &= G_{1n,2} p_{1n}(x) \end{aligned} \right\}, \tag{9.69}$$

where

$$p_{1n}(x) = \exp(\eta_n x), \tag{9.70}$$

and $G_{1n,1}$, $G_{1n,2}$, and η_n are the constants to be determined.

Substitution of (9.69) into (9.63) results in

$$\begin{bmatrix} t_{1n,1} & t_{1n,2} \\ t_{2n,1} & t_{2n,2} \end{bmatrix} \begin{bmatrix} G_{1n,1} \\ G_{1n,2} \end{bmatrix} = 0, \tag{9.71}$$

where

$$t_{1n,1} = t_{2n,2} = \sum_{\substack{k_1 \ge 0, k_2 \ge 0 \\ k_1 + k_2 \le r}} a_{2k_1,2k_2} (-1)^{k_2} \eta_n^{2k_1} \beta_n^{2k_2} + \sum_{\substack{k_1 \ge 0, k_2 \ge 0 \\ k_1 + k_2 \le r-1}} a_{2k_1+1,2k_2} (-1)^{k_2} \eta_n^{2k_1+1} \beta_n^{2k_2}, \tag{9.72}$$

$$t_{1n,2} = \sum_{\substack{k_1 \ge 0, k_2 \ge 0 \\ k_1 + k_2 \le r-1}} a_{2k_1,2k_2+1} (-1)^{k_2} \eta_n^{2k_1} \beta_n^{2k_2+1} + \sum_{\substack{k_1 \ge 0, k_2 \ge 0 \\ k_1 + k_2 \le r-1}} a_{2k_1+1,2k_2+1} (-1)^{k_2} \eta_n^{2k_1+1} \beta_n^{2k_2+1}, \tag{9.73}$$

and

$$t_{2n,1} = \sum_{\substack{k_1 \ge 0, k_2 \ge 0 \\ k_1 + k_2 \le r-1}} a_{2k_1,2k_2+1} (-1)^{k_2+1} \eta_n^{2k_1} \beta_n^{2k_2+1} + \sum_{\substack{k_1 \ge 0, k_2 \ge 0 \\ k_1 + k_2 \le r-1}} a_{2k_1+1,2k_2+1} (-1)^{k_2+1} \eta_n^{2k_1+1} \beta_n^{2k_2+1}. \tag{9.74}$$

Thus, we have derived the characteristic equation (9.71) from which $4r$ characteristic roots $\eta_{n,l}$, $l = 1, 2, \cdots, 4r$, and the corresponding characteristic functions $p_{1n,l}(x)$, $l = 1, 2, \cdots, 4r$, can be obtained.

For $l = 1, 2, \cdots, 4r$, substituting the characteristic roots $\eta_{n,l}$ into (9.71) will lead to the following relations between the coefficients $G_{1n,1}^l$ and $G_{1n,2}^l$

$$\begin{bmatrix} t_{1n,1}(\eta_{n,l}) & t_{1n,2}(\eta_{n,l}) \end{bmatrix} \begin{bmatrix} G_{1n,1}^l \\ G_{1n,2}^l \end{bmatrix} = 0. \tag{9.75}$$

Let

$$\mathbf{a}_{1n,1}^{\mathrm{T}} = [G_{1n,1}^1 \quad G_{1n,1}^2 \quad \cdots \quad G_{1n,1}^{4r}], \tag{9.76}$$

$$\mathbf{a}_{1n,2}^{\mathrm{T}} = [G_{1n,2}^1 \quad G_{1n,2}^2 \quad \cdots \quad G_{1n,2}^{4r}], \tag{9.77}$$

then (9.75) can be rewritten as

$$\begin{bmatrix} \mathbf{a}_{1n,1} \\ \mathbf{a}_{1n,2} \end{bmatrix} = \mathbf{T}_{1n}\mathbf{a}_{1n,1}, \tag{9.78}$$

where $\mathbf{T}_{1n}$ is the transformation matrix.

Denote

$$\mathbf{p}_{1n}^{\mathrm{T}}(x) = [p_{1n,1}(x) \quad p_{1n,2}(x) \quad \cdots \quad p_{1n,4r}(x)], \tag{9.79}$$

the coefficient functions $\xi_{1n}(x)$ and $\xi_{2n}(x)$ can then be expressed as

$$\xi_{1n}(x) = \mathbf{p}_{1n}^{\mathrm{T}}(x) \cdot \mathbf{a}_{1n,1}, \tag{9.80}$$

$$\xi_{2n}(x) = \mathbf{p}_{1n}^{\mathrm{T}}(x) \cdot \mathbf{a}_{1n,2}. \tag{9.81}$$

From (8.50) and (8.51) we have

$$\mathbf{R}_{1n}\mathbf{a}_{1n,1} = \mathbf{q}_{1n,1}, \tag{9.82}$$

$$\mathbf{R}_{1n}\mathbf{a}_{1n,2} = \mathbf{q}_{1n,2}, \tag{9.83}$$

where the matrix $\mathbf{R}_{1n}$, and the sub-vectors (of boundary Fourier coefficients) $\mathbf{q}_{1n,1}$ and $\mathbf{q}_{1n,2}$ are previously defined in (8.52) and (8.53) in Chapter 8. However, it's worthy of noting that the dimensions of $\mathbf{R}_{1n}$ have been adjusted from $2r \times 2r$ to $2r \times 4r$.

By combining (9.82) and (9.83) as

$$\mathbf{S}_{1n} \begin{bmatrix} \mathbf{a}_{1n,1} \\ \mathbf{a}_{1n,2} \end{bmatrix} = \begin{bmatrix} \mathbf{q}_{1n,1} \\ \mathbf{q}_{1n,2} \end{bmatrix}, \tag{9.84}$$

where

$$\mathbf{S}_{1n} = \begin{bmatrix} \mathbf{R}_{1n} & \mathbf{0} \\ \mathbf{0} & \mathbf{R}_{1n} \end{bmatrix}, \tag{9.85}$$

we can rewrite (9.78) into

$$\mathbf{S}_{1n}\mathbf{T}_{1n}\mathbf{a}_{1n,1} = \begin{bmatrix} \mathbf{q}_{1n,1} \\ \mathbf{q}_{1n,2} \end{bmatrix} \tag{9.86}$$

or

$$\mathbf{a}_{1n,1} = (\mathbf{S}_{1n}\mathbf{T}_{1n})^{-1} \begin{bmatrix} \mathbf{q}_{1n,1} \\ \mathbf{q}_{1n,2} \end{bmatrix}. \tag{9.87}$$

Therefore, the one-dimensional functions $\xi_{1n}(x)$ and $\xi_{2n}(x)$ can be finally given as

$$\begin{bmatrix} \xi_{1n}(x) \\ \xi_{2n}(x) \end{bmatrix} = \mathbf{p}_{1n,R}^{\mathrm{T}}(x) \cdot \begin{bmatrix} \mathbf{q}_{1n,1} \\ \mathbf{q}_{1n,2} \end{bmatrix}, \tag{9.88}$$

where

$$\mathbf{p}_{1n,R}^{\mathrm{T}}(x) = \begin{bmatrix} \mathbf{p}_{1n}^{\mathrm{T}}(x) & \mathbf{0} \\ \mathbf{0} & \mathbf{p}_{1n}^{\mathrm{T}}(x) \end{bmatrix} \cdot \mathbf{T}_{1n}(\mathbf{S}_{1n}\mathbf{T}_{1n})^{-1}. \tag{9.89}$$

Accordingly, the vector of basis functions are defined as

$$\Phi_{1n}^{\mathrm{T}}(x,y) = \begin{bmatrix} \Phi_{1n,1}^{\mathrm{T}}(x,y) & \Phi_{1n,2}^{\mathrm{T}}(x,y) \end{bmatrix} = \mathbf{H}_{1n}(y) \cdot \mathbf{p}_{1n,R}^{\mathrm{T}}(x), \tag{9.90}$$

where

$$\mathbf{H}_{1n}(y) = \begin{bmatrix} \cos(\beta_n y) & \sin(\beta_n y) \end{bmatrix}. \tag{9.91}$$

The corresponding vectors of the basis functions for the higher order partial derivatives can be directly obtained:

$$\Phi_{1n}^{(k_1,k_2)\mathrm{T}}(x,y) = \begin{bmatrix} \Phi_{1n,1}^{(k_1,k_2)\mathrm{T}}(x,y) & \Phi_{1n,2}^{(k_1,k_2)\mathrm{T}}(x,y) \end{bmatrix} = \mathbf{H}_{1n}^{(k_2)}(y) \cdot \mathbf{p}_{1n,R}^{(k_1)\mathrm{T}}(x), \tag{9.92}$$

where

$$\mathbf{p}_{1n,R}^{(k_1)\mathrm{T}}(x) = \begin{bmatrix} \mathbf{p}_{1n}^{(k_1)\mathrm{T}}(x) & \mathbf{0} \\ \mathbf{0} & \mathbf{p}_{1n}^{(k_1)\mathrm{T}}(x) \end{bmatrix} \cdot \mathbf{T}_{1n}(\mathbf{S}_{1n}\mathbf{T}_{1n})^{-1} \tag{9.93}$$

and

$$\mathbf{H}_{1n}^{(k_2)}(y) = \left[\quad [\cos(\beta_n y)]^{(k_2)} \quad [\sin(\beta_n y)]^{(k_2)} \quad \right]. \tag{9.94}$$

In the above derivations, the functions $\mathbf{p}_{1n}^{\mathrm{T}}(x)$ is made relevant to the specific problem on hand by explicitly enforcing them to satisfy (9.40). As a matter of fact, if a set of linearly independent homogeneous solutions, $p_{1nl,H}(x,y)$, $l = 1, 2, \cdots, 4r$, is already available somehow, then we can directly construct the basis functions as

$$\xi_{1n}(x)\cos(\beta_n y) + \xi_{2n}(x)\sin(\beta_n y) = \mathbf{p}_{1n,H}^{\mathrm{T}}(x,y) \cdot \mathbf{a}_{1n}, \tag{9.95}$$

where

$$\mathbf{p}_{1n,H}^{\mathrm{T}}(x,y) = [p_{1n1,H}(x,y) \quad p_{1n2,H}(x,y) \quad \cdots \quad p_{1n(4r),H}(x,y)], \tag{9.96}$$

and

$$\mathbf{a}_{1n}^{\mathrm{T}} = [G_{1n}^1 \quad G_{1n}^2 \quad \cdots \quad G_{1n}^{4r}]. \tag{9.97}$$

Inserting $y = 0$ and $y = b/2n$ successively in (9.95), we obtain

$$\xi_{1n}(x) = \mathbf{p}_{1n,H}^{\mathrm{T}}(x,0) \cdot \mathbf{a}_{1n} \tag{9.98}$$

and

$$\xi_{2n}(x) = \mathbf{p}_{1n,H}^{\mathrm{T}}(x,b/2n) \cdot \mathbf{a}_{1n}. \tag{9.99}$$

Differentiating both sides of (9.98) and (9.99) up to $2r - 1$ times with respect to x leads to

$$\xi_{1n}^{(k_1)}(x) = \mathbf{p}_{1n,H}^{(k_1,0)\mathrm{T}}(x,0) \cdot \mathbf{a}_{1n}, k_1 = 0, 1, \cdots, 2r-1, \tag{9.100}$$

and

$$\xi_{2n}^{(k_1)}(x) = \mathbf{p}_{1n,H}^{(k_1,0)\mathrm{T}}(x,b/2n) \cdot \mathbf{a}_{1n}, k_1 = 0, 1, \cdots, 2r-1. \tag{9.101}$$

Then, by substituting $x = a$ and $x = -a$ successively in (9.100) and (9.101), and subtracting the resulting equations, we yield

$$\xi_{1n}^{(k_1)}(a) - \xi_{1n}^{(k_1)}(-a) = [\mathbf{p}_{1n,H}^{(k_1,0)\mathrm{T}}(a,0) - \mathbf{p}_{1n,H}^{(k_1,0)\mathrm{T}}(-a,0)] \cdot \mathbf{a}_{1n}, k_1 = 0, 1, \cdots, 2r-1, \tag{9.102}$$

and

$$\xi_{2n}^{(k_1)}(a) - \xi_{2n}^{(k_1)}(-a) = [\mathbf{p}_{1n,H}^{(k_1,0)\mathrm{T}}(a,b/2n) - \mathbf{p}_{1n,H}^{(k_1,0)\mathrm{T}}(-a,b/2n)] \cdot \mathbf{a}_{1n}, k_1 = 0, 1, \cdots, 2r-1. \tag{9.103}$$

Therefore, referring to (8.43) and (8.44), we obtain

$$\mathbf{R}_{1n,H}\mathbf{a}_{1n} = \begin{bmatrix} \mathbf{q}_{1n,1} \\ \mathbf{q}_{1n,2} \end{bmatrix} \tag{9.104}$$

where the matrices

$$\mathbf{R}_{1n,H} = \begin{bmatrix} \mathbf{R}_{1n1,H} \\ \mathbf{R}_{1n2,H} \end{bmatrix}, \tag{9.105}$$

$$\mathbf{R}_{1n1,H} = \begin{bmatrix} \mathbf{p}_{1n,H}^{(0,0)\mathrm{T}}(a,0) - \mathbf{p}_{1n,H}^{(0,0)\mathrm{T}}(-a,0) \\ \mathbf{p}_{1n,H}^{(1,0)\mathrm{T}}(a,0) - \mathbf{p}_{1n,H}^{(1,0)\mathrm{T}}(-a,0) \\ \vdots \\ \mathbf{p}_{1n,H}^{(2r-1,0)\mathrm{T}}(a,0) - \mathbf{p}_{1n,H}^{(2r-1,0)\mathrm{T}}(-a,0) \end{bmatrix}, \tag{9.106}$$

$$\mathbf{R}_{1n2,H} = \begin{bmatrix} \mathbf{p}_{1n,H}^{(0,0)\mathrm{T}}(a,b/2n) - \mathbf{p}_{1n,H}^{(0,0)\mathrm{T}}(-a,b/2n) \\ \mathbf{p}_{1n,H}^{(1,0)\mathrm{T}}(a,b/2n) - \mathbf{p}_{1n,H}^{(1,0)\mathrm{T}}(-a,b/2n) \\ \vdots \\ \mathbf{p}_{1n,H}^{(2r-1,0)\mathrm{T}}(a,b/2n) - \mathbf{p}_{1n,H}^{(2r-1,0)\mathrm{T}}(-a,b/2n) \end{bmatrix}, \tag{9.107}$$

and the sub-vectors of boundary Fourier coefficients $\mathbf{q}_{1n,1}$ and $\mathbf{q}_{1n,2}$ are given by (8.53).

Accordingly, we can define the vector of basis functions

$$\Phi_{1n}^{\mathrm{T}}(x,y) = \begin{bmatrix} \Phi_{1n,1}^{\mathrm{T}}(x,y) & \Phi_{1n,2}^{\mathrm{T}}(x,y) \end{bmatrix} = \mathbf{p}_{1n,H}^{\mathrm{T}}(x,y) \cdot \mathbf{R}_{1n,H}^{-1}, \tag{9.108}$$

and the corresponding vectors of higher order partial derivatives of basis functions

$$\Phi_{1n}^{(k_1,k_2)\mathrm{T}}(x,y) = \begin{bmatrix} \Phi_{1n,1}^{(k_1,k_2)\mathrm{T}}(x,y) & \Phi_{1n,2}^{(k_1,k_2)\mathrm{T}}(x,y) \end{bmatrix} = \mathbf{p}_{1n,H}^{(k_1,k_2)\mathrm{T}}(x,y) \cdot \mathbf{R}_{1n,H}^{-1}, \tag{9.109}$$

where k_1 and k_2 are nonnegative integers.

Now, we proceed to the solution of (9.67)

$$\xi_{10}(x) = G_{10,1}p_{10}(x), \tag{9.110}$$

where

$$p_{10}(x) = \exp(\eta_0 x) \tag{9.111}$$

and $G_{10,1}$ and η_0 are the constants to be determined.

Substitution of (9.110) into (9.67) leads to

$$\sum_{k_1=0}^{2r} a_{k_1,0}\eta_0^{k_1} = 0, \tag{9.112}$$

from which we are able to obtain $2r$ characteristic roots $\eta_{0,l}$ $(l=1, 2, \cdots, 2r)$ and the corresponding homogeneous solutions $p_{10,l}(x)$ $(l=1, 2, \cdots, 2r)$.

Thus, the solution can be expressed as

$$\xi_{10}(x) = \mathbf{p}_{10}^{\mathrm{T}}(x) \cdot \mathbf{a}_{10,1}, \tag{9.113}$$

where

$$\mathbf{p}_{10}^{\mathrm{T}}(x) = [p_{10,1}(x) \quad p_{10,2}(x) \quad \cdots \quad p_{10,2r}(x)], \tag{9.114}$$

and

$$\mathbf{a}_{10,1}^{\mathrm{T}} = [G_{10,1}^1 \quad G_{10,1}^2 \quad \cdots \quad G_{10,1}^{2r}]. \tag{9.115}$$

Therefore, from (8.50) we have

$$\mathbf{R}_{10}\mathbf{a}_{10,1} = \mathbf{q}_{10,1}, \tag{9.116}$$

where the matrix $\mathbf{R}_{10}$ and the sub-vector of boundary Fourier coefficients $\mathbf{q}_{10,1}$ are given by (8.52) and (8.53), respectively.

Finally, the function $\xi_{10}(x)$ can be written as

$$\xi_{10}(x) = \mathbf{p}_{10,R}^{\mathrm{T}}(x) \cdot \mathbf{q}_{10,1}, \tag{9.117}$$

where

$$\mathbf{p}_{10,R}^{\mathrm{T}}(x) = \mathbf{p}_{10}^{\mathrm{T}}(x) \cdot \mathbf{R}_{10}^{-1}. \tag{9.118}$$

The sub-vector of the basis functions can be accordingly defined as

$$\Phi_{10,1}^{\mathrm{T}}(x, y) = \mathbf{H}_{10}(y) \cdot \mathbf{p}_{10,R}^{\mathrm{T}}(x), \tag{9.119}$$

where

$$\mathbf{H}_{10}(y) = \begin{bmatrix} 1 \\ 2 \end{bmatrix}. \tag{9.120}$$

The basis function vectors for higher order partial derivatives are readily available

$$\Phi_{10,1}^{(k_1,k_2)\mathrm{T}}(x, y) = \mathbf{H}_{10}^{(k_2)}(y) \cdot \mathbf{p}_{10,R}^{(k_1)\mathrm{T}}(x), \tag{9.121}$$

where k_1 and k_2 are nonnegative integers.

Based on the procedures described above, the boundary function $\varphi_1(x,y)$ is readily obtained as

$$\varphi_1(x,y) = \mathbf{\Phi}_1^T(x,y) \cdot \mathbf{q}_1. \tag{9.122}$$

The other boundary function $\varphi_2(x,y)$ can be similarly constructed without difficulties.

9.5 EQUIVALENT TRANSFORMATION OF THE SOLUTION

The generalized Fourier series was first introduced as a generic means to expand a sufficiently smooth function and its derivatives defined on a compact domain. In the previous sections, it was further developed into a general method for solving linear differential equations with constant coefficients.

In the context of solving linear differential equations, it appears from Sec. 9.1 and 9.2 that the involved Fourier coefficients and expansion constants can be fully determined from forcing the solution to satisfy the differential equation and the boundary conditions such as (9.3) or (9.26). However, this may encounter difficulties with the use of a variational method, where the pre-satisfaction of the displacement type boundary condition is often required (for convenience, the term "displacement" is here simply used to denote the solution function even though it may actually represent other physically-different variables such as pressure, temperature, etc.). Therefore, an equivalent solution transformation scheme will be derived in this section for the generalized Fourier series solution defined on the domain $[-a,a] \times [-b,b]$.

Suppose that the displacement type boundary conditions at $x = a$ are prescribed as

$$\mathbf{C}u = \bar{\mathbf{u}}, \tag{9.123}$$

where $\bar{\mathbf{u}}$ is a vector of functions corresponding to the prescribed distributions along $x = a$, and the vector of differential operators is

$$\mathbf{C}^T = [\mathcal{C}_1 \quad \mathcal{C}_2 \quad \cdots \quad \mathcal{C}_r]. \tag{9.124}$$

The elements of vector $\mathbf{C}$ in (9.124) are given as

$$\mathcal{C}_l = \sum_{\substack{k_1 \geq 0, k_2 \geq 0 \\ k_1 + k_2 \leq 2r-1}} c_{k_1,k_2}^l \frac{\partial^{k_1+k_2}}{\partial x^{k_1} \partial y^{k_2}} \quad (l = 1, 2, \cdots, r), \tag{9.125}$$

where c_{k_1,k_2}^l are constant coefficients.

By substituting the generalized Fourier series solution (8.91) into (9.123), we have

$$\mathbf{\Gamma}(a,y) \cdot \mathbf{q} = \bar{\mathbf{u}}(a,y), \tag{9.126}$$

where

$$
\Gamma = \begin{bmatrix} C_1 \Phi_0^{\mathrm{T}} & C_1 \Phi_1^{\mathrm{T}} & C_1 \Phi_2^{\mathrm{T}} & C_1 \Phi_3^{\mathrm{T}} \\ C_2 \Phi_0^{\mathrm{T}} & C_2 \Phi_1^{\mathrm{T}} & C_2 \Phi_2^{\mathrm{T}} & C_2 \Phi_3^{\mathrm{T}} \\ \vdots & \vdots & \vdots & \vdots \\ C_r \Phi_0^{\mathrm{T}} & C_r \Phi_1^{\mathrm{T}} & C_r \Phi_2^{\mathrm{T}} & C_r \Phi_3^{\mathrm{T}} \end{bmatrix},
\tag{9.127}
$$

and

$$
\mathbf{q}^{\mathrm{T}} = [\mathbf{q}_0^{\mathrm{T}} \quad \mathbf{q}_1^{\mathrm{T}} \quad \mathbf{q}_2^{\mathrm{T}} \quad \mathbf{q}_3^{\mathrm{T}}].
\tag{9.128}
$$

Expanding $\bar{\mathbf{u}}(a, y)$ and each element of the matrix $\Gamma(a, y)$ into full-range Fourier series over the interval $[-b, b]$, and comparing the like terms on the both sides of (9.126) will lead to

$$
\mathbf{q}_{b,a+} = \mathbf{R}_{b,a+} \cdot \mathbf{q},
\tag{9.129}
$$

where the vector of expansion coefficients

$$
\mathbf{q}_{b,a+}^{\mathrm{T}} = [\mathbf{q}_{b1,a+}^{\mathrm{T}} \quad \mathbf{q}_{b2,a+}^{\mathrm{T}} \quad \cdots \quad \mathbf{q}_{br,a+}^{\mathrm{T}}]
\tag{9.130}
$$

with $\mathbf{q}_{b1,a+}^{\mathrm{T}}$, $\mathbf{q}_{b2,a+}^{\mathrm{T}}$, ... and $\mathbf{q}_{br,a+}^{\mathrm{T}}$ being the vectors of Fourier coefficients corresponding to the elements of the first, second, ..., and r-th rows of $\bar{\mathbf{u}}(a, y)$, and $\mathbf{R}_{b,a+}$ being the Fourier coefficients matrix corresponding to the matrix $\Gamma(a, y)$.

Similar relationships can be obtained at other boundaries: $x = -a$, $y = b$, and $y = -b$, and they can be combined with (9.129) into

$$
\mathbf{q}_b = \mathbf{R}_b \cdot \mathbf{q},
\tag{9.131}
$$

where

$$
\mathbf{R}_b = \begin{bmatrix} \mathbf{R}_{b,a+} \\ \mathbf{R}_{b,a-} \\ \mathbf{R}_{b,b+} \\ \mathbf{R}_{b,b-} \end{bmatrix}
\tag{9.132}
$$

and

$$
\mathbf{q}_b^{\mathrm{T}} = [\mathbf{q}_{b,a+}^{\mathrm{T}} \quad \mathbf{q}_{b,a-}^{\mathrm{T}} \quad \mathbf{q}_{b,b+}^{\mathrm{T}} \quad \mathbf{q}_{b,b-}^{\mathrm{T}}].
\tag{9.133}
$$

If the vector $\mathbf{q}$ and the coefficient matrix $\mathbf{R}_b$ are rearranged into

$$
\mathbf{q}_{03}^{\mathrm{T}} = [\mathbf{q}_0^{\mathrm{T}} \quad \mathbf{q}_3^{\mathrm{T}}],
\tag{9.134}
$$

$$
\mathbf{q}_{12}^{\mathrm{T}} = [\mathbf{q}_1^{\mathrm{T}} \quad \mathbf{q}_2^{\mathrm{T}}],
\tag{9.135}
$$

and

$$\mathbf{R}_b = [\mathbf{R}_{b,03} \quad \mathbf{R}_{b,12}], \tag{9.136}$$

then (9.131) can be accordingly rewritten as

$$\mathbf{q}_b = \mathbf{R}_{b,03} \cdot \mathbf{q}_{03} + \mathbf{R}_{b,12} \cdot \mathbf{q}_{12}, \tag{9.137}$$

or

$$\mathbf{q}_{12} = -\mathbf{R}_{b,12}^{-1} \mathbf{R}_{b,03} \cdot \mathbf{q}_{03} + \mathbf{R}_{b,12}^{-1} \cdot \mathbf{q}_b. \tag{9.138}$$

Substitution of (9.138) into (8.91) will finally lead to the generalized Fourier series solution in the form of

$$u = \boldsymbol{\Phi}_R^T \cdot \mathbf{q}_R, \tag{9.139}$$

where

$$\boldsymbol{\Phi}_R^T = \begin{bmatrix} \boldsymbol{\Phi}_{03}^T & \boldsymbol{\Phi}_{12}^T \end{bmatrix} \begin{bmatrix} \mathbf{I} & \mathbf{0} \\ -\mathbf{R}_{b,12}^{-1} \mathbf{R}_{b,03} & \mathbf{R}_{b,12}^{-1} \end{bmatrix}, \tag{9.140}$$

$$\boldsymbol{\Phi}_{03}^T = \begin{bmatrix} \boldsymbol{\Phi}_0^T & \boldsymbol{\Phi}_3^T \end{bmatrix}, \tag{9.141}$$

$$\boldsymbol{\Phi}_{12}^T = \begin{bmatrix} \boldsymbol{\Phi}_1^T & \boldsymbol{\Phi}_2^T \end{bmatrix}, \tag{9.142}$$

and

$$\mathbf{q}_R^T = \begin{bmatrix} \mathbf{q}_{03}^T & \mathbf{q}_b^T \end{bmatrix}. \tag{9.143}$$

Even though the solution given by (9.139) still looks similar to its origin (8.91), the displacement-type boundary conditions have already been explicitly incorporated into the solution. In other words, the original Fourier coefficients associated with the boundary functions, $\varphi_1(x, y)$ and $\varphi_2(x, y)$, have been transformed into a new set of Fourier coefficients which are compatible with the prescribed displacement-type boundary conditions.

9.6   INTRODUCTION OF THE SUPPLEMENTARY SOLUTION

It is intuitively understandable that the characteristic of the external force function f can more or less affect the convergence and accuracy of the generalized Fourier series solution of a linear differential equation. Accordingly, the external force function will be here decomposed into two parts: the coarse scale component f_s and the fine scale component $(f - f_s)$, which are to be treated differently.

For example, consider the two-dimensional full-range series solution over the domain $[-a,a] \times [-b,b]$. We select a suitable form of function φ_s such that

$$\mathcal{L}\varphi_s = f_s. \tag{9.144}$$

Thus, the differential equation (9.24) becomes

$$\mathcal{L}(u - \varphi_s) = f - f_s, (x,y) \in (-a,a) \times (-b,b), \tag{9.145}$$

and the boundary conditions shall be accordingly modified to

$$\mathbf{B}(u - \varphi_s) = \mathbf{g} - \mathbf{B}\varphi_s. \tag{9.146}$$

Solving the differential equation (9.145) together with the boundary conditions (9.146) completely falls into the context of the above discussions, and does not call for any further elaborations.

Without loss of generality, the generalized Fourier series solution of (9.145) is now given as

$$u - \varphi_s = \varphi_0 + \varphi_1 + \varphi_2 + \varphi_3 \tag{9.147}$$

or

$$u = \varphi_0 + \varphi_1 + \varphi_2 + \varphi_3 + \varphi_s. \tag{9.148}$$

Evidently, the supplementary term φ_s is now added to the generalized Fourier series solution which originally consists of the boundary functions, φ_1 and φ_2, and the internal and corner functions, φ_0 and φ_3. Since the general solutions and the particular solution have already been used to label the relevant constituents of the solution of (9.24), the term φ_s will be here referred to as the supplementary solution as a distinction.

9.7 THE MULTISCALE CHARACTERISTIC OF THE SOLUTION

It must be emphasized that the decomposition of the solution into several constituents is carefully formulated to warrant a desired convergence characteristic of the resulting Fourier series solution (of a differential equation or boundary value problem). As mentioned earlier, there is theoretically an infinite number of possible choices for the basis functions. This will give us the leeway to select the ones which better fit the particular problems on hand.

The multiscale capability of a solution method is of special interests to many scientific and engineering applications. As described in Sec. 9.6, the decomposition of the solution into the boundary and internal components permits them to be determined independently. Further, the boundary functions are sought as the general solution of the differential equations or the characteristic functions in the different coordinate directions. Thus, it can be expected that such boundary functions or general solutions shall be able to more appropriately interpret the meaning of the differential equation, and hence better capture the spatial characteristics of the solution in the corresponding coordinate directions. The use of the supplementary solution will be considered beneficial to the solution by removing glitches or irregularities from the forcing function.

With developing the multiscale capability in mind, we would like to evolve the structural decomposition of a solution function into a scale decomposition of the solution space. Take the generalized full-range Fourier series of a solution over the domain $[-a,a] \times [-b,b]$ as an example. In the structural decomposition of the solution, the internal function $\varphi_0(x,y)$ is a two-dimensional Fourier series over the domain and corresponds to the primary scale (or medium scale) of the solution space. The boundary function $\varphi_1(x,y)$ is a one-dimensional Fourier series along the y-direction on the interval $[-b,b]$ and its Fourier coefficients are the homogeneous solutions of the ordinary differential equations with respect to the variable x, thus manifesting itself with two facets: the primary scale in the y-direction and an adaptive scale (or a hidden scale related to the original differential equation) in the x-direction. Similarly, the boundary function $\varphi_2(x,y)$ has the primary scale in the x-direction and an adaptive scale in the y-direction. The corner function $\varphi_3(x,y)$ is a two-dimensional polynomial over the domain $[-a,a] \times [-b,b]$ and corresponds to the global scale (or large scale) of the spatial decomposition scales. The global (large) scale, the primary (medium) scale, and the adaptive (hidden) scale, when combined, represent a multiscale spatial decomposition of the solution of a given differential equation or boundary value problem. Accordingly, the particular solution (sum of the internal function $\varphi_0(x,y)$ and the corner function $\varphi_3(x,y)$) with large and medium scales and the general solution (sum of the boundary functions $\varphi_1(x,y)$ and $\varphi_2(x,y)$) with the adaptive scales shall be able to effectively resolve or capture the multiscale characteristic of the solution of a differential equation or boundary value problem. In the current structural decomposition scheme, small scales have not been explicitly mentioned. However, regardless of whether such small scales are actually related to the adaptive scales or not, the inclusion of the supplementary solution is capable of alleviating this concern to some extent, as illustrated later by examples.

Based on what have been said, the generalized Fourier series solutions (or methods) will also be referred to as the multiscale Fourier series solutions (or methods) to emphasize their multiscale capabilities.

9.8 SOLUTION SCHEMES

By far, the multiscale Fourier series methods have been adequately discussed with respect to their convergence characteristics, approximation errors, and the selections of basis functions. We will now briefly talk about some of the specific techniques used for deriving the multiscale Fourier series solutions.

TABLE 9.1

The Vectors of the Unknown Coefficients Corresponding to Different Solution Schemes

Solution Techniques	The Vectors of the Unknown Coefficients	Multiscale Fourier Series Solution
Fourier coefficient comparison method	q_0, q_1, q_2, q_3	(9.148)
Collocation method	q_0, q_1, q_2, q_3	(9.148)
Minimum potential energy method	q_0, q_b, q_3	(9.139)

The solution techniques can typically be classified into three categories. The first category represents an analytical approach by directly comparing the Fourier coefficients of the like terms on both sides of the final system of equations. The second one is based on the weighted residual methods (for instance, the collocation methods), which still fall into the classification of strong-form solutions. The third one includes the variational methods (for instance, the minimum potential energy method), which lead to the so-called weak-form solutions.

These three approaches will be illustrated separately in the following chapters by solving several boundary value problems of broad concerns in science and engineering. The unknown solution variables corresponding to each category are listed in Table 9.1.

10 Multiscale Fourier Series Method for the Convection-Diffusion-Reaction Equation

The convection-diffusion-reaction equation is one of the most studied subjects in mathematical physics. The convection-diffusion-reaction equation and its special cases (i.e., the convection-diffusion equation, the Helmholtz equation, the convection-reaction equation, and the diffusion-reaction equation) are widely used to describe many physical and chemical processes (Hauke and Garcia-Olivares 2001; Hauke 2002), such as, turbulences, combustions, heat transfers, diffusion of pollutants, propagations of acoustic waves, etc. The understanding of these natural phenomena and chemo-physical processes are often perplexed by, among others, the multiscale characteristics of the physical fields as manifested in, such as, the presence of swift transition between the exponential and the propagation regimes (Onate et al. 2006).

Many multiscale solution methods have been developed for the convection-diffusion-reaction equation, which include the stabilized finite element methods (Huerta and Donea 2002; Huerta et al. 2002), the bubble methods (Brezzi et al. 1998; Franca et al. 1998; Sangalli 2004; Parvazinia and Nassehi 2007), the wavelet finite element methods (Dahmen et al. 1997; Dahmen 2001; Donea and Huerta 2003), the meshless methods (Liu and Gu 2005), and the variational multiscale methods (Hughes and Wells 2005; Song et al. 2010). It is evident from their names that these multiscale methods are primarily based on the traditional mesh-based or other discretization methods by making some suitable modifications, such as: use of stabilization terms, inclusion of different scale groups, decomposition of the solution into coarse and fine scale components, etc. Although these multiscale methods have been applied to a variety of boundary value problems, we are still in the constant struggles with respect to the stability of the solution algorithms, proper selection of computational scales, robustness and effectiveness of the methods, balancing the span of scale groups and solution accuracy, high computational costs, and low numerical accuracy for higher order derivatives of field variables.

The multiscale Fourier series method described earlier potentially represents a strategic shift from the existing framework toward better resolving multiscale problems. In what follows, it will be specifically employed to solve the famous convection-diffusion-reaction equation for a wide spectrum of model variables.

DOI: 10.1201/9781003194859-10

10.1 MULTISCALE FOURIER SERIES SOLUTION FOR ONE-DIMENSIONAL CONVECTION-DIFFUSION-REACTION EQUATION

The one-dimensional convection-diffusion-reaction equation is a second order $(2r = 2)$ linear differential equation with constant coefficients. Therefore, the multiscale Fourier series methods outlined in Chapter 9 are directly applicable to this class of boundary value problems.

10.1.1 DESCRIPTION OF THE PROBLEM

The one-dimensional convection-diffusion-reaction equation on the interval $[-a, a]$ can be written in a dimensionless form as (Hauke and Garcia-Olivares 2001)

$$\mathcal{L}_{cdr,1}\varphi = f, \tag{10.1}$$

where

$$\mathcal{L}_{cdr,1} = -\frac{d^2}{dx^2} + P_e \frac{d}{dx} - P_e D_a, \tag{10.2}$$

$f(x)$ is a given source function, P_e is the dimensionless Peclet number related to the convective velocity, and D_a is the dimensionless Damkohler number.

At the boundary, such as, $x = a$ (actually, an end point of the interval), the boundary condition is specified as:

1. Dirichlet boundary condition (or, D-type, for short)

$$\varphi(a) = \bar{\varphi}_a, \tag{10.3}$$

or
2. Neumann boundary condition (or, N-type, for short)

$$\varphi^{(1)}(a) = \bar{\phi}_a, \tag{10.4}$$

where $\bar{\varphi}_a$ and $\bar{\phi}_a$ are respectively the prescribed values at $x = a$ of the solution and its derivative.

10.1.2 THE GENERAL SOLUTION

Suppose that (10.1) has a homogeneous solution

$$p_H(x) = \exp(\eta x), \tag{10.5}$$

where η is a constant to be determined.

TABLE 10.1

Some Real Constants Related to the Characteristic Roots of Characteristic Equation (10.6)

	α_{10}	α_{20}	α_{30}
$4D_a < P_e$	$-\dfrac{P_e}{2}$	0	$\dfrac{\sqrt{P_e^2 - 4P_e D_a}}{2}$
$4D_a > P_e$	$-\dfrac{P_e}{2}$	$\dfrac{\sqrt{4P_e D_a - P_e^2}}{2}$	0
$4D_a = P_e$	$-\dfrac{P_e}{2}$	0	0

Substituting it into the homogeneous form of (10.1), we have the characteristic equation

$$\eta^2 - P_e \eta + P_e D_a = 0. \tag{10.6}$$

It is well known that (10.6) has two distinct real roots when $4D_a < P_e$, two distinct complex roots when $4D_a > P_e$, and one double (real) root when $4D_a = P_e$.

Denote the characteristic roots as

$$\eta_1 = -\alpha_{10} + \alpha_{30} - i\alpha_{20} \text{ and } \eta_2 = -\alpha_{10} - \alpha_{30} + i\alpha_{20} \quad (i = \sqrt{-1}), \tag{10.7}$$

where the real constants α_{10}, α_{20}, and α_{30} are given in Table 10.1.

Presented in Table 10.2 are the corresponding homogeneous solutions $p_{l,H}(x)$, $l = 1, 2$.

If these homogeneous solutions are chosen as the basis functions

$$\mathbf{p}_1^T(x) = [p_{1,H}(x) \quad p_{2,H}(x)], \tag{10.8}$$

then the general solution of (10.1) can be expressed as

$$\varphi_1(x) = \Phi_1^T(x) \cdot \mathbf{q}_1, \tag{10.9}$$

where the vectors of basis function $\Phi_1^T(x)$ and the undetermined constants $\mathbf{q}_1$ are previously defined by (9.61) in Chapter 9.

TABLE 10.2

Expressions for the Homogeneous Solutions $p_{l,H}(x)$, $l = 1, 2$

	$p_{1,H}(x)$	$p_{2,H}(x)$
$4D_a < P_e$	$\exp[(-\alpha_{10} + \alpha_{30})x]$	$\exp[(-\alpha_{10} - \alpha_{30})x]$
$4D_a > P_e$	$\exp(-\alpha_{10}x)\sin(\alpha_{20}x)$	$\exp(-\alpha_{10}x)\sin[\alpha_{20}(a - x)]$
$4D_a = P_e$	$\exp(-\alpha_{10}x)$	$x\exp(-\alpha_{10}x)$

10.1.3 THE SUPPLEMENTARY SOLUTION

Let N_1^s be a positive integer and $\Upsilon^s = \left\{ x_{n_1}, n_1 = 1, 2, \cdots, N_1^s + 1 \right\}$ be a set of interpolation points uniformly distributed on the interval $[-a, a]$. Now, we use a polynomial function $f_s(x)$ to approximate the distributed source $f(x)$ in the sense that

$$f_s(x_{n_1}) = f(x_{n_1}), n_1 = 1, 2, \cdots, N_1^s + 1. \tag{10.10}$$

For this purpose, define a vector of polynomial functions as

$$\mathbf{p}_{fs}^{\mathrm{T}}(x) = [p_{fs,1}(x) \quad p_{fs,2}(x) \quad \cdots \quad p_{fs, N_1^s + 1}(x)], \tag{10.11}$$

where

$$p_{fs,j}(x) = (x/a)^{j-1}, j = 1, 2, \cdots, N_1^s + 1. \tag{10.12}$$

Then the interpolated source function $f_s(x)$ can be expressed as

$$f_s(x) = \mathbf{p}_{fs}^{\mathrm{T}}(x) \cdot \mathbf{a}_{fs}, \tag{10.13}$$

where the vector of unknown constants

$$\mathbf{a}_{fs}^{\mathrm{T}} = [H_{fs,1} \quad H_{fs,2} \quad \cdots \quad H_{fs, N_1^s + 1}]. \tag{10.14}$$

By substituting (10.13) into (10.10), we yield

$$\mathbf{R}_{fs} \mathbf{a}_{fs} = \mathbf{q}_{fs}, \tag{10.15}$$

where

$$\mathbf{R}_{fs} = \begin{bmatrix} \mathbf{p}_{fs}^{\mathrm{T}}(x_1) \\ \mathbf{p}_{fs}^{\mathrm{T}}(x_2) \\ \vdots \\ \mathbf{p}_{fs}^{\mathrm{T}}(x_{N_1^s + 1}) \end{bmatrix} \tag{10.16}$$

and

$$\mathbf{q}_{fs}^{\mathrm{T}} = [f(x_1) \quad f(x_2) \quad \cdots \quad f(x_{N_1^s + 1})]. \tag{10.17}$$

Thus, the interpolated polynomial function $f_s(x)$ is obtained as

$$f_s(x) = \mathbf{p}_{fs}^{\mathrm{T}}(x) \cdot \mathbf{R}_{fs}^{-1} \mathbf{q}_{fs}. \tag{10.18}$$

The supplementary solution of (10.1) will be accordingly sought in the form of polynomials.

Assume $P_e \neq 0$ and $D_a \neq 0$, we select, without loss of generality, the vector of the interpolation functions as

$$\mathbf{p}_s^T(x) = [p_{s,1}(x) \quad p_{s,2}(x) \quad \cdots \quad p_{s,N_1^s+1}(x)], \quad (10.19)$$

where

$$p_{s,j}(x) = (x/a)^{j-1}, \, j = 1, 2, \cdots, N_1^s + 1, \quad (10.20)$$

and the vector of unknown constants

$$\mathbf{a}_s^T = [G_{s,1} \quad G_{s,2} \quad \cdots \quad G_{s,N_1^s+1}]. \quad (10.21)$$

Thus, the supplementary solution can be expressed as

$$\varphi_s(x) = \mathbf{p}_s^T(x) \cdot \mathbf{a}_s. \quad (10.22)$$

By substituting the supplementary solution into the equation

$$\mathcal{L}_{cdr,1}\varphi_s = f_s \quad (10.23)$$

and equating the coefficients of the like powers on both sides, we are able to obtain

$$\mathbf{R}_s\mathbf{a}_s = \mathbf{a}_{fs}, \quad (10.24)$$

or

$$\mathbf{a}_s = \mathbf{R}_s^{-1}\mathbf{a}_{fs}, \quad (10.25)$$

where the transformation matrix

$$\mathbf{R}_s = \begin{bmatrix} -P_eD_a & P_e/a & -1\cdot2/a^2 & 0 & \cdots & 0 \\ 0 & -P_eD_a & 2P_e/a & -2\cdot3/a^2 & \cdots & 0 \\ \vdots & \vdots & \ddots & \ddots & \ddots & \vdots \\ 0 & 0 & 0 & -P_eD_a & (N_1^s-1)P_e/a & -(N_1^s-1)N_1^s/a^2 \\ 0 & 0 & 0 & \cdots & -P_eD_a & N_1^sP_e/a \\ 0 & 0 & 0 & \cdots & 0 & -P_eD_a \end{bmatrix}. \quad (10.26)$$

Define the vector of basis functions

$$\Phi_s^T(x) = \mathbf{p}_s^T(x) \cdot \mathbf{R}_s^{-1} \mathbf{R}_{fs}^{-1}. \tag{10.27}$$

The supplementary solution $\varphi_s(x)$ can be finally written as

$$\varphi_s(x) = \Phi_s^T(x) \cdot \mathbf{q}_{fs}. \tag{10.28}$$

If $P_e \neq 0$ and $D_a = 0$, the components of $\mathbf{p}_s^T(x)$ in (10.19) will be replaced with

$$p_{s,j}(x) = (x/a)^j, \, j = 1, 2, \cdots, N_1^s + 1, \tag{10.29}$$

and the transformation matrix should be accordingly modified to

$$\mathbf{R}_s = \begin{bmatrix} P_e/a & -1 \cdot 2/a^2 & 0 & \cdots & 0 \\ 0 & 2P_e/a & -2 \cdot 3/a^2 & \cdots & 0 \\ \vdots & \vdots & \ddots & \ddots & \vdots \\ 0 & 0 & \cdots & N_1^s P_e/a & -N_1^s(N_1^s+1)/a^2 \\ 0 & 0 & \cdots & 0 & (N_1^s+1)P_e/a \end{bmatrix}. \tag{10.30}$$

The supplementary solution $\varphi_s(x)$ is still given in the form of (10.28).
If $P_e = 0$ and $D_a = 0$, the elements of $\mathbf{p}_s^T(x)$ shall be replaced with

$$p_{s,j}(x) = (x/a)^{j+1}, \, j = 1, 2, \cdots, N_1^s + 1, \tag{10.31}$$

and the transformation matrix accordingly takes the form of

$$\mathbf{R}_s = \begin{bmatrix} -1 \cdot 2/a^2 & 0 & \cdots & 0 \\ 0 & -2 \cdot 3/a^2 & \cdots & 0 \\ \vdots & \vdots & \ddots & \vdots \\ 0 & 0 & 0 & -(N_1^s+1)(N_1^s+2)/a^2 \end{bmatrix}. \tag{10.32}$$

If the supplementary solution $\varphi_s(x)$ is not desired, we can simply set $N_1^s = 0$ to terminate this selection in the multiscale Fourier series solution.

10.1.4 THE PARTICULAR SOLUTION

Denote the error of the interpolated source function $f_s(x)$ as

$$f_p(x) = f(x) - f_s(x). \tag{10.33}$$

It is evident from (10.10) that

$$f_p(-a) = f_p(a) = 0. \tag{10.34}$$

Expand the error function into Fourier series on the interval $[-a, a]$

$$f_p(x) = \sum_{m=0}^{\infty} \mu_m [V_{fp,1m} \cos(\alpha_m x) + V_{fp,2m} \sin(\alpha_m x)], \tag{10.35}$$

where $\alpha_m = m\pi/a$, $\mu_m = \begin{cases} 1/2 & m = 0 \\ 1 & m > 0 \end{cases}$, and $V_{fp,1m}$ and $V_{fp,2m}$ are the Fourier coefficients.

Because of the condition (10.34), the Fourier series of $f_p(x)$ will be generally better converged than its counterpart for the original source function $f(x)$. This convergence improvement tends to have a positive impact on the subsequent multiscale Fourier series solution, which is another potential benefit for the use of the supplementary solution $\varphi_s(x)$.

In matrix form, (10.35) can be rewritten as

$$f_p(x) = \Phi_0^T(x) \cdot \mathbf{q}_{fp}, \tag{10.36}$$

where the vectors of trigonometric functions $\Phi_0^T(x)$ and the Fourier coefficients $\mathbf{q}_{fp}$ should have become self-explaining.

Suppose that the particular solution of (10.1) is expanded into the Fourier series

$$\varphi_0(x) = \Phi_0^T(x) \cdot \mathbf{q}_0, \tag{10.37}$$

where $\mathbf{q}_0$ is the vector of Fourier coefficients.

By substituting (10.36) and (10.37) into the differential equation

$$\mathcal{L}_{cdr,1} \varphi_0 = f_p, \tag{10.38}$$

we then have

$$[\mathcal{L}_{cdr,1} \Phi_0^T(x)] \cdot \mathbf{q}_0 = \Phi_0^T(x) \cdot \mathbf{q}_{fp}. \tag{10.39}$$

The solution of (10.39) can be simply obtained by using, for instance, the collocation method (CM) as described below.

Let the Fourier series solution $\varphi_0(x)$ be truncated to $m = M$, and $\Upsilon_p = \{x_n, n = 1, 2, \cdots, 2M+1\}$ denote a set of collocation points uniformly distributed on the interval $[-a, a]$. Applying (10.39) sequentially to each of these collocation points leads to

$$\mathbf{R}_{p1} \mathbf{q}_0 = \mathbf{R}_{p2} \mathbf{q}_{fp}, \tag{10.40}$$

where

$$\mathbf{R}_{p1} = \begin{bmatrix} -\boldsymbol{\Phi}_0^{(2)\mathrm{T}}(x_1) + P_e\boldsymbol{\Phi}_0^{(1)\mathrm{T}}(x_1) - P_eD_a\boldsymbol{\Phi}_0^{\mathrm{T}}(x_1) \\ -\boldsymbol{\Phi}_0^{(2)\mathrm{T}}(x_2) + P_e\boldsymbol{\Phi}_0^{(1)\mathrm{T}}(x_2) - P_eD_a\boldsymbol{\Phi}_0^{\mathrm{T}}(x_2) \\ \vdots \\ -\boldsymbol{\Phi}_0^{(2)\mathrm{T}}(x_{2M+1}) + P_e\boldsymbol{\Phi}_0^{(1)\mathrm{T}}(x_{2M+1}) - P_eD_a\boldsymbol{\Phi}_0^{\mathrm{T}}(x_{2M+1}) \end{bmatrix} \tag{10.41}$$

and

$$\mathbf{R}_{p2} = \begin{bmatrix} \boldsymbol{\Phi}_0^{\mathrm{T}}(x_1) \\ \boldsymbol{\Phi}_0^{\mathrm{T}}(x_2) \\ \vdots \\ \boldsymbol{\Phi}_0^{\mathrm{T}}(x_{2M+1}) \end{bmatrix}. \tag{10.42}$$

The Fourier coefficients can then be readily obtained from (10.40)

$$\mathbf{q}_0 = \mathbf{R}_{p1}^{-1}\mathbf{R}_{p2}\mathbf{q}_{fp}. \tag{10.43}$$

Alternatively, the particular solution $\varphi_0(x)$ can be determined from the Fourier coefficient comparison method (FCCM). Again, let M be the truncated number for the Fourier series.

For $m = 0$, we simply have

$$-P_eD_aV_{10} = V_{fp,10}, \tag{10.44}$$

that is,

$$V_{10} = -\frac{1}{P_eD_a} \cdot V_{fp,10}. \tag{10.45}$$

For $m = 1, 2, \cdots, M$, comparing the like terms on the both sides of (10.39) results in

$$\begin{bmatrix} \alpha_m^2 - P_eD_a & P_e\alpha_m \\ -P_e\alpha_m & \alpha_m^2 - P_eD_a \end{bmatrix} \begin{bmatrix} V_{1m} \\ V_{2m} \end{bmatrix} = \begin{bmatrix} V_{fp,1m} \\ V_{fp,2m} \end{bmatrix} \tag{10.46}$$

or

$$\begin{bmatrix} V_{1m} \\ V_{2m} \end{bmatrix} = \begin{bmatrix} \alpha_m^2 - P_eD_a & P_e\alpha_m \\ -P_e\alpha_m & \alpha_m^2 - P_eD_a \end{bmatrix}^{-1} \begin{bmatrix} V_{fp,1m} \\ V_{fp,2m} \end{bmatrix}. \tag{10.47}$$

10.1.5   THE MULTISCALE FOURIER SERIES SOLUTION

With all three constituents, (10.37), (10.9), and (10.28), being available, the multiscale Fourier series solution of the one-dimensional convection-diffusion-reaction equation can now be expressed as

$$
\begin{aligned}
\varphi(x) &= \varphi_0(x) + \varphi_1(x) + \varphi_s(x) \\
&= \boldsymbol{\Phi}_0^{\mathrm{T}}(x) \cdot \mathbf{q}_0 + \boldsymbol{\Phi}_1^{\mathrm{T}}(x) \cdot \mathbf{q}_1 + \boldsymbol{\Phi}_s^{\mathrm{T}}(x) \cdot \mathbf{q}_{fs},
\end{aligned}
\tag{10.48}
$$

where the vectors $\mathbf{q}_0$ and $\mathbf{q}_{fs}$ can be directly calculated from the source functions $f(x)$ and its corresponding interpolation function $f_s(x)$; however, the vector $\mathbf{q}_1$ needs to be determined from the prescribed boundary conditions.

To be specific, we assume that the boundary conditions are as below

$$
\mathbf{q}_b = [\varphi(-a) \quad \varphi(a)]^{\mathrm{T}}.
\tag{10.49}
$$

Substituting (10.48) into (10.49) leads to

$$
\mathbf{q}_1 = \mathbf{R}_f^{-1} \mathbf{q}_b - \mathbf{R}_f^{-1}
\begin{bmatrix}
\boldsymbol{\Phi}_0^{\mathrm{T}}(-a) \cdot \mathbf{q}_0 + \boldsymbol{\Phi}_s^{\mathrm{T}}(-a) \cdot \mathbf{q}_{fs} \\
\boldsymbol{\Phi}_0^{\mathrm{T}}(a) \cdot \mathbf{q}_0 + \boldsymbol{\Phi}_s^{\mathrm{T}}(a) \cdot \mathbf{q}_{fs}
\end{bmatrix},
\tag{10.50}
$$

where

$$
\mathbf{R}_f =
\begin{bmatrix}
\boldsymbol{\Phi}_1^{\mathrm{T}}(-a) \\
\boldsymbol{\Phi}_1^{\mathrm{T}}(a)
\end{bmatrix}.
\tag{10.51}
$$

Making use of (10.50), (10.48) can be rewritten as

$$
\varphi(x) = \boldsymbol{\Phi}_1^{\mathrm{T}}(x) \cdot \mathbf{R}_f^{-1} \mathbf{q}_b + \boldsymbol{\Phi}_0^{\mathrm{T}}(x) \cdot \mathbf{q}_0 + \boldsymbol{\Phi}_s^{\mathrm{T}}(x) \cdot \mathbf{q}_{fs}
$$

$$
- \boldsymbol{\Phi}_1^{\mathrm{T}}(x) \cdot \mathbf{R}_f^{-1}
\begin{bmatrix}
\boldsymbol{\Phi}_0^{\mathrm{T}}(-a) \cdot \mathbf{q}_0 + \boldsymbol{\Phi}_s^{\mathrm{T}}(-a) \cdot \mathbf{q}_{fs} \\
\boldsymbol{\Phi}_0^{\mathrm{T}}(a) \cdot \mathbf{q}_0 + \boldsymbol{\Phi}_s^{\mathrm{T}}(a) \cdot \mathbf{q}_{fs}
\end{bmatrix}.
\tag{10.52}
$$

Denote

$$
\varphi_b(x) = \boldsymbol{\Phi}_b^{\mathrm{T}}(x) \cdot \mathbf{q}_b
\tag{10.53}
$$

and

$$
\varphi_f(x) = \boldsymbol{\Phi}_0^{\mathrm{T}}(x) \cdot \mathbf{q}_0 + \boldsymbol{\Phi}_s^{\mathrm{T}}(x) \cdot \mathbf{q}_{fs} - \boldsymbol{\Phi}_b^{\mathrm{T}}(x) \cdot
\begin{bmatrix}
\boldsymbol{\Phi}_0^{\mathrm{T}}(-a) \cdot \mathbf{q}_0 + \boldsymbol{\Phi}_s^{\mathrm{T}}(-a) \cdot \mathbf{q}_{fs} \\
\boldsymbol{\Phi}_0^{\mathrm{T}}(a) \cdot \mathbf{q}_0 + \boldsymbol{\Phi}_s^{\mathrm{T}}(a) \cdot \mathbf{q}_{fs}
\end{bmatrix},
\tag{10.54}
$$

where

$$\Phi_b^T(x) = \Phi_1^T(x) \cdot \mathbf{R}_f^{-1}. \tag{10.55}$$

The multiscale Fourier series solution of the one-dimensional convection-diffusion-reaction equation can be finally obtained as

$$\varphi(x) = \varphi_b(x) + \varphi_f(x). \tag{10.56}$$

10.2   ONE-DIMENSIONAL NUMERICAL EXAMPLES

The convection-diffusion-reaction equation will now be solved for various combinations of model parameters and boundary conditions. Table 10.3 shows the parameter setups in each of the test groups specifically used to evaluate the convergence characteristics and accuracies of the multiscale Fourier series solutions. These parameters are typically related to either the physical process itself (such as those directly specified via the convection-diffusion-reaction equation and the boundary conditions), or the solution procedures (such as, the order of the interpolation polynomials, the approaches taken in deriving the discrete system equations, etc.). The tests in the first group are specifically used to investigate if and how the convergence characteristics of the multiscale Fourier series solutions are affected by the mathematical processes (e.g., the FCCM and the CM) employed in deriving the final systems. In the second test group, our attentions are focused

---

TABLE 10.3

The Division of Test Groups According to the Model Parameters and Solution Options

Test Groups	Cases	Boundary Conditions	Model Parameters	Orders of Interpolation Polynomials	Solution Schemes
1	a	DD	$P_e = 3, D_a = 90$	$N_1^s = 0$	FCCM
	b	$\varphi(-a) = 1, \varphi(a) = 0$			CM
2	a	DD	$P_e = 3, D_a = 90$	$N_1^s = 0$	FCCM
	b	$\varphi(-a) = 1, \varphi(a) = 0$		$N_1^s = 1$	
	c			$N_1^s = 2$	
3	a	DD	$P_e = 3, D_a = 90$	$N_1^s = 0$	FCCM
	b	$\varphi(-a) = 1, \varphi(a) = 0$	$P_e = 1, D_a = 30$		
	c		$P_e = 30, D_a = 1$		
	d		$P_e = 200, D_a = -1$		
4	a	DD	$P_e = 3, D_a = 90$	$N_1^s = 0$	FCCM
		$\varphi(-a) = 1, \varphi(a) = 0$			
	b	DN			
		$\varphi(-a) = 1, \varphi^{(1)}(a) = 1$			

on the effects of the orders of the interpolation polynomials used to approximate the source function. The third test group is dedicated to studying the impacts on the convergence characteristics of the model parameters (i.e., P_e and D_a) of the convection-diffusion-reaction equation when the corresponding physical states vary from the strong reaction type to the reaction-dominated type, then to the convection-dominated type, and eventually to the strong convection type. Finally, in the fourth test group, we will consider the effects of the boundary conditions on the convergence characteristics of the multiscale Fourier series solutions.

In all these tests, the source function remains to be the same, and is given by the following third order polynomial

$$f(x) = 10^3 + 2 \times 10^3 \frac{x}{a} + 5 \times 10^3 \left(\frac{x}{a}\right)^2 + 10^4 \left(\frac{x}{a}\right)^3, x \in [-a, a]. \qquad (10.57)$$

Thus, when the order of the interpolation polynomial of the source function is set equal to 3, the corresponding vector of the Fourier coefficients

$$\mathbf{q}_0 = \mathbf{0}, \qquad (10.58)$$

and the multiscale Fourier series solution given by (10.56) simply "degenerates" into the exact solution (of the convection-diffusion-reaction equation), which will be subsequently used as the reference in calculating the numerical errors associated with the solutions obtained in other ways.

10.2.1 CONVERGENCE CHARACTERISTICS

The first case in group 1 is considered as the baseline in the convergence studies. The internal and boundary approximation errors of the function $\varphi(x)$ and its first and second order derivatives are displayed in Figure 10.1 against the truncation number: $M = 2, 3, 5, 10, 20, 30,$ and 40.

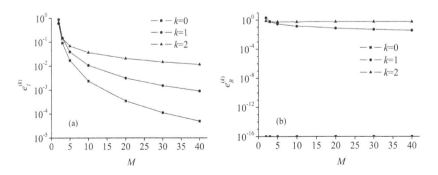

FIGURE 10.1 The convergence characteristics of multiscale Fourier series solutions of one-dimensional convection-diffusion-reaction equation: (a) $e_I^{(k)}(\varphi_M)$-M curves and (b) $e_B^{(k)}(\varphi_M)$-M curves.

With the increase of the number of terms retained in the multiscale Fourier series solution, the internal error of $\varphi(x)$ decreases at an impressive speed. Take the error index $e_I^{(0)}(\varphi_M)$ at $M = 10$ for example. Each time the truncation number M is doubled, the error index is reduced approximately by an order of magnitude. In the meantime, it is interesting to notice that the boundary approximation $e_B^{(0)}(\varphi_M)$ is error-free or extremely small (below 1.0E–16), even for small truncation numbers.

The internal approximation of $\varphi^{(1)}(x)$ also shows a decent improvement as the truncation number increases. Even though the convergence rate of $\varphi^{(1)}(x)$ has slowed down noticeably as compared with $\varphi(x)$, $e_I^{(1)}(\varphi_{40})$ is still smaller than $e_I^{(1)}(\varphi_{10})$ by an order of magnitude. The internal approximation of $\varphi^{(2)}(x)$ deteriorates furthermore, but still displays a satisfactory trending as the truncation number increases.

The boundary error of $\varphi^{(1)}(x)$ reduces slowly with the truncation number: the error index is approximately equal to 1.0E–1 for $M = 10$, and keeps a trend of sustained decreasing to about 1.0E–2 for $M = 40$. In contrast, the boundary error with $\varphi^{(2)}(x)$ is almost unaffected by the truncation number, indicating a possible failure of the multiscale Fourier series in approximating the boundary values of the second order derivative. However, this shall not be automatically perceived as problematic because the boundary conditions only involve with the derivatives of orders lower than that of the differential equation (i.e., $2r = 2$ in the current problem); thus, examining the boundary error of the second order derivative perhaps represents an over-reach of the current solution.

The first test group is designated to compare the efficacy of the approaches, the FCCM and the CM, which are adopted in deriving the final system of discrete equations as discussed in Sec. 9.8. The convergence characteristics of the resulting multiscale Fourier series solutions are compared in Figure 10.2 (the FCCM results are actually the reproductions of those previously shown in Figure 10.1). It is observed that, while the convergence trends (or more explicitly, the convergence rates) for both of them are very similar, the internal approximation errors with the FCCM solution are noticeably much smaller than their CM counterparts. In the meantime, the boundary errors are roughly the same for both solutions. Based on these observations, the FCCM-based solution method is here considered the preferred choice, and will be invariably used in the subsequent testing cases.

Now, we turn to the cases in the second test group to examine the effects of the orders of the polynomials used to interpolate the source function. The convergence characteristics of the corresponding multiscale Fourier series solutions are presented in Figure 10.3.

It is seen that by using the interpolated source functions, the corresponding multiscale Fourier series solutions all have demonstrated remarkable improvements over their counterparts obtained directly from the original source function over the entire solution interval (internal plus boundaries). In particular, the internal errors are considerably reduced with the increase of the order of the interpolation polynomials. While we don't wish to speculate too much in regards to the possible reasons, the use of the interpolation polynomials will certainly help

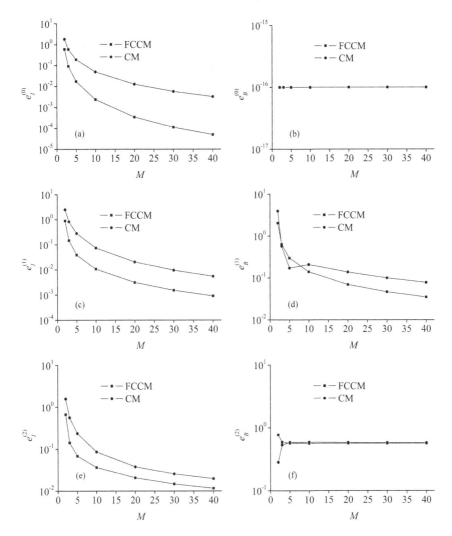

FIGURE 10.2 Convergence comparison of the multiscale Fourier series solutions with different approaches in deriving the discrete system of equations (FCCM and CM): (a) $e_I^{(0)}(\varphi_M)$-M curves; (b) $e_B^{(0)}(\varphi_M)$ -M curves; (c) $e_I^{(1)}(\varphi_M)$-M curves; (d) $e_B^{(1)}(\varphi_M)$ -M curves; (e) $e_I^{(2)}(\varphi_M)$-M curves; and (f) $e_B^{(2)}(\varphi_M)$-M curves.

improve the continuity of the (periodically extended) source function f_p (refer to (10.34)), and hence the convergence of the Fourier series (10.35) and the final solution (10.56).

Perhaps, the significance of using the interpolated source function is more meaningfully manifested in the substantial reduction of the approximation errors both in the interior and boundary regions. Also, the degree of the improvement is seen to increase with the orders of the interpolation polynomials. Even though

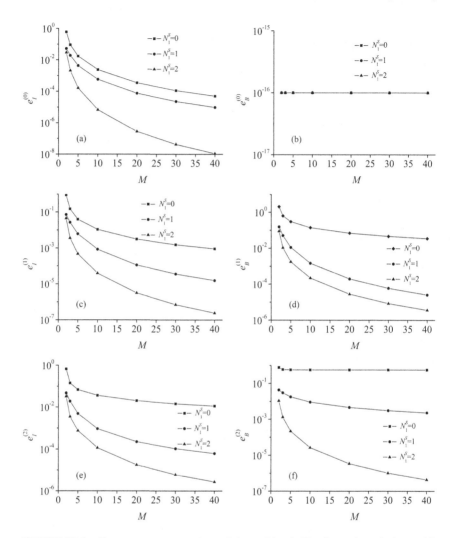

FIGURE 10.3 Convergence comparison of the multiscale Fourier series solutions with different orders of interpolation polynomials of source function: (a) $e_I^{(0)}(\varphi_M)$-M curves; (b) $e_B^{(0)}(\varphi_M)$-M curves; (c) $e_I^{(1)}(\varphi_M)$-M curves; (d) $e_B^{(1)}(\varphi_M)$-M curves; (e) $e_I^{(2)}(\varphi_M)$-M curves; and (f) $e_B^{(2)}(\varphi_M)$-M curves.

the interpolation polynomials are originally introduced to better capture the multiscale characteristic of a boundary value problem, we've already had a pleasant surprise here.

The next group of tests involves investigating the behaviors of the multiscale Fourier series solutions under various physical states: the strong reaction type ($P_e = 3$, $D_a = 90$), the reaction-dominated type ($P_e = 1$, $D_a = 30$), the convection-dominated type ($P_e = 30$, $D_a = 1$), and the strong convection type ($P_e = 200$, $D_a = -1$).

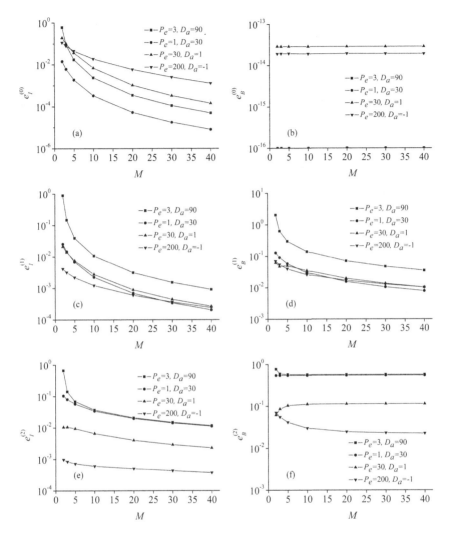

FIGURE 10.4 Convergence comparison of the multiscale Fourier series solutions with different model parameters P_e and D_a: (a) $e_I^{(0)}(\varphi_M)$-M curves; (b) $e_B^{(0)}(\varphi_M)$-M curves; (c) $e_I^{(1)}(\varphi_M)$-M curves; (d) $e_B^{(1)}(\varphi_M)$-M curves; (e) $e_I^{(2)}(\varphi_M)$-M curves; and (f) $e_B^{(2)}(\varphi_M)$-M curves.

As shown in Figure 10.4, the remarkable convergence rates (the slopes) of the multiscale Fourier series solutions $\varphi(x)$ and $\varphi^{(1)}(x)$ are almost unaffected by the model parameters. In comparison, the approximation errors can vary considerably, even by orders of magnitude, as the parameters change.

The convergence of the multiscale Fourier series of $\varphi(x)$ and $\varphi^{(1)}(x)$ are considered very satisfactory at the boundaries of the solution domain. Moreover, the approximation errors of $\varphi(x)$ are extremely small (all below 1.0E–13) regardless of the values of the truncation numbers. However, the same cannot be said about the multiscale Fourier series solution $\varphi^{(2)}(x)$, even though the situation has improved for the case of $P_e = 200$ and $D_a = -1$.

The final group of tests is focused on the effects of the boundary conditions. The convergence characteristics of the multiscale Fourier series solution are demonstrated in Figure 10.5 for two different types of boundary conditions. It is seen that when the boundary condition at $x = a$ is changed from $\varphi(a) = 0$ to $\varphi^{(1)}(a) = 1$,

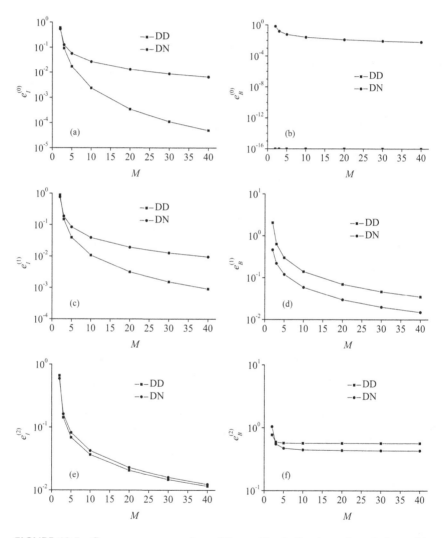

FIGURE 10.5 Convergence comparison of the multiscale Fourier series solutions with different boundary conditions: (a) $e_I^{(0)}(\varphi_M)$-M curves; (b) $e_B^{(0)}(\varphi_M)$-M curves; (c) $e_I^{(1)}(\varphi_M)$-M curves; (d) $e_B^{(1)}(\varphi_M)$-M curves; (e) $e_I^{(2)}(\varphi_M)$-M curves; and (f) $e_B^{(2)}(\varphi_M)$-M curves.

the convergence of the multiscale Fourier series solution $\varphi(x)$ has become noticeably deteriorated over the entire solution domain. In particular, the boundary error $e_B^{(0)}(\varphi_M)$ increases from 1.0E–16 to 1.0E–2 at $M = 40$, but a trending of converging is still clearly presented. A similar deterioration is also observed for $\varphi^{(1)}(x)$. In the meantime, the convergence characteristics and approximation errors of $\varphi^{(2)}(x)$ are kind of insensitive to the specified boundary conditions throughout the solution domain; we can speculate, if we have to, that the "error" with $\varphi^{(1)}(a) = 1$ has indiscriminately spilled over the entire solution domain.

10.2.2 COMPUTATIONAL EFFICIENCY

To check on the computational efficiency, we will compare the multiscale Fourier series method with the bubble function method (Franca and Russo 1997) by solving this one-dimensional convection-diffusion-reaction problem. The third test group in Table 10.3 is used for the comparative studies. Two types of source functions are considered:

1. Constant distribution

$$f(x) = 10^3, x \in [-a,a]; \tag{10.59}$$

2. Linear distribution

$$f(x) = 10^3 + 2 \times 10^3 \frac{x}{a}, x \in [-a,a]. \tag{10.60}$$

Therefore, this is a total of eight different cases considered here.

In all the following calculations, the multiscale Fourier series solution is truncated to $M = 40$, and the solution domain is divided into 100 uniform meshes in the bubble function method. Thus, the number of unknowns is 83 in the multiscale Fourier series method, and 101 in the bubble function method; that is, the final system in the multiscale Fourier series method is slightly smaller than that in the bubble function method. For convenience, the model parameters and the corresponding results are summarized in Table 10.4. It should be noted that the errors for both solution methods are calculated at the 10001 sampling points evenly distributed in the solution domain.

For each of the 8 cases, the CPU time is 2 seconds for the multiscale Fourier series solution in comparison with 49 seconds for the bubble function solution. It is true that the CPU times can actually be influenced by the programming skills and some other known/unknown factors. However, the increase of computational efficiency by almost 25 times should be considered meaningful by any measure.

In the meantime, the numerical accuracies of the multiscale Fourier series solutions are also observed far better, by orders of magnitude, than the reference (the bubble function) solutions. Besides, the multiscale Fourier series method has exhibited an excellent robustness and unity in dealing with different physical

TABLE 10.4

Comparison of the Computational Efficiencies of the Multiscale Fourier Series Method and the Bubble Function Method

Method	Number of Unknowns	CPU Time(s)	Source Function	Model Parameters	Solution Errors $\varphi(x)$	$\varphi^{(1)}(x)$	$\varphi^{(2)}(x)$
Multiscale Fourier series method	83	2	Constant	$P_e = 3, D_a = 90$	1.0849E–13	9.8395E–14	1.1072E–13
				$P_e = 1, D_a = 30$	1.8423E–13	1.9183E–13	1.9945E–13
				$P_e = 30, D_a = 1$	1.6526E–13	2.0002E–14	1.2077E–14
				$P_e = 200, D_a = -1$	1.8946E–13	3.1784E–15	1.6337E–15
Multiscale Fourier series method	83	2	Linear	$P_e = 3, D_a = 90$	3.1502E–05	8.6717E–04	1.2883E–02
				$P_e = 1, D_a = 30$	3.4250E–06	1.3947E–04	7.7467E–03
				$P_e = 30, D_a = 1$	4.1961E–05	9.9038E–05	9.5799E–04
				$P_e = 200, D_a = -1$	5.6963E–04	1.1687E–04	1.8354E–04
Bubble function method	101	49	Constant	$P_e = 3, D_a = 90$	2.2078E–04	3.8535E–03	1.0692E–01
				$P_e = 1, D_a = 30$	1.0400E–04	8.3772E–03	7.0050E–01
				$P_e = 30, D_a = 1$	1.6512E–04	1.4376E–03	1.8817E–02
				$P_e = 200, D_a = -1$	2.2520E–03	3.3312E–03	6.9497E–03
Bubble function method	101	49	Linear	$P_e = 3, D_a = 90$	2.9823E–02	5.9332E–02	4.0030E–01
				$P_e = 1, D_a = 30$	3.9407E–02	1.2020E–01	2.6545E–01
				$P_e = 30, D_a = 1$		Failure	
				$P_e = 200, D_a = -1$		Failure	

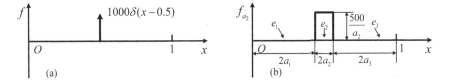

FIGURE 10.6 The Green's function type of source function: (a) theoretical description and (b) approximate description.

states as dictated by the model parameters and/or source functions, in view of the fact that the bubble function method has actually failed in the last two cases. Thus, it is fair to claim that the multiscale Fourier series method has convincingly outperformed the bubble function method with respect to the computational efficiency and numerical accuracy.

10.2.3 MULTISCALE CHARACTERISTICS

In this section, we use a Green's function type of source to amplify the multiscale behavior of the one-dimensional convection-diffusion-reaction equation and to examine if the multiscale Fourier series solution can react responsively and appropriately.

As shown in Figure 10.6(a), the source function is theoretically specified as

$$f(x) = 1000\delta(x - 0.5), \ x \in [0,1]. \tag{10.61}$$

In the numerical calculations, this Green's function type of source function is approximated by a piecewise function. For instance, the source function can be treated as a rectangular impulse of width $2a_2$ and height $500/a_2$, and being zero elsewhere, as illustrated in Figure 10.6(b). Accordingly, as listed in Table 10.5, we will here consider two different solution schemes: the whole interval solution and the subinterval solution. In the whole interval solution scheme, the

---

TABLE 10.5
Two Different Solution Schemes Used to Solve the One-Dimensional Convection-Diffusion-Reaction Equation with the Green's Function Type of Source Function

Description of the Problem		Solution Schemes	
Boundary Conditions	**Model Parameters**	**Whole Interval Solution**	**Subinterval Solution**
DD	$P_e = 3, D_a = 90$	$N_1^s = 0$	$N_1^s = 1$
$\varphi(0) = 1, \varphi(1) = 0$	$P_e = 1, D_a = 30$	FCCM	$q_0 = 0$
	$P_e = 30, D_a = 1$		
	$P_e = 200, D_a = -1$		

solution interval is first mapped from $[0,1]$ onto $[-1/2,1/2]$ and the convection-diffusion-reaction equation is then solved directly by setting $a = 1/2$ in the multiscale Fourier series method without using the interpolation polynomial of the source function. In the subinterval solution scheme, the convection-diffusion-reaction equation is solved sequentially on each of the subintervals, $e_1 = [0, 2a_1]$, $e_2 = [2a_1, 2a_1 + 2a_2]$, and $e_3 = [2a_1 + 2a_2, 1]$, by using the multiscale Fourier series method with the introduction of the first order interpolation polynomial of the source function. Evidently, this will lead to an accurate solution in each subinterval. It is clear from Figure 10.6(b) that the given Green's function type of source function is the limiting case of the functions $f_{a_2}(x)$, as the parameter a_2 approaches 0 from above.

In the subinterval solution scheme, the end points of the subintervals are globally labeled as 1, 2, 3, and 4, and the values of the solutions at these end points as φ_1, φ_2, φ_3, and φ_4 accordingly. According to (10.56), the multiscale Fourier series solutions within the subintervals e_1, e_2, and e_3 are then respectively given, in terms of the local coordinates, as

$$\varphi(x) = \Phi_{b,e_1}^{T}(x)[\varphi_1 \; \varphi_2]^{T} + \varphi_{f,e_1}(x), x \in [-a_1, a_1], \qquad (10.62)$$

$$\varphi(x) = \Phi_{b,e_2}^{T}(x)[\varphi_2 \; \varphi_3]^{T} + \varphi_{f,e_2}(x), x \in [-a_2, a_2], \qquad (10.63)$$

and

$$\varphi(x) = \Phi_{b,e_3}^{T}(x)[\varphi_3 \; \varphi_4]^{T} + \varphi_{f,e_3}(x), x \in [-a_3, a_3]; \qquad (10.64)$$

their first order derivatives are as follows:

$$\varphi^{(1)}(x) = \Phi_{b,e_1}^{(1)T}(x)[\varphi_1 \; \varphi_2]^{T} + \varphi_{f,e_1}^{(1)}(x), x \in [-a_1, a_1], \qquad (10.65)$$

$$\varphi^{(1)}(x) = \Phi_{b,e_2}^{(1)T}(x)[\varphi_2 \; \varphi_3]^{T} + \varphi_{f,e_2}^{(1)}(x), x \in [-a_2, a_2], \qquad (10.66)$$

and

$$\varphi^{(1)}(x) = \Phi_{b,e_3}^{(1)T}(x)[\varphi_3 \; \varphi_4]^{T} + \varphi_{f,e_3}^{(1)}(x), x \in [-a_3, a_3]. \qquad (10.67)$$

The values of the solution at the first and fourth end points are actually the specified boundary conditions and those at the second and third end points shall satisfy the compatibility conditions, namely,

$$\left. \begin{array}{c} \varphi_1 = 1 \\ \Phi_{b,e_1}^{(1)T}(a_1)[\varphi_1 \; \varphi_2]^{T} + \varphi_{f,e_1}^{(1)}(a_1) = \Phi_{b,e_2}^{(1)T}(-a_2)[\varphi_2 \; \varphi_3]^{T} + \varphi_{f,e_2}^{(1)}(-a_2) \\ \Phi_{b,e_2}^{(1)T}(a_2)[\varphi_2 \; \varphi_3]^{T} + \varphi_{f,e_2}^{(1)}(a_2) = \Phi_{b,e_3}^{(1)T}(-a_3)[\varphi_3 \; \varphi_4]^{T} + \varphi_{f,e_3}^{(1)}(-a_3) \\ \varphi_4 = 0 \end{array} \right\}. \qquad (10.68)$$

Hence, we determine φ_2 and φ_3, the values of the solution at the intermediate end points, and eventually the solution function on each subinterval (or the entire interval [0,1]) by using (10.62)–(10.64).

In order to converge to the original solution for the Green's function type of source distribution, the whole interval and subinterval solutions will potentially involve two different limit processes of $M \to \infty$ and $a_2 \to 0^+$, respectively.

The convergence characteristics of the solutions corresponding to these two limit processes are displayed in Figures 10.7–10.10 for different sets of computational parameters P_e and D_a. The multiscale behaviors of the solutions are here exhibited in the forms of, such as, glitches, knife edges, sharp gradients, and other irregularities in the solution interval, especially at/near the center where the source is located.

For instance, in the case of $P_e = 3$ and $D_a = 90$, a small glitch appears with $\varphi^{(1)}(x)$ at the action point ($x = 0.5$) of the source function. For $P_e = 1$ and $D_a = 30$, a peak and a steep gradient are respectively observed for $\varphi(x)$ and $\varphi^{(1)}(x)$ near

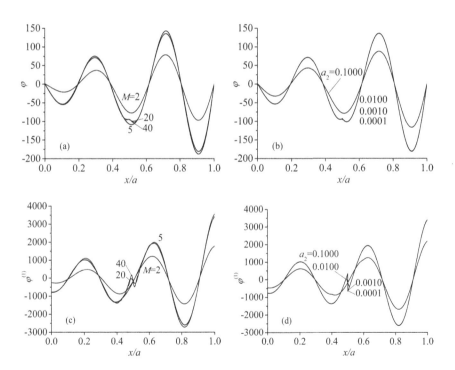

FIGURE 10.7 Convergence processes for the whole interval solution and the subinterval solution with model parameters $P_e = 3$ and $D_a = 90$: (a) φ curves for the whole interval solution with different terms of M; (b) φ curves for the subinterval solution with different values of a_2; (c) $\varphi^{(1)}$ curves for the whole interval solution with different terms of M; and (d) $\varphi^{(1)}$ curves for the subinterval solution with different values of a_2.

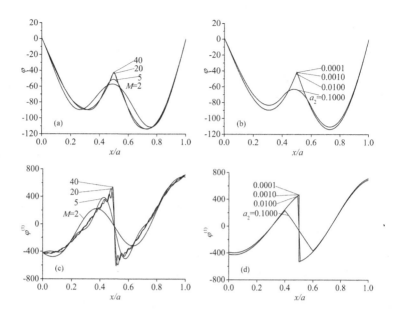

FIGURE 10.8 Convergence processes for the whole interval solution and the subinterval solution with computational parameters $P_e = 1$ and $D_a = 30$: (a) φ curves for the whole interval solution with different terms of M; (b) φ curves for the subinterval solution with different values of a_2; (c) $\varphi^{(1)}$ curves for the whole interval solution with different terms of M; and (d) $\varphi^{(1)}$ curves for the subinterval solution with different values of a_2.

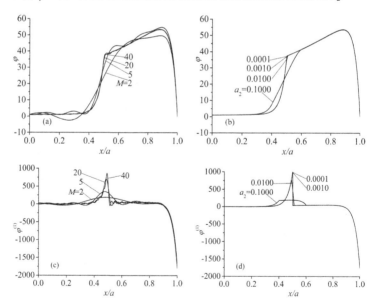

FIGURE 10.9 Convergence processes for the whole interval solution and the subinterval solution with computational parameters $P_e = 30$ and $D_a = 1$: (a) φ curves for the whole interval solution with different terms of M; (b) φ curves for the subinterval solution with different values of a_2; (c) $\varphi^{(1)}$ curves for the whole interval solution with different terms of M; and (d) $\varphi^{(1)}$ curves for the subinterval solution with different values of a_2.

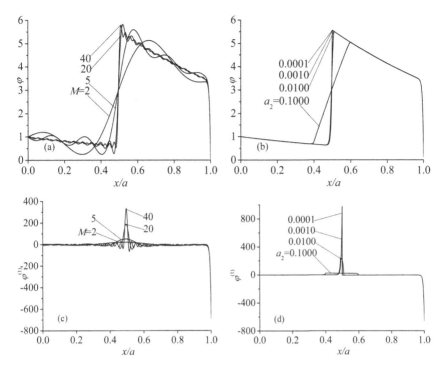

FIGURE 10.10 Convergence processes for the whole interval solution and the subinterval solution with computational parameters $P_e = 200$ and $D_a = -1$: (a) φ curves for the whole interval solution with different terms of M; (b) φ curves for the subinterval solution with different values of a_2; (c) $\varphi^{(1)}$ curves for the whole interval solution with different terms of M; and (d) $\varphi^{(1)}$ curves for the subinterval solution with different values of a_2.

the action point of the source. For $(P_e = 30, D_a = 1)$ and $(P_e = 200, D_a = -1)$, the multiscale characteristics of the solutions are manifested in the sharp peaks with $\varphi^{(1)}(x)$ at the source location, and the steep gradients with both $\varphi(x)$ and $\varphi^{(1)}(x)$ at the source location and in the close proximity of the end point $(x = 1)$ of the interval as well.

10.3 MULTISCALE FOURIER SERIES SOLUTION FOR TWO-DIMENSIONAL CONVECTION-DIFFUSION-REACTION EQUATION

The two-dimensional convection-diffusion-reaction equation is a second order $(2r = 2)$ linear differential equation with constant coefficients, wherein convection specifically refers to the terms involving the first order partial derivatives of the solution function. In what follows, we will apply the multiscale full-range Fourier series method to this convection-diffusion-reaction problem with necessary details.

10.3.1 DESCRIPTION OF THE PROBLEM

The two-dimensional convection-diffusion-reaction equation in the domain $[-a,a] \times [-b,b]$ can be written in the dimensionless form as (Hauke 2002)

$$\mathcal{L}_{cdr,2}\varphi = f, \tag{10.69}$$

where the differential operator

$$\mathcal{L}_{cdr,2} = -\left(\frac{\partial^2}{\partial x^2} + \frac{\partial^2}{\partial y^2}\right) + P_{e1}\frac{\partial}{\partial x} + P_{e2}\frac{\partial}{\partial y} - P_e D_a, \tag{10.70}$$

$\varphi(x,y)$ is the solution function, $f(x,y)$ is the given source distribution, P_{e1} and P_{e2} are respectively the x- and y-component of the dimensionless Peclet number P_e, and D_a is the dimensionless Damkohler number.

The parameters P_{e1} and P_{e2} are not actually independent:

$$P_{e1} = P_e \cos\theta, \; P_{e2} = P_e \sin\theta, \tag{10.71}$$

where θ is the inflow angle.

At the boundary, such as $x = a$, the boundary condition is prescribed as:

1. Dirichlet boundary condition (or, D type, for short)

$$\varphi(a,y) = \bar{\varphi}_a(y), \tag{10.72}$$

or

2. Neumann boundary condition (or, N type, for short)

$$\varphi^{(1,0)}(a,y) = \bar{\phi}_a(y), \tag{10.73}$$

where $\bar{\varphi}_a(y)$ and $\bar{\phi}_a(y)$ are respectively the prescribed distributions of the solution function and its first order partial derivative at the boundary $x = a$.

10.3.2 THE GENERAL SOLUTION

Suppose that (10.69) has a homogeneous solution

$$p_{1n,H}(x,y) = \exp(\eta_n x)\exp(i\beta_n y), \tag{10.74}$$

where n is a nonnegative integer, η_n is an unknown constant, $\beta_n = n\pi/b$ and $i = \sqrt{-1}$.

Substituting (10.74) into the homogeneous form of (10.69) will lead to the characteristic equation

$$\eta_n^2 - P_{e1}\eta_n + (-\beta_n^2 + P_e D_a - iP_{e2}\beta_n) = 0. \tag{10.75}$$

Set

$$\gamma_{1n} = -P_{e1}^2 - 4\beta_n^2 + 4P_e D_a, \gamma_{2n} = -4P_{e2}\beta_n,$$

$$\alpha_{1n} = -\frac{P_{e1}}{2}, \text{ and } \alpha_{2n} + i\alpha_{3n} = \frac{1}{2}\sqrt{\gamma_{1n} + i\gamma_{2n}}. \tag{10.76}$$

Then the characteristic roots are given as

$$\eta_{n,1} = (-\alpha_{1n} + \alpha_{3n}) - i\alpha_{2n} \text{ and } \eta_{n,2} = (-\alpha_{1n} - \alpha_{3n}) + i\alpha_{2n}. \tag{10.77}$$

For $n = 0$, (10.75) has a double real root if $\gamma_{1n} = 0$, two distinct complex roots if $\gamma_{1n} > 0$, or two distinct real roots if $\gamma_{1n} < 0$. For $n > 0$, (10.75) has one double real root if $\gamma_{1n} = 0$ and $\gamma_{2n} = 0$, or two distinct complex/real roots otherwise.

The corresponding homogeneous solutions are summarized in Table 10.6 for $n = 0$, and in Table 10.7 for $n > 0$.

Referring back to Sec. 9.4.2, we can now substitute the original basis functions in (9.96) and (9.114) with the above obtained homogeneous solutions $p_{1nl,H}(x,y)$ and $p_{10l,H}(x,y)$, respectively:

$$\mathbf{p}_{1n,H}^T(x,y) = [p_{1n1,H}(x,y) \quad p_{1n2,H}(x,y) \quad p_{1n3,H}(x,y) \quad p_{1n4,H}(x,y)], \tag{10.78}$$

and

$$\mathbf{p}_{10}^T(x) = [p_{101,H}(x) \quad p_{102,H}(x)]. \tag{10.79}$$

TABLE 10.6

Expressions for the Homogeneous Solutions $p_{10l,H}(x,y)$, $l = 1, 2$

	$\gamma_{1n} = 0$	$\gamma_{1n} > 0$	$\gamma_{1n} < 0$
$p_{101,H}(x,y)$	$\dfrac{\exp(-\alpha_{1n}x)}{\cosh(-\alpha_{1n}a)}$	$\dfrac{\exp(-\alpha_{1n}x)}{\cosh(-\alpha_{1n}a)}\cos(\alpha_{2n}x)$	$\dfrac{\exp[(-\alpha_{1n} + \alpha_{3n})x]}{\cosh[(-\alpha_{1n} + \alpha_{3n})a]}$
$p_{102,H}(x,y)$	$\dfrac{x\ \exp(-\alpha_{1n}x)}{a\ \cosh(-\alpha_{1n}a)}$	$\dfrac{\exp(-\alpha_{1n}x)}{\cosh(-\alpha_{1n}a)}\sin(\alpha_{2n}x)$	$\dfrac{\exp[(-\alpha_{1n} - \alpha_{3n})x]}{\cosh[(-\alpha_{1n} - \alpha_{3n})a]}$

TABLE 10.7

Expressions for the Homogeneous Solutions $p_{1nl,H}(x,y)$, $l = 1, 2, 3, 4$

	$\gamma_{1n} = 0$ and $\gamma_{2n} = 0$	$\gamma_{1n} \neq 0$ or $\gamma_{2n} \neq 0$
$p_{1n1,H}(x,y)$	$\dfrac{\exp(-\alpha_{1n}x)}{\cosh(-\alpha_{1n}a)}\cos(\beta_n y)$	$\dfrac{\exp[(-\alpha_{1n}+\alpha_{3n})x]}{\cosh[(-\alpha_{1n}+\alpha_{3n})a]}\cos(\beta_n y - \alpha_{2n}x)$
$p_{1n2,H}(x,y)$	$\dfrac{x\ \exp(-\alpha_{1n}x)}{a\ \cosh(-\alpha_{1n}a)}\cos(\beta_n y)$	$\dfrac{\exp[(-\alpha_{1n}+\alpha_{3n})x]}{\cosh[(-\alpha_{1n}+\alpha_{3n})a]}\sin(\beta_n y - \alpha_{2n}x)$
$p_{1n3,H}(x,y)$	$\dfrac{\exp(-\alpha_{1n}x)}{\cosh(-\alpha_{1n}a)}\sin(\beta_n y)$	$\dfrac{\exp[(-\alpha_{1n}-\alpha_{3n})x]}{\cosh[(-\alpha_{1n}-\alpha_{3n})a]}\cos(\beta_n y + \alpha_{2n}x)$
$p_{1n4,H}(x,y)$	$\dfrac{x\ \exp(-\alpha_{1n}x)}{a\ \cosh(-\alpha_{1n}a)}\sin(\beta_n y)$	$\dfrac{\exp[(-\alpha_{1n}-\alpha_{3n})x]}{\cosh[(-\alpha_{1n}-\alpha_{3n})a]}\sin(\beta_n y + \alpha_{2n}x)$

Based on the procedures described in Sec. 9.4.2, the general solution corresponding to the boundary functions $\varphi_1(x,y)$ is obtained as

$$\varphi_1(x,y) = \Phi_1^T(x,y)\cdot\mathbf{q}_1, \qquad (10.80)$$

where the vector of basis function $\Phi_1^T(x,y)$ and the vector of boundary Fourier coefficients $\mathbf{q}_1^T$ are defined by (8.62)–(8.64) and (8.68)–(8.70).

In a similar way, the general solution corresponding to the boundary functions $\varphi_2(x,y)$ can be written as

$$\varphi_2(x,y) = \Phi_2^T(x,y)\cdot\mathbf{q}_2, \qquad (10.81)$$

where the vector of basis function $\Phi_2^T(x,y)$, and the vector of boundary Fourier coefficients $\mathbf{q}_2^T$ are as discussed in Sec. 8.2.1.

10.3.3 THE SUPPLEMENTARY SOLUTION

Let N_1^s and N_2^s be positive integers and $\Upsilon^s = \left\{(x_{n_1}, y_{n_2}), n_1 = 1, 2, \cdots, N_1^s + 1, n_2 = 1, 2, \cdots, N_2^s + 1\right\}$ be a set of interpolation points uniformly distributed over the domain $[-a,a]\times[-b,b]$. Now, we use a polynomial function $f_s(x,y)$ to approximate the distributed source $f(x,y)$ such that

$$f_s(x_{n_1}, y_{n_2}) = f(x_{n_1}, y_{n_2}), n_1 = 1, 2, \cdots, N_1^s + 1, n_2 = 1, 2, \cdots, N_2^s + 1. \qquad (10.82)$$

Correspondingly, the vector of polynomial functions can be selected as

$$\mathbf{p}_{fs}^T(x,y) = [p_{fs,11}(x,y) \quad \cdots \quad p_{fs,jl}(x,y) \quad \cdots \quad], \qquad (10.83)$$

where

$$p_{fs,jl}(x,y) = (x/a)^{j-1}(y/b)^{l-1}, \ j = 1, 2, \cdots, N_1^s + 1, l = 1, 2, \cdots, N_2^s + 1. \quad (10.84)$$

Then the interpolated source function $f_s(x,y)$ can be expressed as

$$f_s(x,y) = \mathbf{p}_{fs}^{\mathrm{T}}(x,y) \cdot \mathbf{a}_{fs}, \quad (10.85)$$

where

$$\mathbf{a}_{fs}^{\mathrm{T}} = [H_{fs,11} \quad \cdots \quad H_{fs,jl} \quad \cdots \quad] \quad (10.86)$$

is the vector of unknown constants.

Substitution of (10.85) into (10.82) results in

$$\mathbf{R}_{fs}\mathbf{a}_{fs} = \mathbf{q}_{fs}, \quad (10.87)$$

where

$$\mathbf{R}_{fs} = \begin{bmatrix} \mathbf{p}_{fs}^{\mathrm{T}}(x_1, y_1) \\ \vdots \\ \mathbf{p}_{fs}^{\mathrm{T}}(x_{n_1}, y_{n_2}) \\ \vdots \\ \mathbf{p}_{fs}^{\mathrm{T}}\left(x_{N_1^s+1}, y_{N_2^s+1}\right) \end{bmatrix} \quad (10.88)$$

and

$$\mathbf{q}_{fs}^{\mathrm{T}} = \left[f(x_1, y_1) \quad \cdots \quad f(x_{n_1}, y_{n_2}) \quad \cdots \quad f\left(x_{N_1^s+1}, y_{N_2^s+1}\right) \right]. \quad (10.89)$$

Thus, the polynomial function $f_s(x,y)$ is given as

$$f_s(x,y) = \mathbf{p}_{fs}^{\mathrm{T}}(x,y) \cdot \mathbf{R}_{fs}^{-1}\mathbf{q}_{fs}. \quad (10.90)$$

Take the case of $P_e \neq 0$ and $D_a \neq 0$ for example. The supplementary solution of (10.69) will also be sought in the form of polynomials.

First, select the vector of functions as

$$\mathbf{p}_s^{\mathrm{T}}(x,y) = [p_{s,11}(x,y) \quad \cdots \quad p_{s,jl}(x,y) \quad \cdots \quad], \quad (10.91)$$

where

$$p_{s,jl}(x,y) = (x/a)^{j-1}(y/b)^{l-1}, \ j = 1, 2, \cdots, N_1^s + 1, l = 1, 2, \cdots, N_2^s + 1. \quad (10.92)$$

The supplementary solution is then constructed as

$$\varphi_s(x,y) = \mathbf{p}_s^{\mathrm{T}}(x,y) \cdot \mathbf{a}_s, \quad (10.93)$$

where

$$\mathbf{a}_s^T = [G_{s,11} \quad \cdots \quad G_{s,jl} \quad \cdots \quad]. \tag{10.94}$$

By letting the supplementary solution satisfy the differential equation

$$\mathcal{L}_{cdr,2}\varphi_s = f_s \tag{10.95}$$

specifically at each of the grid points in Υ^s, we then have

$$\mathbf{R}_s\mathbf{a}_s = \mathbf{q}_{fs}, \tag{10.96}$$

or equivalently,

$$\mathbf{a}_s = \mathbf{R}_s^{-1}\mathbf{q}_{fs}, \tag{10.97}$$

where

$$\mathbf{R}_s = \begin{bmatrix} \mathcal{L}_{cdr,2}\mathbf{p}_s^T(x_1, y_1) \\ \vdots \\ \mathcal{L}_{cdr,2}\mathbf{p}_s^T(x_{m_1}, y_{n_2}) \\ \vdots \\ \mathcal{L}_{cdr,2}\mathbf{p}_s^T\left(x_{N_1^s+1}, y_{N_2^s+1}\right) \end{bmatrix}. \tag{10.98}$$

Define the vector of basis functions

$$\Phi_s^T(x, y) = \mathbf{p}_s^T(x, y) \cdot \mathbf{R}_s^{-1}, \tag{10.99}$$

the supplementary solution $\varphi_s(x, y)$ can be expressed as

$$\varphi_s(x, y) = \Phi_s^T(x, y) \cdot \mathbf{q}_{fs}. \tag{10.100}$$

10.3.4 THE PARTICULAR SOLUTION

When the interpolation polynomial $f_s(x, y)$ is used to approximate the source function $f(x, y)$, the error function is accordingly defined as

$$f_p(x, y) = f(x, y) - f_s(x, y). \tag{10.101}$$

In light of the interpolation conditions (10.82), we known that

$$f_p(-a, -b) = 0, \, f_p(-a, b) = 0, \, f_p(a, -b) = 0, \, f_p(a, b) = 0. \tag{10.102}$$

Thus, the function $f_p(x, y)$ can be expanded into a Fourier series, with a better convergence rate than that of $f(x, y)$, over domain $[-a, a] \times [-b, b]$:

$$f_p(x, y) = \sum_{m=0}^{\infty} \sum_{n=0}^{\infty} \lambda_{mn} [V_{fp,1mn} \cos(\alpha_m x) \cos(\beta_n y) + V_{fp,2mn} \sin(\alpha_m x) \cos(\beta_n y)$$

$$+ V_{fp,3mn} \cos(\alpha_m x) \sin(\beta_n y) + V_{fp,4mn} \sin(\alpha_m x) \sin(\beta_n y)], \quad (10.103)$$

where $\alpha_m = m\pi/a$, $\beta_n = n\pi/b$, $\lambda_{mn} = \begin{cases} 1/4 & m = 0, n = 0 \\ 1/2 & m = 1, 2, \cdots, n = 0 \\ 1/2 & m = 0, n = 1, 2, \cdots \\ 1 & m, n = 1, 2, \cdots \end{cases}$,

and $V_{fp,1mn}$, $V_{fp,2mn}$, $V_{fp,3mn}$, and $V_{fp,4mn}$ are the Fourier coefficients of $f_p(x, y)$.

Equation (10.103) is rewritten in matrix form as

$$f_p(x, y) = \Phi_0^T(x, y) \cdot \mathbf{q}_{fp}, \quad (10.104)$$

where $\Phi_0^T(x, y)$ denotes the vector of the trigonometric functions and $\mathbf{q}_{fp}$ the vector of the Fourier coefficients.

The particular solution of (10.69) is also sought in the form of Fourier series

$$\varphi_0(x, y) = \Phi_0^T(x, y) \cdot \mathbf{q}_0, \quad (10.105)$$

where $\mathbf{q}_0$ is the vector of Fourier coefficients yet to be determined.

Substituting (10.104) and (10.105) into the differential equation

$$\mathcal{L}_{cdr,2}(\varphi_0 + \varphi_3) = f_p, \quad (10.106)$$

we have

$$[\mathcal{L}_{cdr,2} \Phi_0^T(x, y)] \cdot \mathbf{q}_0 = \Phi_0^T(x, y) \cdot \mathbf{q}_{fp} - [\mathcal{L}_{cdr,2} \Phi_3^T(x, y)] \cdot \mathbf{q}_3, \quad (10.107)$$

where Φ_3^T and $\mathbf{q}_3$ are the vectors of the basis functions and the unknown coefficients as defined in (8.40) and (8.38).

Since (10.69) is a second order differential equation, each of the vectors Φ_3^T and $\mathbf{q}_3$ actually consists of only one element

$$\Phi_3^T = [\varphi_{3,00}(x, y)] \quad (10.108)$$

and

$$\mathbf{q}_3^T = [q_{3,00}], \quad (10.109)$$

where the function

$$\varphi_{3,00}(x, y) = \frac{xy}{4ab}, \quad (10.110)$$

and $q_{3,00}$ is an undetermined constant.

By substituting (10.104), (10.105), and (8.41) into (10.107), expanding all the involved functions into Fourier series over the domain $[-a,a] \times [-b,b]$, and comparing Fourier coefficients of the like terms on both sides of (10.107), we will be able to establish the relationships among the vectors $\mathbf{q}_0$, $\mathbf{q}_{fp}$ and $\mathbf{q}_3$:

1. For $m = 0$ and $n = 0$,

$$-P_e D_a V_{100} = V_{fp,100};\qquad(10.111)$$

2. For $m > 0$ and $n = 0$,

$$
\begin{bmatrix} \alpha_m^2 - P_e D_a & P_{e1}\alpha_m \\ -P_{e1}\alpha_m & \alpha_m^2 - P_e D_a \end{bmatrix}
\begin{bmatrix} V_{1m0} \\ V_{2m0} \end{bmatrix}
=
\begin{bmatrix} V_{fp,1m0} \\ V_{fp,2m0} \end{bmatrix}
+
\begin{bmatrix} 0 \\ \dfrac{P_{e2}(-1)^m}{bm\pi} \end{bmatrix} q_{3,00};
$$

$$(10.112)$$

3. For $m = 0$ and $n > 0$,

$$
\begin{bmatrix} \beta_n^2 - P_e D_a & P_{e2}\beta_n \\ -P_{e2}\beta_n & \beta_n^2 - P_e D_a \end{bmatrix}
\begin{bmatrix} V_{10n} \\ V_{30n} \end{bmatrix}
=
\begin{bmatrix} V_{fp,10n} \\ V_{fp,30n} \end{bmatrix}
+
\begin{bmatrix} 0 \\ \dfrac{P_{e1}(-1)^n}{an\pi} \end{bmatrix} q_{3,00};\qquad(10.113)
$$

4. For $m > 0$ and $n > 0$,

$$
\begin{bmatrix} \alpha_m^2 + \beta_n^2 - P_e D_a & P_{e1}\alpha_m & P_{e2}\beta_n & 0 \\ -P_{e1}\alpha_m & \alpha_m^2 + \beta_n^2 - P_e D_a & 0 & P_{e2}\beta_n \\ -P_{e2}\beta_n & 0 & \alpha_m^2 + \beta_n^2 - P_e D_a & P_{e1}\alpha_m \\ 0 & -P_{e2}\beta_n & -P_{e1}\alpha_m & \alpha_m^2 + \beta_n^2 - P_e D_a \end{bmatrix}
\begin{bmatrix} V_{1mn} \\ V_{2mn} \\ V_{3mn} \\ V_{4mn} \end{bmatrix}
$$

$$
=
\begin{bmatrix} V_{fp,1mn} \\ V_{fp,2mn} \\ V_{fp,3mn} \\ V_{fp,4mn} \end{bmatrix}
+
\begin{bmatrix} 0 \\ 0 \\ 0 \\ \dfrac{P_e D_a(-1)^{m+n}}{mn\pi^2} \end{bmatrix} q_{3,00}.\qquad(10.114)
$$

10.3.5 THE MULTISCALE FOURIER SERIES SOLUTION

After all four constituents, (10.80), (10.81), (10.100), and (10.105), become available, the multiscale Fourier series solution of the two-dimensional convection-diffusion-reaction equation can be formally expressed as

$$\varphi(x,y) = \varphi_0(x,y) + \varphi_1(x,y) + \varphi_2(x,y) + \varphi_3(x,y) + \varphi_s(x,y)$$
$$= \mathbf{\Phi}_0^{\mathrm{T}}(x,y) \cdot \mathbf{q}_0 + \mathbf{\Phi}_1^{\mathrm{T}}(x,y) \cdot \mathbf{q}_1 + \mathbf{\Phi}_2^{\mathrm{T}}(x,y) \cdot \mathbf{q}_2 + \mathbf{\Phi}_3^{\mathrm{T}}(x,y) \cdot \mathbf{q}_3 + \mathbf{\Phi}_s^{\mathrm{T}}(x,y) \cdot \mathbf{q}_{fs},$$

$$(10.115)$$

where the vectors of unknown constants $\mathbf{q}_0$ and $\mathbf{q}_{fs}$ are directly related to the source function $f(x,y)$ and its interpolation polynomial $f_s(x,y)$, as per (10.111)–(10.114) and (10.89). However, the vectors of unknown constants $\mathbf{q}_1$ and $\mathbf{q}_2$ will have to be determined from the prescribed boundary conditions.

10.4 TWO-DIMENSIONAL NUMERICAL EXAMPLES

In the following discussions, we will adopt an inverse process to validate the multiscale Fourier series solutions of the two-dimensional convection-diffusion-reaction problems. Accordingly, a reference solution $\varphi_{ref}(x,y)$ will first be constructed, and then substituted into (10.69) to determine a compatible source function. The boundary values of the solution and its relevant derivatives are also calculated at the boundaries in accordance with the types of boundary conditions to be prescribed. With such information available, the solutions of such two-dimensional convection-diffusion-reaction problems can be simply obtained by following the standard procedures as presented in the previous section.

As an example, the reference solution will be simply given as a linear combination of homogeneous solutions $p_{1nl,H}(x,y)$, $l = 1, 2, 3, 4$:

$$\varphi_{ref}(x,y) = \sum_{l=1}^{4} p_{1nl,H}(x,y), \tag{10.116}$$

wherein the parameter $\beta_n = n\pi/b$ is replaced with $\beta_{ref} = \pi/2b$.

It is easy to verify that the source function now vanishes identically

$$f_{ref}(x,y) = 0. \tag{10.117}$$

Accordingly, the order of the interpolation polynomial of the source function is set equal to zero. FCCM will be used to derive the final system of equations.

As presented in Table 10.8, four different test groups are designated for better understanding the impacts on the multiscale Fourier series solution of the key variables such as the model parameters, inflow angle, boundary conditions, and aspect ratio. For instance, in the first group of tests, we will focus on the influences of the model parameters P_e and D_a, which measure the relative importance of the convection and reaction terms in the differential equation.

TABLE 10.8

Four Groups of Tests Involving Different Model and Solution Parameters

Test Groups	Cases	Boundary Conditions	Model Parameters	Inflow Angle θ	Aspect Ratio a/b
1	a	DDDD	$P_e = 3, D_a = 90$	$\pi/3$	1.0
	b		$P_e = 1, D_a = 30$		
	c		$P_e = 30, D_a = 1$		
	d		$P_e = 200, D_a = -1$		
2	a	DDDD	$P_e = 3, D_a = 90$	$\pi/3$	1.0
	b			$\pi/4$	
	c			$\pi/6$	
	d			0	
3	a	DDDD	$P_e = 3, D_a = 90$	$\pi/3$	1.0
	b	DDND			
	c	DNND			
4	a	DDDD	$P_e = 3, D_a = 90$	$\pi/3$	1.0
	b				0.67
	c				0.50
	d				1.25
	e				2.0

The intentions of other test groups are quite clear since only one variable is singled out in each group. The Dirichlet type of boundary condition refers to the situation when the value of the solution itself is directly prescribed along a boundary line. In contrast, the Neumann type specifies the distribution of its normal derivative, such as $\varphi^{(1,0)}(a, y)$ along $x = a$. In the third test group, three different boundary conditions, DDDD, DDND, and DNND, are taken into account (refer to Figure 10.11). It should be noted that the first case in each group actually represents the same test which is used as the reference.

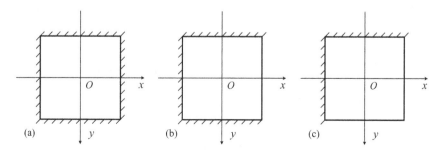

FIGURE 10.11   The types of boundary conditions considered: (a) DDDD boundary condition; (b) DDND boundary condition; and (c) DNND boundary condition.

10.4.1 CONVERGENCE CHARACTERISTICS

As before, the convergence characteristics of the multiscale Fourier series solutions are investigated by focusing on how the overall and regional (internal, boundary, and corner) errors of the solution $\varphi(x, y)$ and its first order partial derivatives are affected by the selected truncation numbers: $M = 2, 3, 5, 10, 20, 30$, and 40.

For the reference or baseline configuration, Figure 10.12 demonstrates that the multiscale Fourier series of $\varphi(x, y)$ and its first order partial derivatives converge nicely both inside the solution domain, and on the boundaries. Although boundary error index for $\varphi^{(0,1)}(x, y)$ appears to be considerably larger than that for $\varphi^{(1,0)}(x, y)$, their convergence rates are actually very similar, as indicated by the fact that the difference between these two error curves is almost unchanged. If the convergence of $\varphi(x, y)$ is considered acceptable at the corners, the first order partial derivatives have clearly failed the convergence test.

It is evident from Figures 10.13 and 10.14 that the model variables can affect the multiscale Fourier series solutions in various ways. While the model parameters

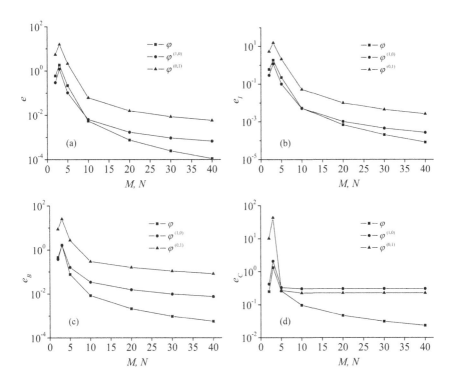

FIGURE 10.12 Convergence characteristics of the multiscale Fourier series solution of two-dimensional convection-diffusion-reaction equation (the reference configuration): (a) $e^{(k_1, k_2)}(\varphi_{M,N})$-$M$, N curves; (b) $e_I^{(k_1, k_2)}(\varphi_{M,N})$-$M$, N curves; (c) $e_B^{(k_1, k_2)}(\varphi_{M,N})$-$M$, N curves; and (d) $e_C^{(k_1, k_2)}(\varphi_{M,N})$-$M$, N curves.

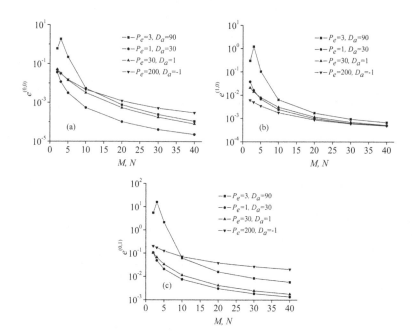

FIGURE 10.13 Convergence comparison of the multiscale Fourier series solutions with different model parameters P_e and D_a: (a) $e^{(0,0)}(\varphi_{M,N})$-M, N curves; (b) $e^{(1,0)}(\varphi_{M,N})$-M, N curves; and (c) $e^{(0,1)}(\varphi_{M,N})$-M, N curves.

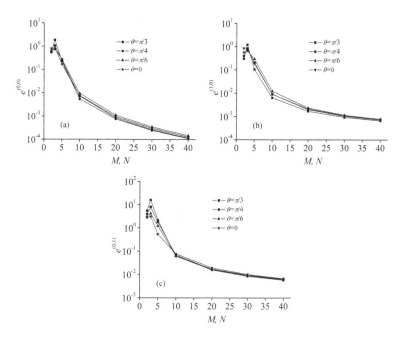

FIGURE 10.14 Convergence comparison of the multiscale Fourier series solutions with different inflow angles: (a) $e^{(0,0)}(\varphi_{M,N})$-M, N curves; (b) $e^{(1,0)}(\varphi_{M,N})$-M, N curves; and (c) $e^{(0,1)}(\varphi_{M,N})$-M, N curves.

P_e and D_a clearly have the marked effects on the convergence behaviors and approximation errors of the solutions, the impacts of the inflow angles are kind of considered intangible.

It is illustrated in Figure 10.15 that in terms of the numerical accuracy, the multiscale Fourier series solution is better suited for the DDDD boundary condition than the other two boundary conditions involving Neumann type specified, at least, along one boundary line.

Figure 10.16 shows the convergence behaviors of the multiscale Fourier series solutions as the aspect ratio a/b varies from 0.5 to 2.0. It is observed that the solution errors are considerably larger for the two smaller aspect ratios. An immediate reaction to this is to increase the number of terms retained in y-direction so that the series expansion approximately has an equal spatial resolution in both x- and y-direction (e.g., setting $N/M = 2$ for $a/b = 0.50$). Indeed, this has immediately led to a significant reduction of the solution errors, as illustrated in Figure 10.17. However, this cannot fully explain why the solution behaves satisfactory for the case of $a/b = 2.0$ in Figure 10.16. Consequently, we may have to attribute it to the disparity of the inflow angle $(= \pi/3)$ which has resulted in the biased distributions in the x- and y-direction.

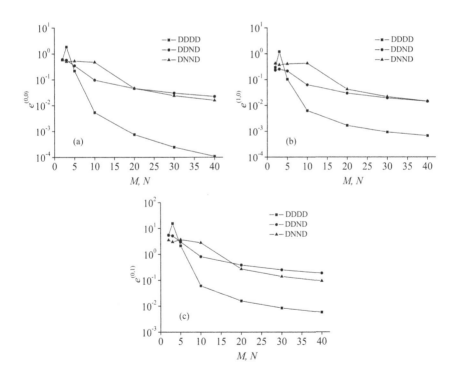

FIGURE 10.15 Convergence comparison of the multiscale Fourier series solutions with different boundary conditions: (a) $e^{(0,0)}(\varphi_{M,N})$-M, N curves; (b) $e^{(1,0)}(\varphi_{M,N})$-M, N curves; and (c) $e^{(0,1)}(\varphi_{M,N})$-M, N curves.

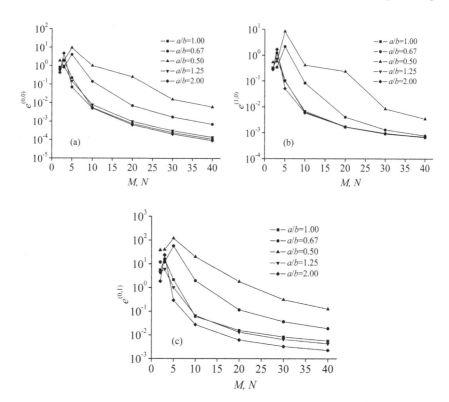

FIGURE 10.16 Convergence comparison of the multiscale Fourier series solutions with different aspect ratios: (a) $e^{(0,0)}(\varphi_{M,N})$-M, N curves; (b) $e^{(1,0)}(\varphi_{M,N})$-M, N curves; and (c) $e^{(0,1)}(\varphi_{M,N})$-M, N curves.

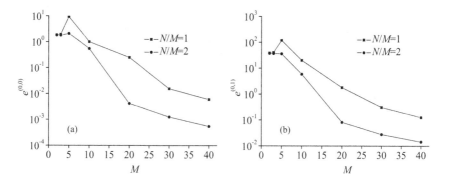

FIGURE 10.17 Convergence comparison of the multiscale Fourier series solutions with different values of N/M ($a/b = 0.5$): (a) $e^{(0,0)}(\varphi_{M,N})$-M curves and (b) $e^{(0,1)}(\varphi_{M,N})$-M curves.

10.4.2 MULTISCALE CHARACTERISTICS

Based on (10.116) and Table 10.7, it can be expected that the (reference) solution will exhibit some peculiar behaviors when, by properly administering the model parameters such as P_e and D_a, the convection-diffusion-reaction equation accordingly goes through the transition from the strong reaction type to the strong convection type. Take the cases in test group 1 as example. The spatial distributions of the reference solutions and the multiscale Fourier series solutions (obtained by setting $M = N = 40$) are displayed in Figures 10.18–10.21 sequentially for each

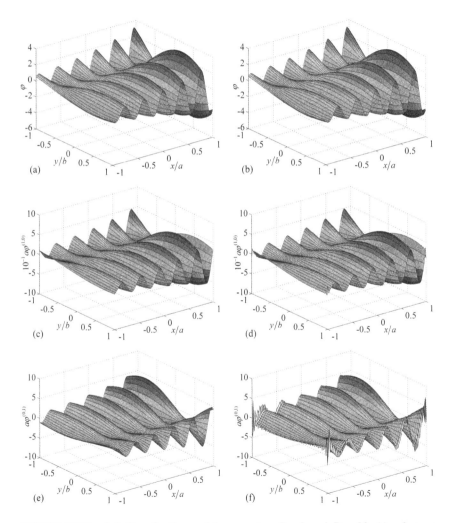

FIGURE 10.18 Solutions for the model parameters $P_e = 3$ and $D_a = 90$: (a) reference solution $\varphi(x, y)$; (b) multiscale Fourier series solution $\varphi(x, y)$; (c) $\varphi^{(1,0)}(x, y)$ (reference); (d) $\varphi^{(1,0)}(x, y)$ (calculated); (e) $\varphi^{(0,1)}(x, y)$ (reference); and (f) $\varphi^{(0,1)}(x, y)$ (calculated).

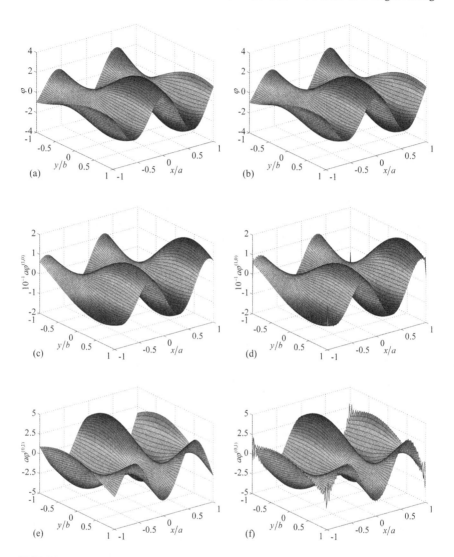

FIGURE 10.19 Solutions for the model parameters $P_e = 1$ and $D_a = 30$: (a) reference solution $\varphi(x, y)$; (b) multiscale Fourier series solution $\varphi(x, y)$; (c) $\varphi^{(1,0)}(x, y)$ (reference); (d) $\varphi^{(1,0)}(x, y)$ (calculated); (e) $\varphi^{(0,1)}(x, y)$ (reference); and (f) $\varphi^{(0,1)}(x, y)$ (calculated).

case. It is seen from these plots that with the changes of the physical states, the corresponding solution varies from simple oscillations to exponential decay, and eventually to the forming of boundary layers, a typical multiscale characteristic with a strong convection phenomenon.

In each of these cases, the excellent convergence and numerical accuracy of the multiscale Fourier series solution are repeatedly demonstrated by its close

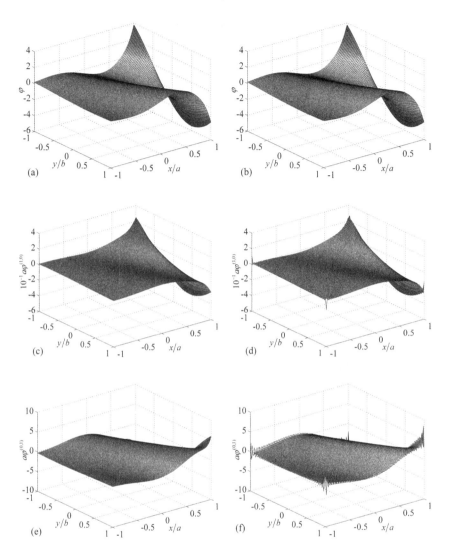

FIGURE 10.20 Solutions for the model parameters $P_e = 30$ and $D_a = 1$: (a) reference solution $\varphi(x, y)$; (b) multiscale Fourier series solution $\varphi(x, y)$; (c) $\varphi^{(1,0)}(x, y)$ (reference); (d) $\varphi^{(1,0)}(x, y)$ (calculated); (e) $\varphi^{(0,1)}(x, y)$ (reference); and (f) $\varphi^{(0,1)}(x, y)$ (calculated).

similarity to the reference in regards to the spatial distribution over the solution domain. In comparison, the accuracies for the partial derivatives are not equally good, as predominantly indicated by presence of the glitches or spikes at or near the edges and corners.

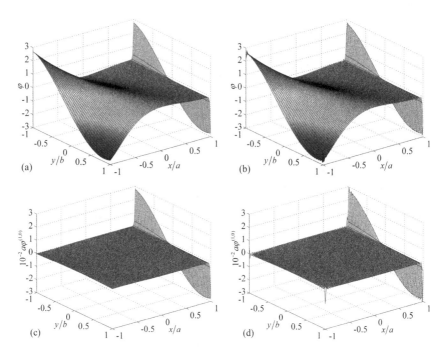

FIGURE 10.21   Solutions for the model parameters $P_e = 200$ and $D_a = -1$: (a) reference solution $\varphi(x, y)$; (b) multiscale Fourier series solution $\varphi(x, y)$; (c) $\varphi^{(1,0)}(x, y)$ (reference); and (d) $\varphi^{(1,0)}(x, y)$ (calculated).

REFERENCES

Brezzi, F., L. P. Franca, and A. Russo. 1998. Further considerations on residual-free bubbles for advective-diffusive equations. *Computer Methods in Applied Mechanics and Engineering* 166: 25–33.

Dahmen, W. 2001. Wavelet methods for PDEs-some recent developments. *Journal of Computational and Applied Mathematics* 128: 133–185.

Dahmen, W., A. J. Kurdila, and P. Oswald. 1997. *Multiscale Wavelet Methods for Partial Differential Equations*. San Diego: Academic Press.

Donea, J., and A. Huerta. 2003. *Finite Element Methods for Flow Problems*. Chichester: John Wiley & Sons Ltd.

Franca, L. P., and A. Russo. 1997. Mass lumping emanating from residual-free bubbles. *Computer Methods in Applied Mechanics and Engineering* 142: 353–360.

Franca, L. P., A. Nesliturk, and M. Stynes. 1998. On the stability of residual-free bubbles for convection-diffusion problems and their approximation by a two-level finite element method. *Computer Methods in Applied Mechanics and Engineering* 166: 35–49.

Hauke, G. 2002. A simple subgrid scale stabilized method for the advection-diffusion-reaction equation. *Computer Methods in Applied Mechanics and Engineering* 191: 2925–2947.

Hauke, G., and A. Garcia-Olivares. 2001. Variational subgrid scale formulations for the advection-diffusion-reaction equation. *Computer Methods in Applied Mechanics and Engineering* 190: 6847–6865.

Huerta, A., and J. Donea. 2002. Time-accurate solution of stabilized convection-diffusion-reaction equations: I – Time and space discretization. *Communications in Numerical Methods in Engineering* 18: 565–573.

Huerta, A., B. Roig, and J. Donea. 2002. Time-accurate solution of stabilized convection-diffusion-reaction equations: II – Accuracy analysis and examples. *Communications in Numerical Methods in Engineering* 18: 575–584.

Hughes, T. J. R., and G. N. Wells. 2005. Conservation properties for the Galerkin and stabilised forms of the advection-diffusion and incompressible Navier-Stokes equations. *Computer Methods in Applied Mechanics and Engineering* 194: 1141–1159.

Liu, G. R., and Y. T. Gu. 2005. *An Introduction to Meshfree Methods and Their Programming*. Dordrecht: Springer.

Onate, E., J. Miquel, and G. Hauke. 2006. Stabilized formulation for the advection-diffusion-absorption equation using finite calculus and linear finite elements. *Computer Methods in Applied Mechanics and Engineering* 195: 3926–3946.

Parvazinia, M., and V. Nassehi. 2007. Multiscale finite element modeling of diffusion-reaction equation using bubble functions with bilinear and triangular elements. *Computer Methods in Applied Mechanics and Engineering* 196: 1095–1107.

Sangalli, G. 2004. A discontinuous residual-free bubble method for advection-diffusion problems. *Journal of Engineering Mathematics* 49: 149–162.

Song, L. N., Y. R. Hou, and H. B. Zheng. 2010. A variational multiscale method based on bubble functions for convection-dominated convection-diffusion equation. *Applied Mathematics and Computation* 217: 2226–2237.

11 Bending of Thick Plates on Elastic Foundations

Solid mechanics, Elasticity, and Continuum Mechanics are the typical titles for the college courses and technical books in which the deformations of elastic bodies under various loading conditions are of primary concerns. This is perhaps the oldest scientific and engineering discipline, and has given rise to many classical boundary value problems such as the vibrations of stretched strings and membranes, and deformations of elastic bars, beams, and plates. In the literature, there is a rich pool of studies about the elastic bending of rectangular plates with various complicating factors. In this chapter, we will focus on one special class of plate problems: bending of thick plates resting on an elastic foundation. This topic is of not only practical interests to many scientific and engineering fields, but also technical challenges to mathematical physics methods (Henwood et al. 1982; Shi et al. 1994; Rashed et al. 1999; Sun et al. 1999; Wen 2008) due to the intrinsic complexities (e.g., multiscale characteristics) of the subject matters.

11.1 DESCRIPTION OF THE PROBLEM

In this section, we consider a thick plate of length a, width b and thickness h, resting on an elastic foundation. The deformations of the plate are described by the deflection of the mid-surface w, and its rotations β_x and β_y about y- and x-axis, respectively.

The upper surface of the plate is subjected to a transverse load q, and the lower surface to a reactive force q_e from the foundation. A perfect contact (or no gap) is assumed to exist between the plate and the foundation, and the reactive force $q_e(x, y)$ is given in the bi-parametric form as

$$q_e = kw - G_p\nabla^2 w, \tag{11.1}$$

where k is the modulus of the substrate, G_p is the shear modulus of the foundation, and the Laplace differential operator

$$\nabla^2 = \frac{\partial^2}{\partial x^2} + \frac{\partial^2}{\partial y^2}. \tag{11.2}$$

According to Reissner's thick plate theory (He and Shen 1993), the transverse displacement w satisfies the following differential equation

$$\left. \begin{array}{r} \mathcal{L}_p w = (q - q_e) - \dfrac{(2-\mu)h^2}{10(1-\mu)} \nabla^2 (q - q_e) \\[4mm] \mathcal{L}_\psi \psi = 0 \end{array} \right\}, \tag{11.3}$$

where $\psi(x, y)$ is a stress function, and the differential operators

$$\mathcal{L}_p = D\nabla^2 \nabla^2 \tag{11.4}$$

and

$$\mathcal{L}_\psi = \nabla^2 - \frac{10}{h^2}. \tag{11.5}$$

The moments and shear forces are related to the deformations:

$$\left[\begin{array}{c} M_x \\ M_y \\ M_{xy} \end{array} \right] = D \left[\begin{array}{ccc} 1 & \mu & 0 \\ \mu & 1 & 0 \\ 0 & 0 & \dfrac{1-\mu}{2} \end{array} \right] \left[\begin{array}{c} \dfrac{\partial \beta_x}{\partial x} \\[3mm] \dfrac{\partial \beta_y}{\partial y} \\[3mm] \dfrac{\partial \beta_x}{\partial y} + \dfrac{\partial \beta_y}{\partial x} \end{array} \right] \tag{11.6}$$

and

$$\left[\begin{array}{c} Q_x \\ Q_y \end{array} \right] = \left[\begin{array}{cc} C_s & 0 \\ 0 & C_s \end{array} \right] \left[\begin{array}{c} \beta_x + \dfrac{\partial w}{\partial x} \\[3mm] \beta_y + \dfrac{\partial w}{\partial y} \end{array} \right], \tag{11.7}$$

where M_x, M_y, and M_{xy} are the bending and torsional moments, Q_x and Q_y are the transverse shear forces, $D = Eh^3 / 12(1 - \mu^2)$ is the flexural rigidity of the plate, and $C_s = 5Gh / 6$ is a shear constant.

In addition, there is the relationship between the transverse shear forces, Q_x and Q_y, and the stress function ψ

$$\left. \begin{array}{l} Q_x = \dfrac{\partial \psi}{\partial y} - D\dfrac{\partial}{\partial x}(\nabla^2 w) \\[4mm] Q_y = -\dfrac{\partial \psi}{\partial x} - D\dfrac{\partial}{\partial y}(\nabla^2 w) \end{array} \right\}. \tag{11.8}$$

Three kinds of boundary conditions are typically prescribed along an edge, such as, $x = a$:

1. Generalized clamped edge (C type)

$$w(a,y) = \overline{w}(a,y), \beta_x(a,y) = \overline{\beta}_x(a,y), \beta_y(a,y) = \overline{\beta}_y(a,y), \qquad (11.9)$$

2. Generalized simply supported edge (S type)

$$w(a,y) = \overline{w}(a,y), M_x(a,y) = \overline{M}_x(a,y), M_{xy}(a,y) = \overline{M}_{xy}(a,y), \qquad (11.10)$$

3. Generalized free edge (F type)

$$Q_x(a,y) = \overline{Q}_x(a,y), M_x(a,y) = \overline{M}_x(a,y), M_{xy}(a,y) = \overline{M}_{xy}(a,y), \qquad (11.11)$$

where $\overline{w}$, $\overline{\beta}_x$, $\overline{\beta}_y$, $\overline{M}_x$, $\overline{M}_y$, $\overline{M}_{xy}$, and $\overline{Q}_x$, $\overline{Q}_y$ are respectively the prescribed displacement, rotations, bending moments, torsional moments and shear force at the edge $x = a$. The term "generalized" is used here to indicate that the prescribed boundary distributions can be any non-zero values.

11.2 THE MULTISCALE FOURIER SERIES SOLUTIONS

The bending deformations of a thick plate resting on an elastic foundation are described by a fourth order ($2r_w = 4$) linear differential equation for the transverse displacement and a second order ($2r_\psi = 2$) linear differential equation for the stress function. Moreover, since the system of equations includes only the solution functions and their partial derivatives of even orders, the transverse displacement $w(x,y)$ and the stress function $\psi(x,y)$ will be here sought in the forms of the half-range sine-sine series and the half-range cosine-cosine series, respectively.

11.2.1 THE GENERAL SOLUTION OF THE TRANSVERSE DISPLACEMENT

Denote the differential operators

$$\mathcal{L}_{pf} = D_h \nabla^2 \nabla^2 - G_{ph} \nabla^2 + k \qquad (11.12)$$

and

$$\mathcal{L}_q = 1 - \frac{(2-\mu)h^2}{10(1-\mu)} \nabla^2, \qquad (11.13)$$

where

$$D_h = D + \frac{(2-\mu)h^2}{10(1-\mu)} G_p$$

and

$$G_{ph} = G_p + \frac{(2-\mu)h^2}{10(1-\mu)} k.$$

Substitution of (11.1) into the first equation in (11.3) leads to

$$\mathcal{L}_{pf} w = \mathcal{L}_q q. \tag{11.14}$$

Suppose that the homogeneous solution of (11.14) is given as

$$p_{1n,Hw}(x,y) = \exp(\eta_{nw} x)\exp(i\beta_n y), \tag{11.15}$$

where n is a positive integer, η_{nw} is an undetermined constant, $\beta_n = n\pi / b$ and $i = \sqrt{-1}$.

Substituting (11.15) into the homogeneous form of (11.14), we yield the characteristics equation

$$D_h \eta_{nw}^4 - (2D_h \beta_n^2 + G_{ph})\eta_{nw}^2 + D_h \beta_n^4 + G_{ph}\beta_n^2 + k = 0. \tag{11.16}$$

If $\Delta_h = G_{ph}^2 - 4D_h k > 0$, (11.16) has four distinct real roots:

$$\eta_{nw,1} = \alpha_{1n}, \eta_{nw,2} = -\alpha_{1n}, \eta_{nw,3} = \alpha_{2n}, \text{ and } \eta_{nw,4} = -\alpha_{2n}, \tag{11.17}$$

where

$$\alpha_{1n} = \sqrt{\beta_n^2 + \frac{1}{2D_h}\left[G_{ph} + \sqrt{G_{ph}^2 - 4D_h k}\right]}$$

and

$$\alpha_{2n} = \sqrt{\beta_n^2 + \frac{1}{2D_h}\left[G_{ph} - \sqrt{G_{ph}^2 - 4D_h k}\right]}.$$

If $\Delta_h = G_{ph}^2 - 4D_h k = 0$, (11.16) has two distinct double real roots:

$$\eta_{nw,1} = \eta_{nw,2} = \alpha_{3n} \text{ and } \eta_{nw,3} = \eta_{nw,4} = -\alpha_{3n}, \tag{11.18}$$

where

$$\alpha_{3n} = \sqrt{\beta_n^2 + \frac{G_{ph}}{2D_h}}.$$

If $\Delta_h = G_{ph}^2 - 4D_h k < 0$, (11.16) has four distinct complex roots:

$$\eta_{nw,1,2,3,4} = \pm\alpha_{5n} \pm i\alpha_{6n}, \tag{11.19}$$

TABLE 11.1

Expressions for Characteristic Functions $p_{1nl,w}(x)$, $l = 1, 2, 3, 4$.

	$\Delta_h > 0$	$\Delta_h = 0$	$\Delta_h < 0$
$p_{1n1,w}(x)$	$\dfrac{\sinh(\alpha_{1n}x)}{\sinh(\alpha_{1n}a)}$	$\dfrac{\sinh(\alpha_{3n}x)}{\sinh(\alpha_{3n}a)}$	$\dfrac{\sinh(\alpha_{5n}x)\sin(\alpha_{6n}x)}{\sinh(\alpha_{5n}a)\sin(\alpha_{6n}a)}$
$p_{1n2,w}(x)$	$\dfrac{\sinh[\alpha_{1n}(a-x)]}{\sinh(\alpha_{1n}a)}$	$\dfrac{x\sinh(\alpha_{3n}x)}{a\sinh(\alpha_{3n}a)}$	$\dfrac{\sinh(\alpha_{5n}x)\sin[\alpha_{6n}(a-x)]}{\sinh(\alpha_{5n}a)\sin(\alpha_{6n}a)}$
$p_{1n3,w}(x)$	$\dfrac{\sinh(\alpha_{2n}x)}{\sinh(\alpha_{2n}a)}$	$\dfrac{\sinh[\alpha_{3n}(a-x)]}{\sinh(\alpha_{3n}a)}$	$\dfrac{\sinh[\alpha_{5n}(a-x)]\sin(\alpha_{6n}x)}{\sinh(\alpha_{5n}a)\sin(\alpha_{6n}a)}$
$p_{1n4,w}(x)$	$\dfrac{\sinh[\alpha_{2n}(a-x)]}{\sinh(\alpha_{2n}a)}$	$\dfrac{(a-x)\sinh[\alpha_{3n}(a-x)]}{a\sinh(\alpha_{3n}a)}$	$\dfrac{\sinh[\alpha_{5n}(a-x)]\sin[\alpha_{6n}(a-x)]}{\sinh(\alpha_{5n}a)\sin(\alpha_{6n}a)}$

where $\alpha_{5n} > 0$, $\alpha_{6n} > 0$ and

$$\alpha_{5n} + i\alpha_{6n} = \sqrt{\beta_n^2 + \frac{1}{2D_h}\left[G_{ph} + i\sqrt{4D_hk - G_{ph}^2}\right]}.$$

The corresponding characteristic functions $p_{1nl,w}(x)$, $l = 1, 2, 3, 4$, are presented in Table 11.1.

Denote

$$\mathbf{p}_{1n,Hw}^{T}(x,y) = [p_{1n1,Hw}(x,y) \quad p_{1n2,Hw}(x,y) \quad p_{1n3,Hw}(x,y) \quad p_{1n4,Hw}(x,y)], \quad (11.20)$$

where

$$p_{1nl,Hw}(x,y) = p_{1nl,w}(x)\sin(\beta_n y), l = 1, 2, 3, 4. \tag{11.21}$$

Then the general solution of (11.14) can be expressed as

$$w_1(x,y) = \mathbf{\Phi}_{1,w}^{T}(x,y)\cdot\mathbf{q}_{1,w}, \tag{11.22}$$

where $\mathbf{\Phi}_{1,w}^{T}(x,y)$ is the vector of the basis functions, and $\mathbf{q}_{1,w}$ is the vector of the undetermined coefficients as previously defined in (9.122) with some minor modifications.

11.2.2 THE GENERAL SOLUTION OF THE STRESS FUNCTION

Similarly, suppose that the second equation in (11.3) has the homogeneous solution

$$p_{1n,H\psi}(x,y) = \exp(\eta_{n\psi}x)\exp(i\beta_n y), \tag{11.23}$$

where n is a positive integer, $\eta_{n\psi}$ is an undetermined constant, and $\beta_n = n\pi/b$.

By substituting (11.23) into the second of the equations (11.3), we yield the characteristic equation

$$\eta_{n\psi}^2 - \beta_n^2 - \frac{10}{h^2} = 0, \tag{11.24}$$

which has two distinct real roots as

$$\eta_{n\psi,1} = \alpha_{7n}, \text{ and } \eta_{n\psi,2} = -\alpha_{7n}, \tag{11.25}$$

where

$$\alpha_{7n} = \sqrt{\beta_n^2 + \frac{10}{h^2}}.$$

Thus, the two homogeneous solutions are obtained as

$$p_{1nl,H\psi}(x,y) = p_{1nl,\psi}(x)\cos(\beta_n y) \quad (l = 1,2), \tag{11.26}$$

where

$$p_{1n1,\psi}(x) = \frac{\sinh(\alpha_{7n}x)}{\sinh(\alpha_{7n}a)} \tag{11.27}$$

and

$$p_{1n2,\psi}(x) = \frac{\sinh[(\alpha_{7n}(a-x)]}{\sinh(\alpha_{7n}a)}. \tag{11.28}$$

By defining the sub-vector of the basis functions

$$\mathbf{p}_{1n,H\psi}^T(x,y) = [p_{1n1,H\psi}(x,y) \quad p_{1n2,H\psi}(x,y)], \tag{11.29}$$

the general solution of the second equation in (11.3) can be expressed as

$$\psi_1(x,y) = \mathbf{\Phi}_{1,\psi}^T(x,y) \cdot \mathbf{q}_{1,\psi}, \tag{11.30}$$

where $\mathbf{\Phi}_{1,\psi}^T(x,y)$ is the vector of the basis functions, and $\mathbf{q}_{1,\psi}$ is the vector of the undetermined coefficients, similar to those in (9.122).

11.2.3 EXPRESSIONS OF THE MULTISCALE FOURIER SERIES SOLUTIONS

The transverse displacement function $w(x,y)$ is governed by the fourth order linear differential equation with constant coefficients, and will be here expanded into the multiscale Fourier sine-sine series over the solution domain $[0,a] \times [0,b]$ as

$$w(x,y) = w_0(x,y) + w_1(x,y) + w_2(x,y) + w_3(x,y)$$
$$= \mathbf{\Phi}_{0,w}^T(x,y) \cdot \mathbf{q}_{0,w} + \mathbf{\Phi}_{1,w}^T(x,y) \cdot \mathbf{q}_{1,w} + \mathbf{\Phi}_{2,w}^T(x,y) \cdot \mathbf{q}_{2,w} + \mathbf{\Phi}_{3,w}^T(x,y) \cdot \mathbf{q}_{3,w}. \tag{11.31}$$

More concisely, by defining the vector of the basis functions

$$\boldsymbol{\Phi}_w^\mathrm{T}(x,y) = \left[\boldsymbol{\Phi}_{0,w}^\mathrm{T}(x,y) \quad \boldsymbol{\Phi}_{1,w}^\mathrm{T}(x,y) \quad \boldsymbol{\Phi}_{2,w}^\mathrm{T}(x,y) \quad \boldsymbol{\Phi}_{3,w}^\mathrm{T}(x,y) \right] \quad (11.32)$$

and the vector of the undetermined constants

$$\mathbf{q}_w^\mathrm{T} = \left[\mathbf{q}_{0,w}^\mathrm{T} \quad \mathbf{q}_{1,w}^\mathrm{T} \quad \mathbf{q}_{2,w}^\mathrm{T} \quad \mathbf{q}_{3,w}^\mathrm{T} \right], \quad (11.33)$$

the displacement function can be written as

$$w(x,y) = \boldsymbol{\Phi}_w^\mathrm{T}(x,y) \cdot \mathbf{q}_w. \quad (11.34)$$

The stress function $\psi(x,y)$ is governed by second order linear differential equation with constant coefficients, and will be expanded into the multiscale Fourier cosine-cosine series over the domain $[0,a] \times [0,b]$ as

$$\begin{aligned} \psi(x,y) &= \psi_0(x,y) + \psi_1(x,y) + \psi_2(x,y) \\ &= \boldsymbol{\Phi}_{0,\psi}^\mathrm{T}(x,y) \cdot \mathbf{q}_{0,\psi} + \boldsymbol{\Phi}_{1,\psi}^\mathrm{T}(x,y) \cdot \mathbf{q}_{1,\psi} + \boldsymbol{\Phi}_{2,\psi}^\mathrm{T}(x,y) \cdot \mathbf{q}_{2,\psi}. \end{aligned} \quad (11.35)$$

It is worthy of note that since $2r_\psi = 2$, the corner function or corner conditions are no longer the prerequisites (see Theorem 6.4 in Sec. 6.4) for this series solution $\psi(x,y)$ to be termwise differentiable for the partial derivatives of up to the second order.

Because the second equation of (11.3) is a homogeneous equation, we immediately have

$$\mathbf{q}_{0,\psi} = \mathbf{0}. \quad (11.36)$$

Define the vector of the basis functions

$$\boldsymbol{\Phi}_\psi^\mathrm{T}(x,y) = \left[\boldsymbol{\Phi}_{1,\psi}^\mathrm{T}(x,y) \quad \boldsymbol{\Phi}_{2,\psi}^\mathrm{T}(x,y) \right], \quad (11.37)$$

and the vector of the undetermined constants

$$\mathbf{q}_\psi^\mathrm{T} = \left[\mathbf{q}_{1,\psi}^\mathrm{T} \quad \mathbf{q}_{2,\psi}^\mathrm{T} \right], \quad (11.38)$$

the stress function can be rewritten as

$$\psi(x,y) = \boldsymbol{\Phi}_\psi^\mathrm{T}(x,y) \cdot \mathbf{q}_\psi. \quad (11.39)$$

By combining (11.34) and (11.39), we finally have the multiscale Fourier series solution for the flexural deformation of a thick plate resting on an elastic foundation

$$\begin{bmatrix} w \\ \psi \end{bmatrix} = \boldsymbol{\Phi} \cdot \mathbf{q}, \quad (11.40)$$

where the matrix of the basis functions

$$\Phi = \begin{bmatrix} \Phi_w^T & 0 \\ 0 & \Phi_\psi^T \end{bmatrix} \tag{11.41}$$

and the vector of the undetermined constants

$$\mathbf{q}^T = \begin{bmatrix} \mathbf{q}_w^T & \mathbf{q}_\psi^T \end{bmatrix}. \tag{11.42}$$

11.2.4 Equivalent Transformation of the Solutions

By substituting the multiscale Fourier series solution (11.40) into (11.7) and (11.8), we are able to obtain

$$\begin{bmatrix} w \\ \beta_x \\ \beta_y \end{bmatrix} = \Gamma \cdot \mathbf{q}, \tag{11.43}$$

where the transformation matrix

$$\Gamma = \begin{bmatrix} \Phi_w^T & 0 \\ -\dfrac{D}{C_s}\Phi_w^{(3,0)T} - \dfrac{D}{C_s}\Phi_w^{(1,2)T} - \Phi_w^{(1,0)T} & \dfrac{1}{C_s}\Phi_\psi^{(0,1)T} \\ -\dfrac{D}{C_s}\Phi_w^{(2,1)T} - \dfrac{D}{C_s}\Phi_w^{(0,3)T} - \Phi_w^{(0,1)T} & -\dfrac{1}{C_s}\Phi_\psi^{(1,0)T} \end{bmatrix}. \tag{11.44}$$

Specifically, along the edge $x = a$, (11.43) becomes

$$\begin{bmatrix} w(a,y) \\ \beta_x(a,y) \\ \beta_y(a,y) \end{bmatrix} = \Gamma(a,y) \cdot \mathbf{q}. \tag{11.45}$$

The displacement $w(a,y)$, the rotations $\beta_x(a,y)$, and $\beta_y(a,y)$(and accordingly, the elements in the first, second, and third rows of the transformation matrix $\Gamma(a,y)$) are then respectively expanded into the Fourier sine series, Fourier sine series, and Fourier cosine series over the interval $[0,b]$. By comparing the Fourier coefficients of the like-terms on both sides of (11.45), we have

$$\begin{bmatrix} \mathbf{q}_{w,1a} \\ \mathbf{q}_{\beta x,1a} \\ \mathbf{q}_{\beta y,1a} \end{bmatrix} = \mathbf{R}_{pf,1a} \cdot \mathbf{q},$$
(11.46)

where $\mathbf{q}_{w,1a}$, $\mathbf{q}_{\beta x,1a}$, and $\mathbf{q}_{\beta y,1a}$ are the sub-vectors of the Fourier coefficients corresponding to $w(a,y)$, $\beta_x(a,y)$, and $\beta_y(a,y)$, and $\mathbf{R}_{pf,1a}$ is the sub-matrix of Fourier coefficients corresponding to the transformation matrix $\Gamma(a,y)$.

Similar relationships can be derived from the displacement type boundary conditions which are specified along other edges at $x = 0$, $y = b$, and $y = 0$.

Accordingly, the final system can be written as

$$\mathbf{q}_b = \mathbf{R}_{pf} \cdot \mathbf{q}$$
(11.47)

with

$$\mathbf{R}_{pf} = \begin{bmatrix} \mathbf{R}_{pf,1a} \\ \mathbf{R}_{pf,10} \\ \mathbf{R}_{pf,2b} \\ \mathbf{R}_{pf,20} \end{bmatrix}$$
(11.48)

and

$$\mathbf{q}_b = \begin{bmatrix} \mathbf{q}_{b,1a} \\ \mathbf{q}_{b,10} \\ \mathbf{q}_{b,2b} \\ \mathbf{q}_{b,20} \end{bmatrix},$$
(11.49)

where the sub-vectors of the boundary Fourier coefficients are defined as

$$\mathbf{q}_{b,1a} = \begin{bmatrix} \mathbf{q}_{w,1a} \\ \mathbf{q}_{\beta x,1a} \\ \mathbf{q}_{\beta y,1a} \end{bmatrix}, \mathbf{q}_{b,10} = \begin{bmatrix} \mathbf{q}_{w,10} \\ \mathbf{q}_{\beta x,10} \\ \mathbf{q}_{\beta y,10} \end{bmatrix},$$

$$\mathbf{q}_{b,2b} = \begin{bmatrix} \mathbf{q}_{w,2b} \\ \mathbf{q}_{\beta x,2b} \\ \mathbf{q}_{\beta y,2b} \end{bmatrix}, \text{ and } \mathbf{q}_{b,20} = \begin{bmatrix} \mathbf{q}_{w,20} \\ \mathbf{q}_{\beta x,20} \\ \mathbf{q}_{\beta y,20} \end{bmatrix}.$$
(11.50)

Let

$$\mathbf{q}_{03}^{\mathrm{T}} = \begin{bmatrix} \mathbf{q}_{0,w}^{\mathrm{T}} & \mathbf{q}_{3,w}^{\mathrm{T}} \end{bmatrix},$$
(11.51)

$$\mathbf{q}_{12}^{\mathrm{T}} = \begin{bmatrix} \mathbf{q}_{1,w}^{\mathrm{T}} & \mathbf{q}_{2,w}^{\mathrm{T}} & \mathbf{q}_{1,\psi}^{\mathrm{T}} & \mathbf{q}_{2,\psi}^{\mathrm{T}} \end{bmatrix}, \tag{11.52}$$

and the matrix $\mathbf{R}_{pf}$ be accordingly rearranged as

$$\mathbf{R}_{pf} = [\mathbf{R}_{pf,03} \quad \mathbf{R}_{pf,12}]. \tag{11.53}$$

We can rewrite (11.47) into

$$\mathbf{q}_b = \mathbf{R}_{pf,03} \cdot \mathbf{q}_{03} + \mathbf{R}_{pf,12} \cdot \mathbf{q}_{12} \tag{11.54}$$

or

$$\mathbf{q}_{12} = -\mathbf{R}_{pf,12}^{-1} \mathbf{R}_{pf,03} \cdot \mathbf{q}_{03} + \mathbf{R}_{pf,12}^{-1} \cdot \mathbf{q}_b. \tag{11.55}$$

The multiscale Fourier series solution (11.40) can then be expressed as

$$\begin{bmatrix} w \\ \psi \end{bmatrix} = \mathbf{\Phi}_R \cdot \mathbf{q}_R, \tag{11.56}$$

where the matrix of the basis functions

$$\mathbf{\Phi}_R = \begin{bmatrix} \mathbf{\Phi}_{0,w}^{\mathrm{T}} & \mathbf{\Phi}_{3,w}^{\mathrm{T}} & \mathbf{\Phi}_{1,w}^{\mathrm{T}} & \mathbf{\Phi}_{2,w}^{\mathrm{T}} & \mathbf{0} & \mathbf{0} \\ \mathbf{0} & \mathbf{0} & \mathbf{0} & \mathbf{0} & \mathbf{\Phi}_{1,\psi}^{\mathrm{T}} & \mathbf{\Phi}_{2,\psi}^{\mathrm{T}} \end{bmatrix} \begin{bmatrix} \mathbf{I} & \mathbf{0} \\ -\mathbf{R}_{pf,12}^{-1} \mathbf{R}_{pf,03} & \mathbf{R}_{pf,12}^{-1} \end{bmatrix} \tag{11.57}$$

and the vector of the undetermined constants

$$\mathbf{q}_R^{\mathrm{T}} = [\mathbf{q}_{03}^{\mathrm{T}} \quad \mathbf{q}_b^{\mathrm{T}}]. \tag{11.58}$$

11.2.5 Expressions of Stress Resultants

The matrix $\mathbf{\Phi}_R$ of the basis functions can be partitioned in block form as

$$\mathbf{\Phi}_R = \begin{bmatrix} \mathbf{\Phi}_{Rw} \\ \mathbf{\Phi}_{R\psi} \end{bmatrix}. \tag{11.59}$$

Then the transverse displacement and the rotations of the thick plate can be accordingly expressed as

$$\begin{bmatrix} w \\ \beta_x \\ \beta_y \end{bmatrix} = \mathbf{\Gamma}_{Rd} \cdot \mathbf{q}_R, \tag{11.60}$$

where the matrix

$$\Gamma_{Rd} = \begin{bmatrix} \Gamma_{R,w} \\ \Gamma_{R,\beta_x} \\ \Gamma_{R,\beta_y} \end{bmatrix} = \begin{bmatrix} \Phi_{Rw} \\ -\dfrac{D}{C_s}\Phi_{Rw}^{(3,0)} - \dfrac{D}{C_s}\Phi_{Rw}^{(1,2)} - \Phi_{Rw}^{(1,0)} + \dfrac{1}{C_s}\Phi_{R\psi}^{(0,1)} \\ -\dfrac{D}{C_s}\Phi_{Rw}^{(2,1)} - \dfrac{D}{C_s}\Phi_{Rw}^{(0,3)} - \Phi_{Rw}^{(0,1)} - \dfrac{1}{C_s}\Phi_{R\psi}^{(1,0)} \end{bmatrix}. \tag{11.61}$$

Further, the bending and torsional moments are given as

$$\begin{bmatrix} M_x \\ M_y \\ M_{xy} \end{bmatrix} = \Gamma_{RM} \cdot \mathbf{q}_R, \tag{11.62}$$

where the matrix

$$\Gamma_{RM} = \begin{bmatrix} \Gamma_{R,M_x} \\ \Gamma_{R,M_y} \\ \Gamma_{R,M_{xy}} \end{bmatrix} \tag{11.63}$$

with

$$\Gamma_{R,M_x} = -\dfrac{D^2}{C_s}\Phi_{Rw}^{(4,0)} - \dfrac{D^2(1+\mu)}{C_s}\Phi_{Rw}^{(2,2)} - D\Phi_{Rw}^{(2,0)} - \dfrac{\mu D^2}{C_s}\Phi_{Rw}^{(0,4)} - \mu D\Phi_{Rw}^{(0,2)}$$

$$+ \dfrac{D(1-\mu)}{C_s}\Phi_{R\psi}^{(1,1)}, \quad \Gamma_{R,M_y} = -\dfrac{\mu D^2}{C_s}\Phi_{Rw}^{(4,0)} - \dfrac{D^2(1+\mu)}{C_s}\Phi_{Rw}^{(2,2)} - \mu D\Phi_{Rw}^{(2,0)}$$

$$- \dfrac{D^2}{C_s}\Phi_{Rw}^{(0,4)} - D\Phi_{Rw}^{(0,2)} - \dfrac{D(1-\mu)}{C_s}\Phi_{R\psi}^{(1,1)}, \quad \Gamma_{R,M_{xy}} = -\dfrac{D^2(1-\mu)}{C_s}\Phi_{Rw}^{(3,1)}$$

$$- \dfrac{D^2(1-\mu)}{C_s}\Phi_{Rw}^{(1,3)} - D(1-\mu)\Phi_{Rw}^{(1,1)} - \dfrac{D(1-\mu)}{2C_s}\Phi_{R\psi}^{(2,0)} + \dfrac{D(1-\mu)}{2C_s}\Phi_{R\psi}^{(0,2)}.$$

According to (11.8), the shear forces can be calculated from

$$\begin{bmatrix} Q_x \\ Q_y \end{bmatrix} = \Gamma_{RQ} \cdot \mathbf{q}_R, \tag{11.64}$$

where

$$\Gamma_{RQ} = \begin{bmatrix} \Gamma_{R,Q_x} \\ \Gamma_{R,Q_y} \end{bmatrix} = \begin{bmatrix} -D\Phi_{Rw}^{(3,0)} - D\Phi_{Rw}^{(1,2)} + \Phi_{R\psi}^{(0,1)} \\ -D\Phi_{Rw}^{(2,1)} - D\Phi_{Rw}^{(0,3)} - \Phi_{R\psi}^{(1,0)} \end{bmatrix}. \tag{11.65}$$

11.3   THE SOLUTION OBTAINED FROM THE ENERGY PRINCIPLE

For the elastic system consisting of a thick Reissner plate and an elastic foundation, the total energy is given by (Li and Zhang 2002; Abdalla and Ibrahim 2006)

$$\prod = \prod_p + \prod_f + \prod_q + \prod_\sigma ,$$

(11.66)

where $\prod_p$ is the potential energy of the Reissner plate, $\prod_f$ is the potential energy of the elastic foundation, $\prod_q$ is the potential energy of the transverse load applied to the thick plate, and $\prod_\sigma$ is the total potential energy corresponding to the moments and shear forces acting on the edges.

Specifically, the potential energy of thick Reissner plate is calculated from

$$\prod_p = \frac{1}{2}\int_0^a\int_0^b \left[d_{11}M_x^2 + 2d_{12}M_xM_y + d_{22}M_y^2 + d_{33}M_{xy}^2 + d_{44}Q_x^2 + d_{55}Q_y^2 \right]dxdy,$$

(11.67)

where

$$d_{11} = d_{22} = \frac{1}{D(1-\mu^2)}, d_{12} = -\frac{\mu}{D(1-\mu^2)}, d_{33} = \frac{2}{D(1-\mu)}, \text{ and } d_{44} = d_{55} = \frac{1}{C_s}.$$

The potential energy of the elastic foundation is

$$\prod_p = \frac{1}{2}\int_0^a\int_0^b \left[kw^2 + G_p\left(\frac{\partial w}{\partial x}\right)^2 + G_p\left(\frac{\partial w}{\partial y}\right)^2 \right]dxdy.$$

(11.68)

The potential energy of transverse load acting on the thick plate is defined as

$$\prod_p = -\int_0^a\int_0^b qw\,dxdy.$$

(11.69)

Finally, if all edges of the plate are subjected to the applied bending and torsional moments, and the transverse shear forces, the corresponding potential energy is determined by

$$\begin{aligned}
\prod_\sigma = &-\int_0^b [\overline{Q}_x(a,y)w(a,y) + \overline{M}_x(a,y)\beta_x(a,y) + \overline{M}_{xy}(a,y)\beta_y(a,y)]dy \\
&+ \int_0^b [\overline{Q}_x(0,y)w(0,y) + \overline{M}_x(0,y)\beta_x(0,y) + \overline{M}_{xy}(0,y)\beta_y(0,y)]dy \\
&- \int_0^a [\overline{Q}_y(x,b)w(x,b) + \overline{M}_y(x,b)\beta_y(x,b) + \overline{M}_{xy}(x,b)\beta_x(x,b)]dx \\
&+ \int_0^a [\overline{Q}_y(x,0)w(x,0) + \overline{M}_y(x,0)\beta_y(x,0) + \overline{M}_{xy}(x,0)\beta_x(x,0)]dx.
\end{aligned}$$

(11.70)

Substitution of (11.67)–(11.70) into (11.66) results in

$$\prod = \frac{1}{2}\mathbf{q}_R^{\mathrm{T}}\mathbf{K}_{pf}\mathbf{q}_R - \mathbf{q}_R^{\mathrm{T}}\mathbf{Q}_{pf},\qquad(11.71)$$

where the stiffness matrix

$$\mathbf{K}_{pf} = \int_0^a\int_0^b\Bigg[d_{11}\,\boldsymbol{\Gamma}_{R,M_x}^{\mathrm{T}}\,\boldsymbol{\Gamma}_{R,M_x} + d_{12}\,\boldsymbol{\Gamma}_{R,M_x}^{\mathrm{T}}\,\boldsymbol{\Gamma}_{R,M_y} + d_{12}\,\boldsymbol{\Gamma}_{R,M_y}^{\mathrm{T}}\,\boldsymbol{\Gamma}_{R,M_x} + d_{22}\,\boldsymbol{\Gamma}_{R,M_y}^{\mathrm{T}}\,\boldsymbol{\Gamma}_{R,M_y}$$

$$+ d_{33}\,\boldsymbol{\Gamma}_{R,M_{xy}}^{\mathrm{T}}\,\boldsymbol{\Gamma}_{R,M_{xy}} + d_{44}\,\boldsymbol{\Gamma}_{R,Q_x}^{\mathrm{T}}\,\boldsymbol{\Gamma}_{R,Q_x} + d_{55}\,\boldsymbol{\Gamma}_{R,Q_y}^{\mathrm{T}}\,\boldsymbol{\Gamma}_{R,Q_y} + k\,\boldsymbol{\Gamma}_{R,w}^{\mathrm{T}}\,\boldsymbol{\Gamma}_{R,w}$$

$$+ G_p\left(\frac{1}{C_s}\boldsymbol{\Gamma}_{R,Q_x} - \boldsymbol{\Gamma}_{R,\beta_x}\right)^{\mathrm{T}}\left(\frac{1}{C_s}\boldsymbol{\Gamma}_{R,Q_x} - \boldsymbol{\Gamma}_{R,\beta_x}\right)$$

$$+ G_p\left(\frac{1}{C_s}\boldsymbol{\Gamma}_{R,Q_y} - \boldsymbol{\Gamma}_{R,\beta_y}\right)^{\mathrm{T}}\left(\frac{1}{C_s}\boldsymbol{\Gamma}_{R,Q_y} - \boldsymbol{\Gamma}_{R,\beta_y}\right)\Bigg]dxdy,$$

$$(11.72)$$

and the vector of the resultant forces

$$\mathbf{Q}_{pf} = \int_0^a\int_0^b q\,\boldsymbol{\Gamma}_{R,w}^{\mathrm{T}}$$

$$+ \int_0^b\Big[\overline{Q}_x(a,y)\,\boldsymbol{\Gamma}_{R,w}^{\mathrm{T}}(a,y) + \overline{M}_x(a,y)\,\boldsymbol{\Gamma}_{R,\beta_x}^{\mathrm{T}}(a,y) + \overline{M}_{xy}(a,y)\,\boldsymbol{\Gamma}_{R,\beta_y}^{\mathrm{T}}(a,y)\Big]dy$$

$$- \int_0^b\Big[\overline{Q}_x(0,y)\,\boldsymbol{\Gamma}_{R,w}^{\mathrm{T}}(0,y) + \overline{M}_x(0,y)\,\boldsymbol{\Gamma}_{R,\beta_x}^{\mathrm{T}}(0,y) + \overline{M}_{xy}(0,y)\,\boldsymbol{\Gamma}_{R,\beta_y}^{\mathrm{T}}(0,y)\Big]dy$$

$$+ \int_0^a\Big[\overline{Q}_y(x,b)\,\boldsymbol{\Gamma}_{R,w}^{\mathrm{T}}(x,b) + \overline{M}_y(x,b)\,\boldsymbol{\Gamma}_{R,\beta_y}^{\mathrm{T}}(x,b) + \overline{M}_{xy}(x,b)\,\boldsymbol{\Gamma}_{R,\beta_x}^{\mathrm{T}}(x,b)\Big]dx$$

$$- \int_0^a\Big[\overline{Q}_y(x,0)\,\boldsymbol{\Gamma}_{R,w}^{\mathrm{T}}(x,0) + \overline{M}_y(x,0)\,\boldsymbol{\Gamma}_{R,\beta_y}^{\mathrm{T}}(x,0) + \overline{M}_{xy}(x,0)\,\boldsymbol{\Gamma}_{R,\beta_x}^{\mathrm{T}}(x,0)\Big]dx.$$

$$(11.73)$$

The solution, corresponding to the stationary condition of (11.71), is simply known as

$$\mathbf{K}_{pf}\mathbf{q}_R = \mathbf{Q}_{pf}.\qquad(11.74)$$

In numerical calculations, all the involved series expansions will have to be truncated to contain only a pre-determined number of terms in each direction. Thus,

(11.74) actually represents a finite system of linear algebraic equations about the expansion coefficients of the multiscale Fourier series solutions.

11.4 NUMERICAL EXAMPLES

In this section, the inverse process is again adopted to validate the multiscale Fourier series solutions of the elastic problems concerning a thick plate resting on an elastic foundation. The linear combination of homogeneous solutions $p_{1nl,Hw}(x,y)$ ($l = 1, 2, 3, 4$) derived in Sec. 11.2.1 is taken as the reference (or exact) solution, namely,

$$w_{ref}(x,y) = \sum_{l=1}^{4} 10^{-3} a \cdot p_{1nl,Hw}(x,y) + \frac{D}{ka^3}, \qquad (11.75)$$

where $\beta_n = n\pi / b$ is replaced with $\beta_{ref} = \pi / 2b$.

It is obvious from (11.75) that the plate is also subjected to a uniform load

$$q_{ref}(x,y) = Da^{-3}. \qquad (11.76)$$

The multiscale Fourier series solution will be solved using the energy method.

As listed in Table 11.2, four groups of numerical tests are designated for a comprehensive investigation of the convergence and accuracy of the multiscale Fourier series solution against some of the key model variables, such as, foundation parameters, thickness-to-length ratio, length-to-width ratio, and boundary conditions. In particular, the first group involves with varying the foundation parameters $k_r = ka^4 / D$ and $G_{pr} = G_p a^2 / D$ in the fairly wide ranges from 1 to 10^6 and 1 to 2×10^3, respectively. The second test group is focused on the effects on the solution of the different thickness-to-length ratios, 0.1, 0.01, 0.2, and 0.4, since the plates here are specially considered "thick." In the third group, our attentions are focused on the influences of the length-to-width ratios : 0.5, 0.67, 1.0, 1.25, and 2.0. Finally, the fourth group concerns three different boundary conditions applied to each edge of the plate: the generalized clamped boundary condition (CCCC), the generalized simply supported boundary condition (SSSS), and the generalized free boundary condition (FFFF). In the process of applying the boundary conditions, if the bending moments, the torsional moments, or the shear forces are involved in the prescribed boundary conditions, their distributions can be directly substituted into (11.73) to determine the vector of the resultant forces $\mathbf{Q}_{pf}$; if the transverse deflection or the rotations are involved in the prescribed boundary conditions, their Fourier coefficients will be calculated first and then assigned to the corresponding sub-vectors of the solution vector $\mathbf{q}_R$ in (11.74).

TABLE 11.2
The Testing Setups or Model Parameters in the Numerical Tests.

Groups	Tests	Boundary Conditions	Foundation Parameters (k_r, G_{pr})	Length-to-Width Ratios a/b	Thickness-to-Length Ratios h/a
1	a	CCCC	$(1, 1)$	1.0	0.1
	b		$(100, 10)$		
	c		$(10^4, 100)$		
	d		$(10^6, 2000)$		
2	a	CCCC	$(1, 1)$	1.0	0.1
	b				0.01
	c				0.2
	d				0.4
3	a	CCCC	$(1, 1)$	1.0	0.1
	b			0.67	
	c			0.50	
	d			1.25	
	e			2.0	
4	a	CCCC	$(1, 1)$	1.0	0.1
	b	SSSS			
	c	FFFF			

The first test in each group represents the baseline case: $k_r = 1, G_{pr} = 1, a/b = 1$, $h/a = 0.1$, and the plate is clamped along each edge.

11.4.1 CONVERGENCE CHARACTERISTICS

In the following examples, the convergence and approximation errors are carefully studied by successively truncating the multiscale Fourier series to $M = N = 2, 3, 5, 10, 15,$ and 20.

The convergence characteristics (actually, the error indexes) of the obtained solutions are first presented in Figure 11.1 for the baseline configuration; in particular, both local and global approximation errors with some of the solution variables, such as, the deflection w, the rotations β_x and β_y, and the bending moments M_x and M_y are carefully assessed using the grid (or check) points located in the corresponding regions. It is evident that the displacement converges nicely over the entire solution domain. In comparison, however, the approximation errors have deteriorated significantly for other solution variables like the rotations and the bending moments which involve the derivatives of the displacement function (the rotations and bending moments can be roughly considered to be linearly dependent upon the first and the second order derivatives

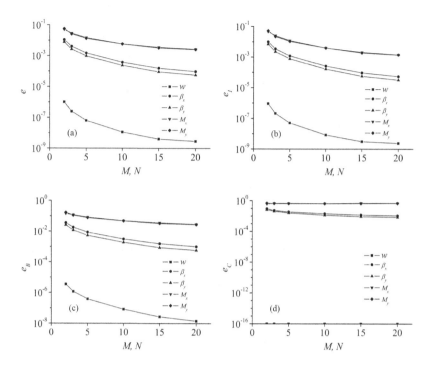

FIGURE 11.1 Convergence characteristics of the multiscale Fourier series solution for the elastic bending of a thick plate resting on the bi-parametric foundation: (a) e-M; N curves; (b) e_I-M, N curves; (c) e_B-M, N curves; and (d) e_C-M, N curves.

of transverse deflection, respectively). Besides, the errors tend to increase as the checking points move from the internal region to the boundaries, or from the boundaries to the corners.

As illustrated in Figure 11.2, the foundation parameters k_r and G_{pr} can influence the solutions substantially, even by orders of magnitude, in terms of the convergence and approximation errors. Accordingly, we may indulge ourselves in speculating that as these two key physical parameters vary considerably, the behavior of the plate-foundation system has experienced some swift transitions from one state to another.

Figures 11.3 and 11.4 shows that except for $h / a = 0.4$, the convergence and accuracy of the solutions are relatively insensitive to the plate geometries, such as, the thickness-to-length and length-to-width ratios. While it is noticeable that the convergence curves for different length-to-width ratios have more or less drifted away from each other (see Figure 11.4), we like to attribute this to the disparity of the spatial resolutions of the series expansion in the x- and y-direction.

In comparison with other model parameters, the adjustment of boundary conditions appears to have more remarked effects on the convergence and accuracy of the multiscale Fourier series solutions. It is seen from Figure 11.5 that the transverse deflections can be satisfactorily calculated with only a small number

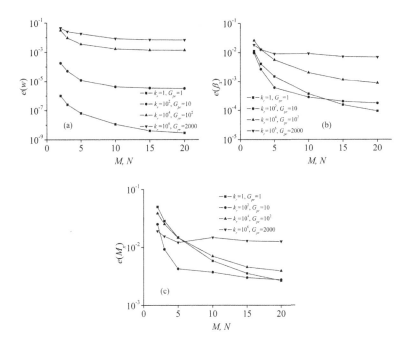

FIGURE 11.2 Convergence comparison of the multiscale Fourier series solutions with different foundation parameters k_r and G_{pr}: (a) $e(w)$-M, N curves; (b) $e(\beta_x)$-M, N curves; and (c) $e(M_x)$-M, N curves.

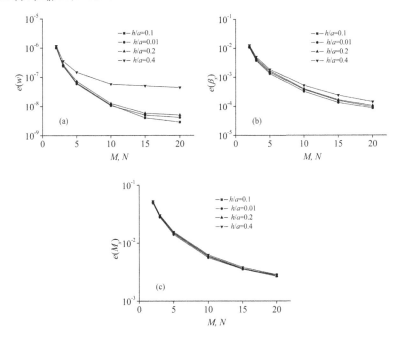

FIGURE 11.3 Convergence comparison of the multiscale Fourier series solutions with different thickness-to-length ratios: (a) $e(w)$-M, N curves; (b) $e(\beta_x)$-M, N curves; and (c) $e(M_x)$-M, N curves.

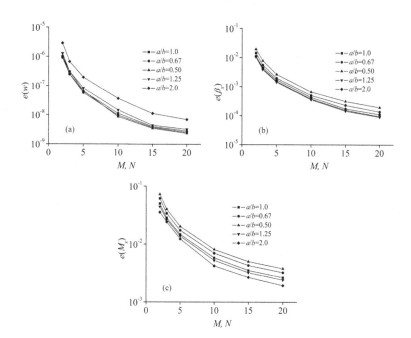

FIGURE 11.4 Convergence comparison of the multiscale Fourier series solutions with different length-to-width ratios: (a) $e(w)$-M, N curves; (b) $e(\beta_x)$-M, N curves; and (c) $e(M_x)$-M, N curves.

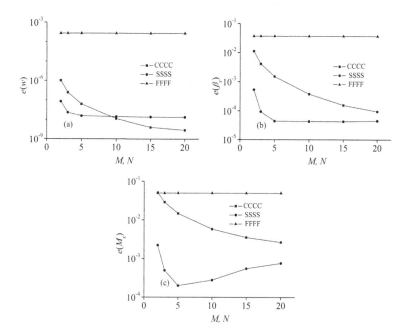

FIGURE 11.5 Convergence comparison of the multiscale Fourier series solutions with different boundary conditions: (a) $e(w)$-M, N curves; (b) $e(\beta_x)$-M, N curves; and (c) $e(M_x)$-M, N curves.

of terms included in the series solutions. The rotation β_x and moment M_x can also be obtained with decent accuracy. While the convergences of the deflection, rotation, and moment solutions are all considered great for the generalized clamped boundary condition (CCCC), the same cannot be said about the other two boundary conditions. Since the rotations and the moments are respectively related to the first and second order derivatives of the transverse deflection, it is understandable that the convergences of the resulting solutions will accordingly deteriorate with the orders of the involved derivatives.

It should be added that for a specific boundary condition the selection of the most suitable form of Fourier series expansion is very important to the quality of the solution. For instance, for the clamped boundary condition, both the transverse displacement and rotations vanish along an edge, the corresponding supplementary terms will thus become unnecessary. However, the use of the Fourier cosine-cosine series will be able to ensure the displacement to have the C^2 continuity (in comparison, only C^1 continuity for the Fourier sine-sine series). As discussed in Chapter 4, for the FFFF boundary conditions, a better performance of the solution can be expected if the displacement and the stress function are expanded into the generalized Fourier cosine-cosine and sine-sine series, respectively.

11.4.2 MULTISCALE CHARACTERISTICS

As shown in Figure 11.2, the convergence and accuracy of the multiscale Fourier series solution are quite sensitive to the foundation parameters k_r and G_{pr}. Among the possible reasons, we are particularly interested in knowing if this elastic system actually exhibits a multiscale characteristic as the model parameters are in the neighborhoods of certain "critical" values. Specifically, by fixing $k_r = 10^4$, we select the several different values for the parameter G_{pr}: 160, 170, 180, 190, and 300. By truncating the series expansions to $M = N = 20$, the corresponding results for some primary solution variables are presented in Figures 11.6–11.10. It is clearly demonstrated that as the model parameter G_{pr} increases, the plate deformations and other relevant physical variables have accordingly undergone a transition from the relatively even wave-like oscillations into the boundary-layer type of multiscale distributions.

While the displacement solution generally compares well with the reference, the physical variables involving the derivatives (especially, those of higher orders) are relatively less accurate as manifested in the spikes at or near the edges or corners. Such approximation errors are considered as the direct results of the reduced degree of smoothness of the corresponding variables, and believed to be more locally confined to the boundary regions. In some cases, this kind of errors may be diminished by artificially boosting the smoothness of the involved solution variables. However, in view of the fact that the smoothness of the primary variable(s) within the solution domain is usually dictated by the physical process itself (or more explicitly, the order of the governing differential equation), this may not always be a viable option; instead, we will resort to increasing the number of the terms included in the series expansions if a better solution is desired.

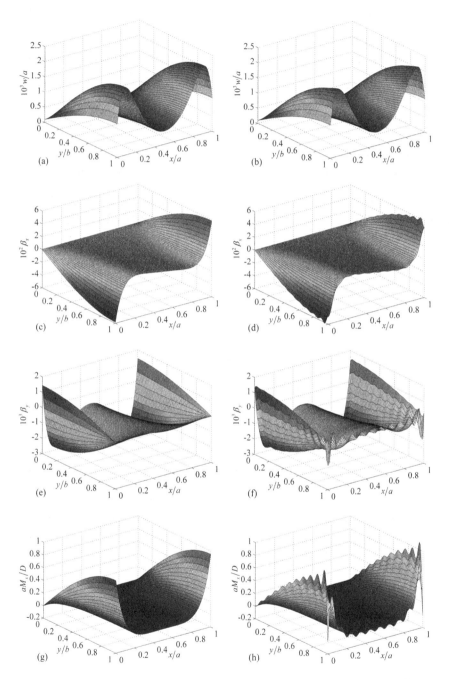

FIGURE 11.6 The deflection, rotation, and moment distributions for $k_r = 10^4$ and $G_{pr} = 160$: (a) w (reference); (b) w (calculated); (c) β_x (reference); (d) β_x (calculated); (e) β_y (reference); (f) β_y (calculated); (g) M_x (reference); and (h) M_x (calculated).

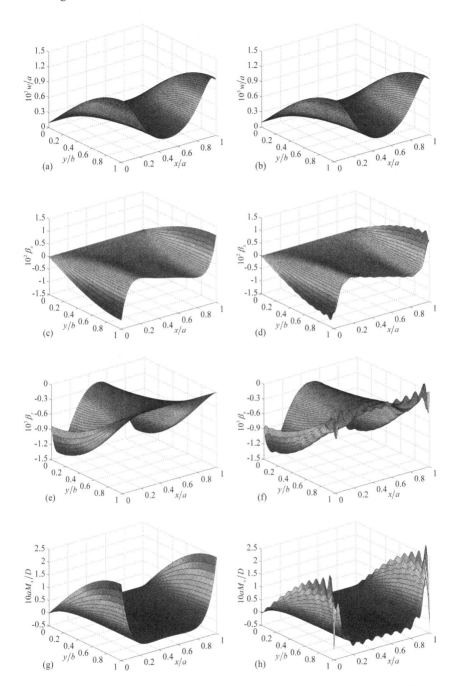

FIGURE 11.7 The deflection, rotation, and moment distributions for $k_r = 10^4$ and $G_{pr} = 170$: (a) w (reference); (b) w (calculated); (c) β_x (reference); (d) β_x (calculated); (e) β_y (reference); (f) β_y (calculated); (g) M_x (reference); and (h) M_x (calculated).

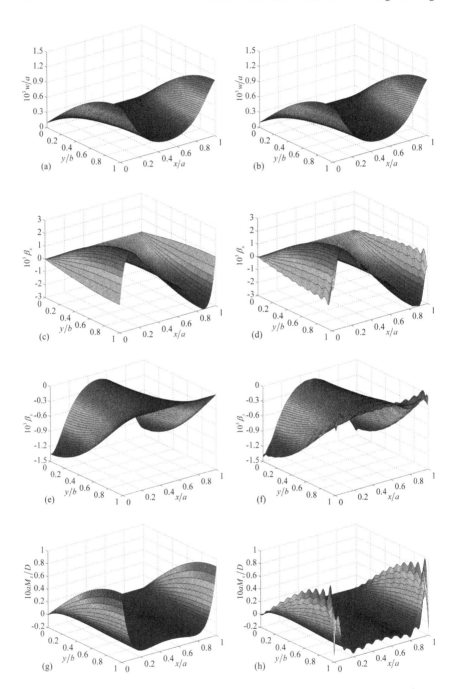

FIGURE 11.8 The deflection, rotation, and moment distributions for $k_r = 10^4$ and $G_{pr} = 180$: (a) w (reference); (b) w (calculated); (c) β_x (reference); (d) β_x (calculated); (e) β_y (reference); (f) β_y (calculated); (g) M_x (reference); and (h) M_x (calculated).

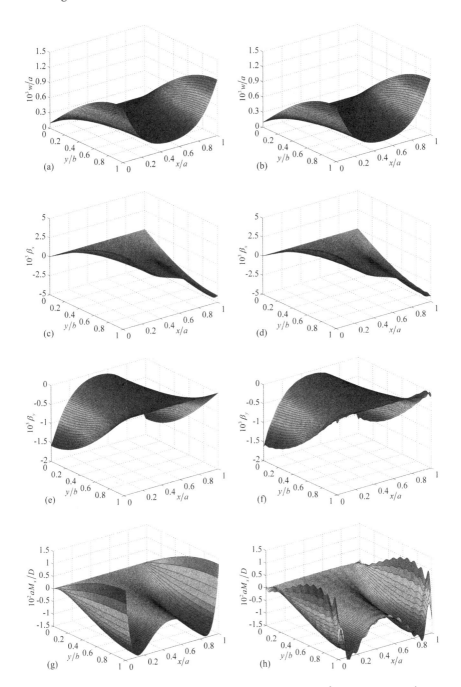

FIGURE 11.9 The deflection, rotation, and moment distributions for $k_r = 10^4$ and $G_{pr} = 190$: (a) w (reference); (b) w (calculated); (c) β_x (reference); (d) β_x (calculated); (e) β_y (reference); (f) β_y (calculated); (g) M_x (reference); and (h) M_x (calculated).

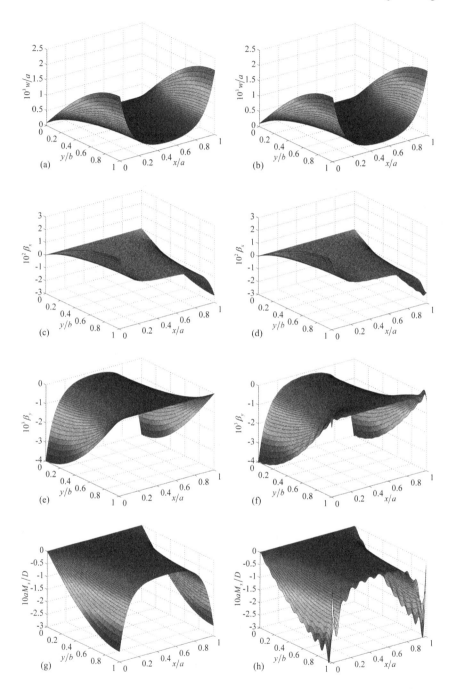

FIGURE 11.10   The deflection, rotation, and moment distributions for $k_r = 10^4$ and $G_{pr} = 300$: (a) w (reference); (b) w (calculated); (c) β_x (reference); (d) β_x (calculated); (e) β_y (reference); (f) β_y (calculated); (g) M_x (reference); and (h) M_x (calculated).

REFERENCES

Abdalla, J. A., and A. M. Ibrahim. 2006. Development of a discrete Reissner-Mindlin element on Winkler foundation. *Finite Elements in Analysis and Design* 42: 740–748.

He, F. B., and Y. P. Shen. 1993. *Theories of Plates and Shells*. Xi'an: Xi'an Jiaotong University Press (in Chinese).

Henwood, D. J., J. R. Whiteman, and A. L. Yettram. 1982. Fourier series solution for a rectangular thick plate with free edges on an elastic foundation. *International Journal for Numerical Methods in Engineering* 18: 1801–1820.

Li, Y. B., and D. C. Zhang. 2002. Middle-thick bending plates on two-parameter elastic foundation by finite element method of lines. *China Civil Engineering Journal* 35(5): 87–92 (in Chinese).

Rashed, Y. F., M. H. Aliabadi, and C. A. Brebbia. 1999. A boundary element formulation for a Reissner plate on a Pasternak foundation. *Computers and Structures* 70: 515–532.

Shi, X. P., S. A. Tan, and T. F. Fwa. 1994. Rectangular thick plate with free edges on Pasternak foundation. *Journal of Engineering Mechanics* 120(5): 971–988.

Sun, W. M., G. S. Yang, and C. Z. Zhang. 1999. General analytic solution for a thick plate on biparametric foundation. *Engineering Mechanics* 16(2): 71–78 (in Chinese).

Wen, P. H. 2008. The fundamental solution of Mindlin plates resting on an elastic foundation in the Laplace domain and its applications. *International Journal of Solids and Structures* 45: 1032–1050.

12 Wave Propagation in Elastic Waveguides

The study of wave propagation in an elastic waveguide represents a specific aspect of the century-old research interest in the vibrations of bars, beams, plates, shells, and other shapes of solids. The waveguides refer to a class of elastic bodies which have neighboring parallel surfaces. The waves propagating in a waveguide will thus interact (in the forms of, such as, reflections, refractions, and diffractions) with the parallel boundaries, causing dispersions, partial mode conversion, or other complicating effects. The waves traveling in a waveguide is typically the superposition of many independent wave modes, and each of them has its unique characteristic and will react to geometrical and physical features or irregularities differently. While this tends to make the process much more complicated and difficult to understand, this phenomenon can potentially reveal a wealth of information about these features for the purpose of detection or interrogation. The wave propagation in waveguides has found broad applications in engineering (Miklowitz 1966) in the fields of, such as, non-destructive testing, vibration control, detection of defects or damages in mechanical systems and civil infrastructures, and in geophysics and seismology (Bullen 1963) for better understanding the earthquake phenomena and nuclear detection problems.

12.1 DESCRIPTION OF THE PROBLEM

Consider an elastic waveguide with a rectangular cross-section $2a \times 2b$, as shown in Figure 12.1.

The wave displacements satisfy the following three-dimensional elastodynamic equations (Achenbach 1973)

$$\left.\begin{array}{l} \rho\dfrac{\partial^2 u}{\partial t^2} = (\lambda + \mu)\dfrac{\partial e}{\partial x} + \mu\nabla^2 u \\[2mm] \rho\dfrac{\partial^2 v}{\partial t^2} = (\lambda + \mu)\dfrac{\partial e}{\partial y} + \mu\nabla^2 v \\[2mm] \rho\dfrac{\partial^2 w}{\partial t^2} = (\lambda + \mu)\dfrac{\partial e}{\partial z} + \mu\nabla^2 w \end{array}\right\}, \tag{12.1}$$

where u, v, and w are the displacements of the elastic body or waveguide; λ and μ are Lame's constants, and ρ is the material density; the volumetric strain

$$e = \frac{\partial u}{\partial x} + \frac{\partial v}{\partial y} + \frac{\partial w}{\partial z} \tag{12.2}$$

DOI: 10.1201/9781003194859-12

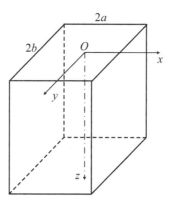

FIGURE 12.1   A rectangular cross-sectioned waveguide.

and the Laplace differential operator

$$\nabla^2 = \frac{\partial^2}{\partial x^2} + \frac{\partial^2}{\partial y^2} + \frac{\partial^2}{\partial z^2}. \tag{12.3}$$

The corresponding constitutive equations are

$$\left.\begin{array}{lll}
\sigma_x = \lambda e + 2\mu\dfrac{\partial u}{\partial x}, & \sigma_y = \lambda e + 2\mu\dfrac{\partial v}{\partial y}, & \sigma_z = \lambda e + 2\mu\dfrac{\partial w}{\partial z} \\[3mm]
\tau_{xy} = \mu\left(\dfrac{\partial u}{\partial y} + \dfrac{\partial v}{\partial x}\right), & \tau_{yz} = \mu\left(\dfrac{\partial v}{\partial z} + \dfrac{\partial w}{\partial y}\right), & \tau_{xz} = \mu\left(\dfrac{\partial w}{\partial x} + \dfrac{\partial u}{\partial z}\right)
\end{array}\right\}. \tag{12.4}$$

The boundary conditions, for instance, at $x = a$ are prescribed as:

1. the clamped edge (C),

$$u(a, y) = 0,\; v(a, y) = 0,\; w(a, y) = 0, \tag{12.5}$$

or

2. the free edge (F),

$$\sigma_x(a, y) = 0,\; \tau_{xy}(a, y) = 0,\; \tau_{xz}(a, y) = 0. \tag{12.6}$$

12.2   THE MULTISCALE FOURIER SERIES SOLUTIONS

Elastic waves propagating in a rectangular waveguide are governed by a system of second order linear differential equations ($2r_u = 2r_v = 2r_w = 2$) with constant coefficients. Thus, the displacements can be expressed as the two-dimensional multiscale Fourier series expansions as described below.

12.2.1 THE DIFFERENTIAL EQUATIONS OF MODAL FUNCTIONS

The discussions will be only focused on the static-state wave propagations in an elastic waveguide. The displacement solutions are accordingly sought in the forms of

$$
\left.
\begin{array}{l}
u = \varphi_u(x,y)\cos(kz - \omega t) \\
v = \varphi_v(x,y)\cos(kz - \omega t) \\
w = \varphi_w(x,y)\sin(kz - \omega t)
\end{array}
\right\}, \tag{12.7}
$$

where φ_u, φ_v, and φ_w are referred to as the modal functions, ω is the angular frequency, and k is the wavenumber which can be real, imaginary, and complex, respectively corresponding to the traveling, near-field, and attenuated waves in the waveguide.

By substituting the displacement solutions (12.7) into (12.1), we will have the differential equations for the modal functions

$$
\mathbf{L}_\varphi
\begin{bmatrix}
\varphi_u \\
\varphi_v \\
\varphi_w
\end{bmatrix} = 0, \tag{12.8}
$$

where

$$
\mathbf{L}_\varphi =
\begin{bmatrix}
\mathbf{L}_{\varphi u} \\
\mathbf{L}_{\varphi v} \\
\mathbf{L}_{\varphi w}
\end{bmatrix} \tag{12.9}
$$

with the differential operators being defined as

$$
\mathbf{L}_{\varphi u} = \left[(\lambda + 2\mu)\frac{\partial^2}{\partial x^2} + \mu\frac{\partial^2}{\partial y^2} + (\rho\omega^2 - \mu k^2) \quad (\lambda + \mu)\frac{\partial^2}{\partial x \partial y} \quad (\lambda + \mu)k\frac{\partial}{\partial x} \right], \tag{12.10}
$$

$$
\mathbf{L}_{\varphi v} = \left[(\lambda + \mu)\frac{\partial^2}{\partial x \partial y} \quad \mu\frac{\partial^2}{\partial x^2} + (\lambda + 2\mu)\frac{\partial^2}{\partial y^2} + (\rho\omega^2 - \mu k^2) \quad (\lambda + \mu)k\frac{\partial}{\partial y} \right], \tag{12.11}
$$

and

$$
\mathbf{L}_{\varphi w} = \left[-(\lambda + \mu)k\frac{\partial}{\partial x} \quad -(\lambda + \mu)k\frac{\partial}{\partial y} \quad \mu\frac{\partial^2}{\partial x^2} + \mu\frac{\partial^2}{\partial y^2} + \rho\omega^2 - (\lambda + 2\mu)k^2 \right]. \tag{12.12}
$$

12.2.2 Structural Decomposition of Modal Functions

Application of the multiscale Fourier series method to solving the current system of second order linear differential equations with constant coefficients should be considered straight-forward by now; the modal functions φ_u, φ_v, and φ_w are correspondingly expressed, over the solution domain $[-\alpha, a] \times [-b, b]$, as:

$$\varphi_u(x,y) = \varphi_{0u}(x,y) + \varphi_{1u}(x,y) + \varphi_{2u}(x,y) + \varphi_{3u}(x,y), \qquad (12.13)$$

$$\varphi_v(x,y) = \varphi_{0v}(x,y) + \varphi_{1v}(x,y) + \varphi_{2v}(x,y) + \varphi_{3v}(x,y), \qquad (12.14)$$

and

$$\varphi_w(x,y) = \varphi_{0w}(x,y) + \varphi_{1w}(x,y) + \varphi_{2w}(x,y) + \varphi_{3w}(x,y), \qquad (12.15)$$

where φ_{0u}, φ_{0v}, φ_{0w} are the internal functions, φ_{1u}, φ_{2u}, φ_{1v}, φ_{2v}, φ_{1w}, φ_{2w} are the boundary functions, and φ_{3u}, φ_{3v}, φ_{3w} are the corner functions, as defined in the previous chapters.

12.2.3 Expressions of Boundary Functions Expanded along the y-Direction

Suppose that (12.8) has the homogeneous solutions:

$$\left.\begin{array}{l} \varphi_{1n,u}(x,y) = \xi_{1n,u}(x)\cos(\beta_n y) + \xi_{2n,u}(x)\sin(\beta_n y) \\ \varphi_{1n,v}(x,y) = \xi_{1n,v}(x)\cos(\beta_n y) + \xi_{2n,v}(x)\sin(\beta_n y) \\ \varphi_{1n,w}(x,y) = \xi_{1n,w}(x)\cos(\beta_n y) + \xi_{2n,w}(x)\sin(\beta_n y) \end{array}\right\}, \qquad (12.16)$$

where $\xi_{1n,u}(x)$, $\xi_{2n,u}(x)$, $\xi_{1n,v}(x)$, $\xi_{2n,v}(x)$, $\xi_{1n,w}(x)$, and $\xi_{2n,w}(x)$ are the one-dimensional functions to be determined, and $\beta_n = n\pi / b$ (n is a nonnegative integer).

Substituting (12.16) into (12.8), we have, for $n > 0$:

the system of equations about the undetermined functions $\xi_{1n,u}(x)$, $\xi_{2n,v}(x)$, and $\xi_{1n,w}(x)$

$$\mathbf{L}_{1n}\begin{bmatrix} \xi_{1n,u} \\ \xi_{2n,v} \\ \xi_{1n,w} \end{bmatrix} = 0, \qquad (12.17)$$

where

$$\mathbf{L}_{1n} = \begin{bmatrix} \mathbf{L}_{1n,u} \\ \mathbf{L}_{1n,v} \\ \mathbf{L}_{1n,w} \end{bmatrix} \tag{12.18}$$

with the differential operators

$$\mathbf{L}_{1n,u} = \left[(\lambda + 2\mu)\frac{d^2}{dx^2} + \rho\omega^2 - \mu\beta_n^2 - \mu k^2 \quad (\lambda + \mu)\beta_n \frac{d}{dx} \quad (\lambda + \mu)k\frac{d}{dx} \right], \tag{12.19}$$

$$\mathbf{L}_{1n,v} = \left[-(\lambda + \mu)\beta_n \frac{d}{dx} \quad \mu\frac{d^2}{dx^2} + \rho\omega^2 - (\lambda + 2\mu)\beta_n^2 - \mu k^2 \quad -(\lambda + \mu)\beta_n k \right], \tag{12.20}$$

and

$$\mathbf{L}_{1n,w} = \left[-(\lambda + \mu)k\frac{d}{dx} \quad -(\lambda + \mu)\beta_n k \quad \mu\frac{d^2}{dx^2} + \rho\omega^2 - \mu\beta_n^2 - (\lambda + 2\mu)k^2 \right]; \tag{12.21}$$

the system of equations about the undetermined functions $\xi_{2n,u}(x)$, $\xi_{1n,v}(x)$, and $\xi_{2n,w}(x)$

$$\mathbf{L}_{2n} \begin{bmatrix} \xi_{2n,u} \\ \xi_{1n,v} \\ \xi_{2n,w} \end{bmatrix} = 0, \tag{12.22}$$

where $\mathbf{L}_{2n}$ can be directly obtained from $\mathbf{L}_{1n}$ by replacing β_n with $-\beta_n$ in (12.19)–(12.21).

For $n = 0$, the system of equations about the undetermined functions $\xi_{10,u}(x)$, $\xi_{10,v}(x)$, and $\xi_{10,w}(x)$ are as follows:

$$\mathbf{L}_{10} \begin{bmatrix} \xi_{10,u} \\ \xi_{10,v} \\ \xi_{10,w} \end{bmatrix} = 0, \tag{12.23}$$

where the differential operators $\mathbf{L}_{10}$ are the same as $\mathbf{L}_{1n}$ except that $\beta_n \equiv 0$.

The solutions of (12.17) are sought in the form of

$$\left.\begin{array}{c} \xi_{1n,u}(x) = G_{1n1,u}\,p_{1n}(x) \\ \xi_{2n,u}(x) = G_{1n2,v}\,p_{1n}(x) \\ \xi_{1n,w}(x) = G_{1n1,w}\,p_{1n}(x) \end{array}\right\}, \qquad (12.24)$$

where

$$p_{1n}(x) = \exp(\eta_n x) \qquad (12.25)$$

and $G_{1n1,u}$, $G_{1n2,v}$, $G_{1n1,w}$, and η_n are the undetermined constants.
Substituting (12.24) into (12.17) leads to the characteristic equation

$$[\mu(-\eta_n^2 + \beta_n^2 + k^2) - \rho\omega^2]^2[(\lambda + 2\mu)(-\eta_n^2 + \beta_n^2 + k^2) - \rho\omega^2] = 0. \qquad (12.26)$$

If k is a real number, and

$$\Delta_1 = \beta_n^2 + k^2 - \frac{\rho\omega^2}{\mu} > 0, \qquad (12.27)$$

then (12.26) has two distinct double real roots

$$\eta_{n,1} = \eta_{n,2} = \alpha_{1n}, \text{ and } \eta_{n,3} = \eta_{n,4} = -\alpha_{1n}, \qquad (12.28)$$

where $\alpha_{1n} = \sqrt{\beta_n^2 + k^2 - \frac{\rho\omega^2}{\mu}}$;
if $\Delta_1 < 0$, then (12.26) has two distinct double imaginary roots as

$$\eta_{n,1} = \eta_{n,2} = i\alpha_{2n}, \text{ and } \eta_{n,3} = \eta_{n,4} = -i\alpha_{2n}, \qquad (12.29)$$

where $\alpha_{2n} = \sqrt{-\left(\beta_n^2 + k^2 - \frac{\rho\omega^2}{\mu}\right)}$.
Similarly, if

$$\Delta_2 = \beta_n^2 + k^2 - \frac{\rho\omega^2}{\lambda + 2\mu} > 0, \qquad (12.30)$$

(12.26) has two more distinct real roots as

$$\eta_{n,5} = \alpha_{3n}, \text{ and } \eta_{n,6} = -\alpha_{3n}, \qquad (12.31)$$

where $\alpha_{3n} = \sqrt{\beta_n^2 + k^2 - \frac{\rho\omega^2}{\lambda + 2\mu}}$;

TABLE 12.1

Expressions for the Basis Functions $p_{1nl}(x)$, $l = 1, 2, 3, 4$.

	$\Delta_1 > 0$	$\Delta_1 < 0$	$\Delta_2 > 0$	$\Delta_2 < 0$
$p_{1n1}(x)$	$\cosh(\alpha_{1n}x)$	$\cos(\alpha_{2n}x)$		
$p_{1n2}(x)$	$\sinh(\alpha_{1n}x)$	$\sin(\alpha_{2n}x)$		
$p_{1n3}(x)$			$\cosh(\alpha_{3n}x)$	$\cos(\alpha_{4n}x)$
$p_{1n4}(x)$			$\sinh(\alpha_{3n}x)$	$\sin(\alpha_{4n}x)$

if $\Delta_2 < 0$, (12.26) has two more distinct imaginary roots as

$$\eta_{n,5} = i\alpha_{4n}, \text{ and } \eta_{n,6} = -i\alpha_{4n}, \tag{12.32}$$

where $\alpha_{4n} = \sqrt{-\left(\beta_n^2 + k^2 - \frac{\rho\omega^2}{\lambda+2\mu}\right)}$.

For clarity, the corresponding basis functions $p_{1nl}(x)$, $l = 1, 2, 3, 4$, are summarized in Table 12.1. Since $\eta_{n,1}$ and $\eta_{n,3}$ are the double roots of (12.26), two more basis functions, $xp_{1n1}(x)$ and $xp_{1n2}(x)$, seemingly need to be included in the list. However, if they are linearly combined with $p_{1nl}(x)$ ($l = 1, 2, 3, 4$) and substituted into (12.17), it can be verified that the combination coefficients for them are identically equal to zero. Therefore, only the basis functions $p_{1nl}(x)$ ($l = 1, 2, 3, 4$) are actually utilized in determining the Fourier coefficients $\xi_{1n,u}(x)$, $\xi_{2n,v}(x)$, and $\xi_{1n,w}(x)$.

Denote

$$\mathbf{p}_{1n}^{T}(x) = [p_{1n1}(x) \quad p_{1n2}(x) \quad p_{1n3}(x) \quad p_{1n4}(x)], \tag{12.33}$$

$$\mathbf{a}_{1n1,u}^{T} = [G_{1n1,u}^{1} \quad G_{1n1,u}^{2} \quad G_{1n1,u}^{3} \quad G_{1n1,u}^{4}], \tag{12.34}$$

$$\mathbf{a}_{1n2,v}^{T} = [G_{1n2,v}^{1} \quad G_{1n2,v}^{2} \quad G_{1n2,v}^{3} \quad G_{1n2,v}^{4}], \tag{12.35}$$

and

$$\mathbf{a}_{1n1,w}^{T} = [G_{1n1,w}^{1} \quad G_{1n1,w}^{2} \quad G_{1n1,w}^{3} \quad G_{1n1,w}^{4}]. \tag{12.36}$$

The Fourier coefficients $\xi_{1n,u}(x)$, $\xi_{2n,v}(x)$, and $\xi_{1n,w}(x)$ can be expressed as

$$\xi_{1n,u}(x) = \mathbf{p}_{1n}^{T}(x) \cdot \mathbf{a}_{1n1,u}, \tag{12.37}$$

$$\xi_{2n,v}(x) = \mathbf{p}_{1n}^{T}(x) \cdot \mathbf{a}_{1n2,v}, \tag{12.38}$$

and

$$\xi_{1n,w}(x) = \mathbf{p}_{1n}^{\mathrm{T}}(x) \cdot \mathbf{a}_{1n1,w}. \tag{12.39}$$

By substituting these expressions into (12.17), we have the following relationships for the undetermined constants:

for $\Delta_1 > 0$,

$$\left.\begin{aligned} \alpha_{1n}G_{1n1,u}^1 + \beta_n G_{1n2,v}^2 + kG_{1n1,w}^2 = 0 \\ \alpha_{1n}G_{1n1,u}^2 + \beta_n G_{1n2,v}^1 + kG_{1n1,w}^1 = 0 \end{aligned}\right\}; \tag{12.40}$$

for $\Delta_1 < 0$,

$$\left.\begin{aligned} -\alpha_{2n}G_{1n1,u}^1 + \beta_n G_{1n2,v}^2 + kG_{1n1,w}^2 = 0 \\ \alpha_{2n}G_{1n1,u}^2 + \beta_n G_{1n2,v}^1 + kG_{1n1,w}^1 = 0 \end{aligned}\right\}; \tag{12.41}$$

for $\Delta_2 > 0$,

$$\left.\begin{aligned} \frac{G_{1n1,u}^3}{-\alpha_{3n}} = \frac{G_{1n2,v}^4}{\beta_n} = \frac{G_{1n1,w}^4}{k} \\ \frac{G_{1n1,u}^4}{-\alpha_{3n}} = \frac{G_{1n2,v}^3}{\beta_n} = \frac{G_{1n1,w}^3}{k} \end{aligned}\right\}; \tag{12.42}$$

for $\Delta_2 < 0$,

$$\left.\begin{aligned} \frac{G_{1n1,u}^3}{-\alpha_{4n}} = \frac{G_{1n2,v}^4}{\beta_n} = \frac{G_{1n1,w}^4}{k} \\ \frac{G_{1n1,u}^4}{\alpha_{4n}} = \frac{G_{1n2,v}^3}{\beta_n} = \frac{G_{1n1,w}^3}{k} \end{aligned}\right\}. \tag{12.43}$$

In light of (12.40)–(12.43), we have

$$\mathbf{a}_{1n1} = \mathbf{T}_{1n1}\mathbf{a}_{1n1,R}, \tag{12.44}$$

where

$$\mathbf{a}_{1n1}^{\mathrm{T}} = [\mathbf{a}_{1n1,u}^{\mathrm{T}} \quad \mathbf{a}_{1n2,v}^{\mathrm{T}} \quad \mathbf{a}_{1n1,w}^{\mathrm{T}}], \tag{12.45}$$

$$\mathbf{a}_{1n1,R}^{\mathrm{T}} = [G_{1n1,u}^1 \quad G_{1n1,u}^2 \quad G_{1n2,v}^3 \quad G_{1n2,v}^4 \quad G_{1n1,w}^1 \quad G_{1n1,w}^2], \tag{12.46}$$

and $\mathbf{T}_{1n1}$ is a transformation matrix defined as

$$\mathbf{T}_{1n1} = \begin{bmatrix} 1 & 0 & 0 & 0 & 0 & 0 \\ 0 & 1 & 0 & 0 & 0 & 0 \\ 0 & 0 & 0 & -\alpha_{3n}/\beta_n & 0 & 0 \\ 0 & 0 & -\alpha_{3n}/\beta_n & 0 & 0 & 0 \\ 0 & -\alpha_{1n}/\beta_n & 0 & 0 & -k/\beta_n & 0 \\ -\alpha_{1n}/\beta_n & 0 & 0 & 0 & 0 & -k/\beta_n \\ 0 & 0 & 1 & 0 & 0 & 0 \\ 0 & 0 & 0 & 1 & 0 & 0 \\ 0 & 0 & 0 & 0 & 1 & 0 \\ 0 & 0 & 0 & 0 & 0 & 1 \\ 0 & 0 & k/\beta_n & 0 & 0 & 0 \\ 0 & 0 & 0 & k/\beta_n & 0 & 0 \end{bmatrix}. \tag{12.47}$$

Basically, (12.44) tells that only six of the 12 undetermined constants are actually independent.

It must be noted that the transformation matrix $\mathbf{T}_{1n1}$ in (12.47) is derived from (12.40) and (12.42) and is specifically suitable for the case of $\Delta_1 > 0$ and $\Delta_2 > 0$. For the other three possible cases, the expressions for the transformation matrix $\mathbf{T}_{1n1}$ will have to be modified accordingly.

In the meantime, referring back to (8.50) and (8.51), we have

$$\mathbf{R}_{1n}\mathbf{a}_{1n1,u} = \mathbf{q}_{1n1,u}, \tag{12.48}$$

$$\mathbf{R}_{1n}\mathbf{a}_{1n2,v} = \mathbf{q}_{1n2,v}, \tag{12.49}$$

and

$$\mathbf{R}_{1n}\mathbf{a}_{1n1,w} = \mathbf{q}_{1n1,w}, \tag{12.50}$$

where

$$\mathbf{R}_{1n} = \begin{bmatrix} \mathbf{p}_{1n}^{(0)\mathrm{T}}(a) - \mathbf{p}_{1n}^{(0)\mathrm{T}}(-a) \\ \mathbf{p}_{1n}^{(1)\mathrm{T}}(a) - \mathbf{p}_{1n}^{(1)\mathrm{T}}(-a) \end{bmatrix}, \tag{12.51}$$

and $\mathbf{q}_{1n1,u}$, $\mathbf{q}_{1n2,v}$, and $\mathbf{q}_{1n1,w}$ are the sub-vectors of the boundary Fourier coefficients, as defined in (8.53).

Combining (12.48)–(12.50) leads to

$$\mathbf{S}_{1n}\mathbf{a}_{1n1} = \mathbf{q}_{1n1}, \tag{12.52}$$

where

$$\mathbf{S}_{1n} = \begin{bmatrix} \mathbf{R}_{1n} & \mathbf{0} & \mathbf{0} \\ \mathbf{0} & \mathbf{R}_{1n} & \mathbf{0} \\ \mathbf{0} & \mathbf{0} & \mathbf{R}_{1n} \end{bmatrix}, \tag{12.53}$$

and

$$\mathbf{q}_{1n1}^{\mathrm{T}} = [\mathbf{q}_{1n1,u}^{\mathrm{T}} \quad \mathbf{q}_{1n2,v}^{\mathrm{T}} \quad \mathbf{q}_{1n1,w}^{\mathrm{T}}]. \tag{12.54}$$

Using (12.44), (12.52) can be rewritten as

$$\mathbf{S}_{1n}\mathbf{T}_{1n1}\mathbf{a}_{1n1,R} = \mathbf{q}_{1n1} \tag{12.55}$$

or

$$\mathbf{a}_{1n1,R} = (\mathbf{S}_{1n}\mathbf{T}_{1n1})^{-1}\mathbf{q}_{1n1}. \tag{12.56}$$

Thus, the Fourier coefficients $\xi_{1n,u}(x)$, $\xi_{2n,v}(x)$, and $\xi_{1n,w}(x)$ can be eventually expressed as

$$\begin{bmatrix} \xi_{1n,u} \\ \xi_{2n,v} \\ \xi_{1n,w} \end{bmatrix} = \mathbf{p}_{1n1,R}^{\mathrm{T}}(x) \cdot \mathbf{q}_{1n1}, \tag{12.57}$$

where

$$\mathbf{p}_{1n1,R}^{\mathrm{T}}(x) = \begin{bmatrix} \mathbf{p}_{1n}^{\mathrm{T}}(x) & \mathbf{0} & \mathbf{0} \\ \mathbf{0} & \mathbf{p}_{1n}^{\mathrm{T}}(x) & \mathbf{0} \\ \mathbf{0} & \mathbf{0} & \mathbf{p}_{1n}^{\mathrm{T}}(x) \end{bmatrix} \cdot \mathbf{T}_{1n1} \cdot (\mathbf{S}_{1n}\mathbf{T}_{1n1})^{-1}. \tag{12.58}$$

The matrix of the basis functions can then be defined as

$$\Phi_{1n,1}^{\mathrm{T}}(x,y) = \mathbf{H}_{1n,1}(y) \cdot \mathbf{p}_{1n1,R}^{\mathrm{T}}(x), \tag{12.59}$$

where

$$\mathbf{H}_{1n,1}(y) = \begin{bmatrix} \cos(\beta_n y) & 0 & 0 \\ 0 & \sin(\beta_n y) & 0 \\ 0 & 0 & \cos(\beta_n y) \end{bmatrix}. \tag{12.60}$$

Evidently, the differentiations to be applied to the displacement solutions can be directly transferred to the basis functions, namely,

$$\Phi_{1n,1}^{(k_1,k_2)\mathrm{T}}(x,y) = \mathbf{H}_{1n,1}^{(k_2)}(y) \cdot \mathbf{p}_{1n1,R}^{(k_1)\mathrm{T}}(x), \tag{12.61}$$

where k_1 and k_2 are nonnegative integers,

$$\mathbf{p}_{1n1,R}^{(k_1)\mathrm{T}}(x) = \begin{bmatrix} \mathbf{p}_{1n}^{(k_1)\mathrm{T}}(x) & \mathbf{0} & \mathbf{0} \\ \mathbf{0} & \mathbf{p}_{1n}^{(k_1)\mathrm{T}}(x) & \mathbf{0} \\ \mathbf{0} & \mathbf{0} & \mathbf{p}_{1n}^{(k_1)\mathrm{T}}(x) \end{bmatrix} \cdot \mathbf{T}_{1n1} \cdot (\mathbf{S}_{1n}\mathbf{T}_{1n1})^{-1} \tag{12.62}$$

and

$$\mathbf{H}_{1n,1}^{(k_2)}(y) = \begin{bmatrix} [\cos(\beta_n y)]^{(k_2)} & 0 & 0 \\ 0 & [\sin(\beta_n y)]^{(k_2)} & 0 \\ 0 & 0 & [\cos(\beta_n y)]^{(k_2)} \end{bmatrix}. \tag{12.63}$$

The Fourier coefficients $\xi_{2n,u}(x)$, $\xi_{1n,v}(x)$, and $\xi_{2n,w}(x)$ can be similarly obtained as

$$\begin{bmatrix} \xi_{2n,u} \\ \xi_{1n,v} \\ \xi_{2n,w} \end{bmatrix} = \mathbf{p}_{1n2,R}^{\mathrm{T}}(x) \cdot \mathbf{q}_{1n2}, \tag{12.64}$$

where

$$\mathbf{q}_{1n2}^{\mathrm{T}} = [\mathbf{q}_{1n2,u}^{\mathrm{T}} \quad \mathbf{q}_{1n1,v}^{\mathrm{T}} \quad \mathbf{q}_{1n2,w}^{\mathrm{T}}] \tag{12.65}$$

and

$$\mathbf{p}_{1n2,R}^{\mathrm{T}}(x) = \begin{bmatrix} \mathbf{p}_{1n}^{\mathrm{T}}(x) & \mathbf{0} & \mathbf{0} \\ \mathbf{0} & \mathbf{p}_{1n}^{\mathrm{T}}(x) & \mathbf{0} \\ \mathbf{0} & \mathbf{0} & \mathbf{p}_{1n}^{\mathrm{T}}(x) \end{bmatrix} \cdot \mathbf{T}_{1n2} \cdot (\mathbf{S}_{1n}\mathbf{T}_{1n2})^{-1}. \tag{12.66}$$

The transformation matrix $\mathbf{T}_{1n2}$ in (12.66) can be obtained from $\mathbf{T}_{1n1}$ given in (12.47) by simply replacing β_n with $-\beta_n$.

The matrix of the basis functions can be accordingly defined as

$$\Phi_{1n,2}^{\mathrm{T}}(x,y) = \mathbf{H}_{1n,2}(y) \cdot \mathbf{p}_{1n2,R}^{\mathrm{T}}(x), \tag{12.67}$$

where

$$\mathbf{H}_{1n,2}(y) = \begin{bmatrix} \sin(\beta_n y) & 0 & 0 \\ 0 & \cos(\beta_n y) & 0 \\ 0 & 0 & \sin(\beta_n y) \end{bmatrix}. \tag{12.68}$$

The matrix of the basis functions is directly differentiable

$$\Phi_{1n,2}^{(k_1,k_2)\mathrm{T}}(x,y) = \mathbf{H}_{1n,2}^{(k_2)}(y) \cdot \mathbf{p}_{1n2,R}^{(k_1)\mathrm{T}}(x), \tag{12.69}$$

where k_1 and k_2 are nonnegative integers.

Now, let's turn to the case of $n = 0$.

By denoting

$$\mathbf{a}_{101,u}^{\mathrm{T}} = [G_{101,u}^1 \quad G_{101,u}^2 \quad G_{101,u}^3 \quad G_{101,u}^4], \tag{12.70}$$

$$\mathbf{a}_{101,v}^{\mathrm{T}} = [G_{101,v}^1 \quad G_{101,v}^2 \quad G_{101,v}^3 \quad G_{101,v}^4], \tag{12.71}$$

and

$$\mathbf{a}_{101,w}^{\mathrm{T}} = [G_{101,w}^1 \quad G_{101,w}^2 \quad G_{101,w}^3 \quad G_{101,w}^4], \tag{12.72}$$

from (12.23) we have:

for $\Delta_1 > 0$,

$$\left.\begin{array}{l} \alpha_{10}G_{101,u}^1 + kG_{101,w}^2 = 0 \\ \alpha_{10}G_{101,u}^2 + kG_{101,w}^1 = 0 \end{array}\right\}; \tag{12.73}$$

for $\Delta_1 < 0$,

$$\left.\begin{array}{l} -\alpha_{20}G_{101,u}^1 + kG_{101,w}^2 = 0 \\ \alpha_{20}G_{101,u}^2 + kG_{101,w}^1 = 0 \end{array}\right\}; \tag{12.74}$$

for $\Delta_2 > 0$,

$$\left.\begin{array}{l} \dfrac{G_{101,u}^3}{-\alpha_{30}} = \dfrac{G_{101,w}^4}{k} \\[2mm] \dfrac{G_{101,u}^4}{-\alpha_{30}} = \dfrac{G_{101,w}^3}{k} \end{array}\right\}; \tag{12.75}$$

for $\Delta_2 < 0$,

$$
\left.
\begin{aligned}
\frac{G_{101,u}^3}{-\alpha_{40}} &= \frac{G_{101,w}^4}{k} \\
\frac{G_{101,u}^4}{\alpha_{40}} &= \frac{G_{101,w}^3}{k}
\end{aligned}
\right\};
\tag{12.76}
$$

$G_{101,v}^1$ and $G_{101,v}^2$ are arbitrary constants, and

$$
G_{101,v}^3 = G_{101,v}^4 = 0.
\tag{12.77}
$$

More concisely, the relationships (12.73)–(12.77) can be rewritten as

$$
\mathbf{a}_{101} = \mathbf{T}_{101}\mathbf{a}_{101,R},
\tag{12.78}
$$

where

$$
\mathbf{a}_{101}^{\mathrm{T}} = [\mathbf{a}_{101,u}^{\mathrm{T}} \quad \mathbf{a}_{101,v}^{\mathrm{T}} \quad \mathbf{a}_{101,w}^{\mathrm{T}}],
\tag{12.79}
$$

$$
\mathbf{a}_{101,R}^{\mathrm{T}} = [G_{101,u}^1 \quad G_{101,u}^2 \quad G_{101,v}^1 \quad G_{101,v}^2 \quad G_{101,w}^3 \quad G_{101,w}^4]
\tag{12.80}
$$

and $\mathbf{T}_{101}$ is the transformation matrix.

The transformation matrix $\mathbf{T}_{101}$ potentially has four different forms, depending upon the signs of Δ_1 and Δ_2. For instance, if $\Delta_1 > 0$ and $\Delta_2 > 0$, it can be explicitly given as

$$
\mathbf{T}_{101} =
\begin{bmatrix}
1 & 0 & 0 & 0 & 0 & 0 \\
0 & 1 & 0 & 0 & 0 & 0 \\
0 & 0 & 0 & 0 & 0 & -\alpha_{30}/k \\
0 & 0 & 0 & 0 & -\alpha_{30}/k & 0 \\
0 & 0 & 1 & 0 & 0 & 0 \\
0 & 0 & 0 & 1 & 0 & 0 \\
0 & 0 & 0 & 0 & 0 & 0 \\
0 & 0 & 0 & 0 & 0 & 0 \\
0 & -\alpha_{10}/k & 0 & 0 & 0 & 0 \\
-\alpha_{10}/k & 0 & 0 & 0 & 0 & 0 \\
0 & 0 & 0 & 0 & 1 & 0 \\
0 & 0 & 0 & 0 & 0 & 1
\end{bmatrix}.
\tag{12.81}
$$

Regardless of the specific expressions for the transformation matrix $\mathbf{T}_{101}$, the Fourier coefficients $\xi_{10,u}(x)$, $\xi_{10,v}(x)$, and $\xi_{10,w}(x)$ can be invariably expressed as

$$
\begin{bmatrix}
\xi_{10,u} \\
\xi_{10,v} \\
\xi_{10,w}
\end{bmatrix}
= \mathbf{p}_{101,R}^{\mathrm{T}}(x) \cdot \mathbf{q}_{101},
\tag{12.82}
$$

where

$$
\mathbf{p}_{101,R}^{\mathrm{T}}(x) =
\begin{bmatrix}
\mathbf{p}_{10}^{\mathrm{T}}(x) & \mathbf{0} & \mathbf{0} \\
\mathbf{0} & \mathbf{p}_{10}^{\mathrm{T}}(x) & \mathbf{0} \\
\mathbf{0} & \mathbf{0} & \mathbf{p}_{10}^{\mathrm{T}}(x)
\end{bmatrix}
\cdot \mathbf{T}_{101} \cdot (\mathbf{S}_{10} \mathbf{T}_{101})^{-1}
\tag{12.83}
$$

and

$$
\mathbf{q}_{101}^{\mathrm{T}} = [\mathbf{q}_{101,u}^{\mathrm{T}} \quad \mathbf{q}_{101,v}^{\mathrm{T}} \quad \mathbf{q}_{101,w}^{\mathrm{T}}].
\tag{12.84}
$$

As before, the matrix of the basis functions is accordingly defined as

$$
\boldsymbol{\Phi}_{10,1}^{\mathrm{T}}(x,y) = \mathbf{H}_{10,1}(y) \cdot \mathbf{p}_{101,R}^{\mathrm{T}}(x),
\tag{12.85}
$$

where

$$
\mathbf{H}_{10,1}(y) =
\begin{bmatrix}
\dfrac{1}{2} & 0 & 0 \\
0 & \dfrac{1}{2} & 0 \\
0 & 0 & \dfrac{1}{2}
\end{bmatrix}.
\tag{12.86}
$$

This matrix is obviously differentiable, that is,

$$
\boldsymbol{\Phi}_{10,1}^{(k_1,k_2)\mathrm{T}}(x,y) = \mathbf{H}_{10,1}^{(k_2)}(y) \cdot \mathbf{p}_{101,R}^{(k_1)\mathrm{T}}(x),
\tag{12.87}
$$

where k_1 and k_2 are nonnegative integers, denoting the orders of differentiations with respect to the x- and y-coordinate.

Define the matrices of the basis functions

$$
\boldsymbol{\Phi}_{1,1}^{\mathrm{T}}(x,y) = [\boldsymbol{\Phi}_{10,1}^{\mathrm{T}}(x,y) \quad \boldsymbol{\Phi}_{11,1}^{\mathrm{T}}(x,y) \quad \cdots \quad \boldsymbol{\Phi}_{1n,1}^{\mathrm{T}}(x,y) \quad \cdots],
\tag{12.88}
$$

$$\Phi_{1,2}^{\mathrm{T}}(x,y) = [\Phi_{11,2}^{\mathrm{T}}(x,y) \quad \Phi_{12,2}^{\mathrm{T}}(x,y) \quad \cdots \quad \Phi_{1n,2}^{\mathrm{T}}(x,y) \quad \cdots], \quad (12.89)$$

$$\Phi_1^{\mathrm{T}}(x,y) = [\Phi_{1,1}^{\mathrm{T}}(x,y) \quad \Phi_{1,2}^{\mathrm{T}}(x,y)], \quad (12.90)$$

and the vectors of the boundary Fourier coefficients

$$\mathbf{q}_{1,1}^{\mathrm{T}} = [\mathbf{q}_{101}^{\mathrm{T}} \quad \mathbf{q}_{111}^{\mathrm{T}} \quad \cdots \quad \mathbf{q}_{1n1}^{\mathrm{T}} \quad \cdots], \quad (12.91)$$

$$\mathbf{q}_{1,2}^{\mathrm{T}} = [\mathbf{q}_{112}^{\mathrm{T}} \quad \mathbf{q}_{122}^{\mathrm{T}} \quad \cdots \quad \mathbf{q}_{1n2}^{\mathrm{T}} \quad \cdots], \quad (12.92)$$

and

$$\mathbf{q}_1^{\mathrm{T}} = [\mathbf{q}_{1,1}^{\mathrm{T}} \quad \mathbf{q}_{1,2}^{\mathrm{T}}]. \quad (12.93)$$

Then the boundary functions, $\varphi_{1u}(x,y)$, $\varphi_{1v}(x,y)$, and $\varphi_{1w}(x,y)$, can be expressed as

$$\begin{bmatrix} \varphi_{1u}(x,y) \\ \varphi_{1v}(x,y) \\ \varphi_{1w}(x,y) \end{bmatrix} = \Phi_1^{\mathrm{T}}(x,y) \cdot \mathbf{q}_1. \quad (12.94)$$

The partial derivatives of the boundary functions, if desired, are directly obtainable from differentiating the matrices of the basis functions accordingly.

12.2.4 EXPRESSIONS OF BOUNDARY FUNCTIONS EXPANDED ALONG THE X-DIRECTION

The other group of the homogeneous solutions of (12.8) is similarly sought as

$$\left. \begin{array}{l} \varphi_{2m,u}(x,y) = \zeta_{1m,u}(y)\cos(\alpha_m x) + \zeta_{2m,u}(y)\sin(\alpha_m x) \\ \varphi_{2m,v}(x,y) = \zeta_{1m,v}(y)\cos(\alpha_m x) + \zeta_{2m,v}(y)\sin(\alpha_m x) \\ \varphi_{2m,w}(x,y) = \zeta_{1m,w}(y)\cos(\alpha_m x) + \zeta_{2m,w}(y)\sin(\alpha_m x) \end{array} \right\}, \quad (12.95)$$

where $\zeta_{1m,u}(y)$, $\zeta_{2m,u}(y)$, $\zeta_{1m,v}(y)$, $\zeta_{2m,v}(y)$, $\zeta_{1m,w}(y)$, and $\zeta_{2m,w}(y)$ are the Fourier coefficients to be determined, and $\alpha_m = m\pi/a$ (m is a nonnegative integer).

Following the same procedures as described above, we will be able to determine the matrices of the basis functions $\Phi_{2m,1}^{\mathrm{T}}(x,y)$, $\Phi_{2m,2}^{\mathrm{T}}(x,y)$, and $\Phi_{20,1}^{\mathrm{T}}(x,y)$. However, due to the inherent symmetry of the homogeneous solutions (12.16) and (12.95), these matrices can be directly obtained from their counterparts, $\Phi_{1n,1}^{\mathrm{T}}(x,y)$, $\Phi_{1n,2}^{\mathrm{T}}(x,y)$, and $\Phi_{10,1}^{\mathrm{T}}(x,y)$, by simply swapping the variables (n, b, x) with (m, a, y) therein. In addition, the 1st row of the resulting matrices shall be switched with the 2nd row, and the 1st and 2nd columns then switched with the 3rd and 4th columns.

Define the matrices of the basis functions

$$\Phi_{2,1}^T(x,y) = [\Phi_{20,1}^T(x,y) \quad \Phi_{21,1}^T(x,y) \quad \cdots \quad \Phi_{2m,1}^T(x,y) \quad \cdots], \quad (12.96)$$

$$\Phi_{2,2}^T(x,y) = [\Phi_{21,2}^T(x,y) \quad \Phi_{22,2}^T(x,y) \quad \cdots \quad \Phi_{2m,2}^T(x,y) \quad \cdots], \quad (12.97)$$

and

$$\Phi_2^T(x,y) = [\Phi_{2,1}^T(x,y) \quad \Phi_{2,2}^T(x,y)], \quad (12.98)$$

and the vectors of boundary Fourier coefficients

$$\mathbf{q}_{2,1}^T = [\mathbf{q}_{201}^T \quad \mathbf{q}_{211}^T \quad \cdots \quad \mathbf{q}_{2m1}^T \quad \cdots], \quad (12.99)$$

$$\mathbf{q}_{2,2}^T = [\mathbf{q}_{212}^T \quad \mathbf{q}_{222}^T \quad \cdots \quad \mathbf{q}_{2m2}^T \quad \cdots], \quad (12.100)$$

and

$$\mathbf{q}_2^T = [\mathbf{q}_{2,1}^T \quad \mathbf{q}_{2,2}^T], \quad (12.101)$$

where

$$\mathbf{q}_{2m1}^T = [\mathbf{q}_{2m2,u}^T \quad \mathbf{q}_{2m1,v}^T \quad \mathbf{q}_{2m1,w}^T], \quad (12.102)$$

$$\mathbf{q}_{2m2}^T = [\mathbf{q}_{2m1,u}^T \quad \mathbf{q}_{2m2,v}^T \quad \mathbf{q}_{2m2,w}^T], \quad (12.103)$$

and

$$\mathbf{q}_{201}^T = [\mathbf{q}_{201,u}^T \quad \mathbf{q}_{201,v}^T \quad \mathbf{q}_{201,w}^T]. \quad (12.104)$$

Thus, the boundary functions $\varphi_{2u}(x,y)$, $\varphi_{2v}(x,y)$, and $\varphi_{2w}(x,y)$ are as follows:

$$\begin{bmatrix} \varphi_{2u}(x,y) \\ \varphi_{2v}(x,y) \\ \varphi_{2w}(x,y) \end{bmatrix} = \Phi_2^T(x,y) \cdot \mathbf{q}_2. \quad (12.105)$$

12.2.5 EXPRESSIONS OF INTERNAL AND CORNER FUNCTIONS

We expand the internal components of the modal functions φ_u, φ_v, and φ_w into the full-range Fourier series over the domain $[-a, a] \times [-b, b]$:

$$\varphi_{0u}(x,y) = \sum_{m=0}^{\infty} \sum_{n=0}^{\infty} \lambda_{mn} [V_{1mn,u} \cos(\alpha_m x)\cos(\beta_n y) + V_{2mn,u} \sin(\alpha_m x)\cos(\beta_n y)$$

$$+ V_{3mn,u} \cos(\alpha_m x)\sin(\beta_n y) + V_{4mn,u} \sin(\alpha_m x)\sin(\beta_n y)],$$

$$(12.106)$$

$$\varphi_{0v}(x,y) = \sum_{m=0}^{\infty}\sum_{n=0}^{\infty}\lambda_{mn}[V_{1mn,v}\cos(\alpha_m x)\cos(\beta_n y) + V_{2mn,v}\sin(\alpha_m x)\cos(\beta_n y)$$

$$+ V_{3mn,v}\cos(\alpha_m x)\sin(\beta_n y) + V_{4mn,v}\sin(\alpha_m x)\sin(\beta_n y)], \qquad (12.107)$$

and

$$\varphi_{0w}(x,y) = \sum_{m=0}^{\infty}\sum_{n=0}^{\infty}\lambda_{mn}[V_{1mn,w}\cos(\alpha_m x)\cos(\beta_n y) + V_{2mn,w}\sin(\alpha_m x)\cos(\beta_n y)$$

$$+ V_{3mn,w}\cos(\alpha_m x)\sin(\beta_n y) + V_{4mn,w}\sin(\alpha_m x)\sin(\beta_n y)], \qquad (12.108)$$

where $\alpha_m = m\pi/a$, $\beta_n = n\pi/b$, $\lambda_{mn} = \begin{cases} 1/4 & m=0, n=0 \\ 1/2 & m=1,2,\cdots, n=0 \\ 1/2 & m=0, n=1,2,\cdots \\ 1 & m,n=1,2,\cdots \end{cases}$, and $V_{smn,u}$, $V_{smn,v}$,

and $V_{smn,w}$ ($s = 1,2,3,4$) are the Fourier coefficients of the functions $\varphi_{0u}(x,y)$, $\varphi_{0v}(x,y)$, and $\varphi_{0w}(x,y)$ respectively.

More concisely, (12.106)–(12.108) can be rewritten in matrix form as

$$\varphi_{0u}(x,y) = \boldsymbol{\Phi}_0^{\mathrm{T}}(x,y)\cdot \mathbf{q}_{0,u}, \qquad (12.109)$$

$$\varphi_{0v}(x,y) = \boldsymbol{\Phi}_0^{\mathrm{T}}(x,y)\cdot \mathbf{q}_{0,v}, \qquad (12.110)$$

and

$$\varphi_{0w}(x,y) = \boldsymbol{\Phi}_0^{\mathrm{T}}(x,y)\cdot \mathbf{q}_{0,w}, \qquad (12.111)$$

where $\boldsymbol{\Phi}_0^{\mathrm{T}}$ is the vector of the basis function, and $\mathbf{q}_{0,u}$, $\mathbf{q}_{0,v}$, and $\mathbf{q}_{0,w}$ are the vectors of Fourier coefficients, as defined in (8.75).

In addition, the corner parts of the modal functions φ_u, φ_v, and φ_w can be expressed as

$$\varphi_{3u}(x,y) = \boldsymbol{\Phi}_3^{\mathrm{T}}(x,y)\cdot \mathbf{q}_{3,u}, \qquad (12.112)$$

$$\varphi_{3v}(x,y) = \boldsymbol{\Phi}_3^{\mathrm{T}}(x,y)\cdot \mathbf{q}_{3,v}, \qquad (12.113)$$

and

$$\varphi_{3w}(x,y) = \boldsymbol{\Phi}_3^{\mathrm{T}}(x,y)\cdot \mathbf{q}_{3,w}, \qquad (12.114)$$

where Φ_3^T is the vector of the basis function and $\mathbf{q}_{3,u}$, $\mathbf{q}_{3,v}$, and $\mathbf{q}_{3,w}$ are the vectors of the undetermined coefficients, as defined in (8.41).

Since (12.8) is a system of second order differential equations, each of the vectors, $\mathbf{q}_{3,u}$, $\mathbf{q}_{3,v}$, $\mathbf{q}_{3,w}$, and Φ_3^T, actually contains only one element, that is,

$$\mathbf{q}_{3,u}^T = [q_{3u,00}], \mathbf{q}_{3,v}^T = [q_{3v,00}], \mathbf{q}_{3,w}^T = [q_{3w,00}], \text{ and } \Phi_3^T = [\varphi_{3,00}(x,y)]. \quad (12.115)$$

where $q_{3u,00}$, $q_{3v,00}$, and $q_{3w,00}$ are constants, and

$$\varphi_{3,00}(x,y) = \frac{xy}{4ab}. \quad (12.116)$$

By substituting the internal and corner functions (12.109)–(12.114) into the differential equations (12.8):

$$\mathbf{L}_\varphi \begin{bmatrix} \varphi_{0u} + \varphi_{3u} \\ \varphi_{0v} + \varphi_{3v} \\ \varphi_{0w} + \varphi_{3w} \end{bmatrix} = 0, \quad (12.117)$$

we have

$$\mathbf{L}_\varphi \begin{bmatrix} \Phi_0^T & 0 & 0 \\ 0 & \Phi_0^T & 0 \\ 0 & 0 & \Phi_0^T \end{bmatrix} \begin{bmatrix} \mathbf{q}_{0,u} \\ \mathbf{q}_{0,v} \\ \mathbf{q}_{0,w} \end{bmatrix} = -\mathbf{L}_\varphi \begin{bmatrix} \Phi_3^T & 0 & 0 \\ 0 & \Phi_3^T & 0 \\ 0 & 0 & \Phi_3^T \end{bmatrix} \begin{bmatrix} \mathbf{q}_{3,u} \\ \mathbf{q}_{3,v} \\ \mathbf{q}_{3,w} \end{bmatrix}. \quad (12.118)$$

In order to solve this equation, we first expand the functions on the right-hand-side into the Fourier series, and then compare the Fourier coefficients of the like terms on the both sides. Suppose that all the series expansions are truncated to $m = M$ and $n = N$. The following relationships can then be established between the two sets of expansion constants, $\{\mathbf{q}_{0,u}^T, \mathbf{q}_{0,v}^T, \mathbf{q}_{0,w}^T\}$ and $\{\mathbf{q}_{3,u}^T, \mathbf{q}_{3,v}^T, \mathbf{q}_{3,w}^T\}$:

for $m = 0$ and $n = 0$,

$$(\rho\omega^2 - \mu k^2)V_{100,u} = -(\lambda + \mu)\frac{1}{ab}q_{3v,00}$$

$$(\rho\omega^2 - \mu k^2)V_{100,v} = -(\lambda + \mu)\frac{1}{ab}q_{3u,00} \quad ; \quad (12.119)$$

$$[(\rho\omega^2 - (\lambda + 2\mu)k^2]V_{100,w} = 0$$

for $m > 0$ and $n = 0$,

$$\left.\begin{array}{l} [-(\lambda+2\mu)\alpha_m^2 + \rho\omega^2 - \mu k^2]V_{1m0,u} + [(\lambda+\mu)k\alpha_m]V_{2m0,w} = 0 \\[4pt] [-(\lambda+2\mu)\alpha_m^2 + \rho\omega^2 - \mu k^2]V_{2m0,u} + [-(\lambda+\mu)k\alpha_m]V_{1m0,w} = 0 \\[4pt] [-\mu\alpha_m^2 + \rho\omega^2 - \mu k^2]V_{1m0,v} = 0 \\[4pt] [-\mu\alpha_m^2 + \rho\omega^2 - \mu k^2]V_{2m0,v} = (\lambda+\mu)k\dfrac{(-1)^m}{bm\pi}q_{3w,00} \\[8pt] [-(\lambda+\mu)k\alpha_m]V_{2m0,u} + [-\mu\alpha_m^2 + \rho\omega^2 - (\lambda+2\mu)k^2]V_{1m0,w} = 0 \\[6pt] [(\lambda+\mu)k\alpha_m]V_{1m0,u} + [-\mu\alpha_m^2 + \rho\omega^2 - (\lambda+2\mu)k^2]V_{2m0,w} \\[6pt] \qquad = (\lambda+\mu)k\dfrac{(-1)^{m+1}}{bm\pi}q_{3v,00} \end{array}\right\}; \qquad (12.120)$$

for $m = 0$ and $n > 0$,

$$\left.\begin{array}{l} [-\mu\beta_n^2 + \rho\omega^2 - \mu k^2]V_{10n,u} = 0 \\[8pt] [-\mu\beta_n^2 + \rho\omega^2 - \mu k^2]V_{30n,u} = (\lambda+\mu)k\dfrac{(-1)^n}{an\pi}q_{3w,00} \\[8pt] [-(\lambda+2\mu)\beta_n^2 + \rho\omega^2 - \mu k^2]V_{10n,v} + [(\lambda+\mu)k\beta_n]V_{30n,w} = 0 \\[6pt] [-(\lambda+2\mu)\beta_n^2 + \rho\omega^2 - \mu k^2]V_{30n,v} + [-(\lambda+\mu)k\beta_n]V_{10n,w} = 0 \\[6pt] [-(\lambda+\mu)k\beta_n]V_{30n,v} + [-\mu\beta_n^2 + \rho\omega^2 - (\lambda+2\mu)k^2]V_{10n,w} = 0 \\[6pt] [(\lambda+\mu)k\beta_n]V_{10n,v} + [-\mu\beta_n^2 + \rho\omega^2 - (\lambda+2\mu)k^2]V_{30n,w} \\[8pt] \qquad = (\lambda+\mu)k\dfrac{(-1)^{n+1}}{an\pi}q_{3u,00} \end{array}\right\}; \qquad (12.121)$$

for $m > 0$ and $n > 0$,

$$\left.\begin{array}{l} [-(\lambda+2\mu)\alpha_m^2 - \mu\beta_n^2 + \rho\omega^2 - \mu k^2]V_{1mn,u} + [(\lambda+\mu)\alpha_m\beta_n]V_{4mn,v} \\[6pt] \quad + [(\lambda+\mu)k\alpha_m]V_{2mn,w} = 0 \\[6pt] [(\lambda+\mu)\alpha_m\beta_n]V_{1mn,u} + [-\mu\alpha_m^2 - (\lambda+2\mu)\beta_n^2 + \rho\omega^2 - \mu k^2]V_{4mn,v} \\[6pt] \quad + [-(\lambda+\mu)k\beta_n]V_{2mn,w} = (\rho\omega^2 - \mu k^2)\dfrac{(-1)^{m+n+1}}{mn\pi^2}q_{3v,00} \\[8pt] [(\lambda+\mu)k\alpha_m]V_{1mn,u} + [-(\lambda+\mu)k\beta_n]V_{4mn,v} \\[6pt] \quad + [-\mu\alpha_m^2 - \mu\beta_n^2 + \rho\omega^2 - (\lambda+2\mu)k^2]V_{2mn,w} = 0 \end{array}\right\}, \qquad (12.122)$$

$$\left.\begin{aligned}
&[-(\lambda+2\mu)\alpha_m^2 - \mu\beta_n^2 + \rho\omega^2 - \mu k^2]V_{2mn,u} + [-(\lambda+\mu)\alpha_m\beta_n]V_{3mn,v} \\
&+[-(\lambda+\mu)k\alpha_m]V_{1mn,w} = 0 \\
&[-(\lambda+\mu)\alpha_m\beta_n]V_{2mn,u} + [-\mu\alpha_m^2 - (\lambda+2\mu)\beta_n^2 + \rho\omega^2 - \mu k^2]V_{3mn,v} \\
&+[-(\lambda+\mu)k\beta_n]V_{1mn,w} = 0 \\
&[-(\lambda+\mu)k\alpha_m]V_{2mn,u} + [-(\lambda+\mu)k\beta_n]V_{3mn,v} \\
&+[-\mu\alpha_m^2 - \mu\beta_n^2 + \rho\omega^2 - (\lambda+2\mu)k^2]V_{1mn,w} = 0
\end{aligned}\right\}, \quad (12.123)$$

$$\left.\begin{aligned}
&[-(\lambda+2\mu)\alpha_m^2 - \mu\beta_n^2 + \rho\omega^2 - \mu k^2]V_{3mn,u} + [-(\lambda+\mu)\alpha_m\beta_n]V_{2mn,v} \\
&+[(\lambda+\mu)k\alpha_m]V_{4mn,w} = 0 \\
&[-(\lambda+\mu)\alpha_m\beta_n]V_{3mn,u} + [-\mu\alpha_m^2 - (\lambda+2\mu)\beta_n^2 + \rho\omega^2 - \mu k^2]V_{2mn,v} \\
&+[(\lambda+\mu)k\beta_n]V_{4mn,w} = 0 \\
&[(\lambda+\mu)k\alpha_m]V_{3mn,u} + [(\lambda+\mu)k\beta_n]V_{2mn,v} \\
&+[-\mu\alpha_m^2 - \mu\beta_n^2 + \rho\omega^2 - (\lambda+2\mu)k^2]V_{4mn,w} \\
&= [\rho\omega^2 - (\lambda+2\mu)k^2]\frac{(-1)^{m+n+1}}{mn\pi^2}q_{3w,00}
\end{aligned}\right\}, \quad (12.124)$$

and

$$\left.\begin{aligned}
&[-(\lambda+2\mu)\alpha_m^2 - \mu\beta_n^2 + \rho\omega^2 - \mu k^2]V_{4mn,u} + [(\lambda+\mu)\alpha_m\beta_n]V_{1mn,v} \\
&+[-(\lambda+\mu)k\alpha_m]V_{3mn,w} = (\rho\omega^2 - \mu k^2)\frac{(-1)^{m+n+1}}{mn\pi^2}q_{3u,00} \\
&[(\lambda+\mu)\alpha_m\beta_n]V_{4mn,u} + [-\mu\alpha_m^2 - (\lambda+2\mu)\beta_n^2 + \rho\omega^2 - \mu k^2]V_{1mn,v} \\
&+[(\lambda+\mu)k\beta_n]V_{3mn,w} = 0 \\
&[-(\lambda+\mu)k\alpha_m]V_{4mn,u} + [(\lambda+\mu)k\beta_n]V_{1mn,v} \\
&+[-\mu\alpha_m^2 - \mu\beta_n^2 + \rho\omega^2 - (\lambda+2\mu)k^2]V_{3mn,w} = 0
\end{aligned}\right\}. \quad (12.125)$$

These equations can be compressed into

$$\mathbf{q}_0 = \mathbf{T}_{03}\mathbf{q}_3, \quad (12.126)$$

where

$$\mathbf{q}_0^{\mathrm{T}} = [\mathbf{q}_{0,u}^{\mathrm{T}} \quad \mathbf{q}_{0,v}^{\mathrm{T}} \quad \mathbf{q}_{0,w}^{\mathrm{T}}], \tag{12.127}$$

$$\mathbf{q}_3^{\mathrm{T}} = [\mathbf{q}_{3,u}^{\mathrm{T}} \quad \mathbf{q}_{3,v}^{\mathrm{T}} \quad \mathbf{q}_{3,w}^{\mathrm{T}}] = [q_{3u,00} \quad q_{3v,00} \quad q_{3w,00}] \tag{12.128}$$

and $\mathbf{T}_{03}$ is the transformation matrix whose expressions will not be explicitly given here for shortness, but can be readily derived from (12.119)–(12.125).

Accordingly, the internal and corner functions can be written as

$$\begin{bmatrix} \varphi_{0u}(x,y) + \varphi_{3u}(x,y) \\ \varphi_{0v}(x,y) + \varphi_{3v}(x,y) \\ \varphi_{0w}(x,y) + \varphi_{3w}(x,y) \end{bmatrix} = \mathbf{\Phi}_{03,R}^{\mathrm{T}}(x,y) \cdot \mathbf{q}_3, \tag{12.129}$$

where

$$\mathbf{\Phi}_{03,R}^{\mathrm{T}}(x,y) = \begin{bmatrix} \Phi_0^{\mathrm{T}} & \mathbf{0} & \mathbf{0} \\ \mathbf{0} & \Phi_0^{\mathrm{T}} & \mathbf{0} \\ \mathbf{0} & \mathbf{0} & \Phi_0^{\mathrm{T}} \end{bmatrix} \cdot \mathbf{T}_{03} + \begin{bmatrix} \Phi_3^{\mathrm{T}} & \mathbf{0} & \mathbf{0} \\ \mathbf{0} & \Phi_3^{\mathrm{T}} & \mathbf{0} \\ \mathbf{0} & \mathbf{0} & \Phi_3^{\mathrm{T}} \end{bmatrix}. \tag{12.130}$$

12.2.6 Expressions of the Multiscale Fourier Series Solution

Denote

$$\mathbf{\Phi}^{\mathrm{T}}(x,y) = [\mathbf{\Phi}_{03,R}^{\mathrm{T}}(x,y) \quad \mathbf{\Phi}_1^{\mathrm{T}}(x,y) \quad \mathbf{\Phi}_2^{\mathrm{T}}(x,y)] \tag{12.131}$$

and

$$\mathbf{q}^{\mathrm{T}} = [\mathbf{q}_3^{\mathrm{T}} \quad \mathbf{q}_1^{\mathrm{T}} \quad \mathbf{q}_2^{\mathrm{T}}]. \tag{12.132}$$

The multiscale Fourier series of the modal functions φ_u, φ_v, and φ_w over the domain $[-a,a] \times [-b,b]$ can be correspondingly given as

$$\begin{bmatrix} \varphi_u(x,y) \\ \varphi_v(x,y) \\ \varphi_w(x,y) \end{bmatrix} = \mathbf{\Phi}^{\mathrm{T}}(x,y) \cdot \mathbf{q}. \tag{12.133}$$

If the matrix $\mathbf{\Phi}^{\mathrm{T}}(x,y)$ is further partitioned into blocks

$$\mathbf{\Phi}^{\mathrm{T}}(x,y) = \begin{bmatrix} \mathbf{\Phi}_u^{\mathrm{T}}(x,y) \\ \mathbf{\Phi}_v^{\mathrm{T}}(x,y) \\ \mathbf{\Phi}_w^{\mathrm{T}}(x,y) \end{bmatrix}, \tag{12.134}$$

the displacement solutions of (12.7) for the wave propagations in the rectangular waveguide can be written as

$$
\left.
\begin{aligned}
u &= [\boldsymbol{\Phi}_u^{\mathrm{T}}(x,y)\cdot \mathbf{q}]\cos(kz-\omega t) \\
v &= [\boldsymbol{\Phi}_v^{\mathrm{T}}(x,y)\cdot \mathbf{q}]\cos(kz-\omega t) \\
w &= [\boldsymbol{\Phi}_w^{\mathrm{T}}(x,y)]\cdot \mathbf{q}]\sin(kz-\omega t)
\end{aligned}
\right\}.
\qquad (12.135)
$$

12.2.7 EXPRESSIONS OF STRESS RESULTANTS

By substituting the multiscale Fourier series solutions (12.135) into the constitutive equations (12.4), the normal stresses σ_x, σ_y, σ_z and shear stresses τ_{xy}, τ_{yz}, τ_{xz} are as follows:

$$
\left.
\begin{aligned}
\sigma_x &= [\boldsymbol{\Gamma}_1 \cdot \mathbf{q}]\cos(kz-\omega t) \\
\sigma_y &= [\boldsymbol{\Gamma}_2 \cdot \mathbf{q}]\cos(kz-\omega t) \\
\sigma_z &= [\boldsymbol{\Gamma}_3 \cdot \mathbf{q}]\cos(kz-\omega t) \\
\tau_{xy} &= [\boldsymbol{\Gamma}_4 \cdot \mathbf{q}]\cos(kz-\omega t) \\
\tau_{yz} &= [\boldsymbol{\Gamma}_5 \cdot \mathbf{q}]\sin(kz-\omega t) \\
\tau_{xz} &= [\boldsymbol{\Gamma}_6 \cdot \mathbf{q}]\sin(kz-\omega t)
\end{aligned}
\right\},
\qquad (12.136)
$$

where

$$
\boldsymbol{\Gamma} =
\begin{bmatrix}
\boldsymbol{\Gamma}_1 \\
\boldsymbol{\Gamma}_2 \\
\boldsymbol{\Gamma}_3 \\
\boldsymbol{\Gamma}_4 \\
\boldsymbol{\Gamma}_5 \\
\boldsymbol{\Gamma}_6
\end{bmatrix}
=
\begin{bmatrix}
(\lambda+2\mu)\boldsymbol{\Phi}_u^{(1,0)\mathrm{T}}(x,y)+\lambda\boldsymbol{\Phi}_v^{(0,1)\mathrm{T}}(x,y)+\lambda k\boldsymbol{\Phi}_w^{\mathrm{T}}(x,y) \\
\lambda\boldsymbol{\Phi}_u^{(1,0)\mathrm{T}}(x,y)+(\lambda+2\mu)\boldsymbol{\Phi}_v^{(0,1)\mathrm{T}}(x,y)+\lambda k\boldsymbol{\Phi}_w^{\mathrm{T}}(x,y) \\
\lambda\boldsymbol{\Phi}_u^{(1,0)\mathrm{T}}(x,y)+\lambda\boldsymbol{\Phi}_v^{(0,1)\mathrm{T}}(x,y)+(\lambda+2\mu)k\boldsymbol{\Phi}_w^{\mathrm{T}}(x,y) \\
\mu\boldsymbol{\Phi}_u^{(0,1)\mathrm{T}}(x,y)+\mu\boldsymbol{\Phi}_v^{(1,0)\mathrm{T}}(x,y) \\
-\mu k\boldsymbol{\Phi}_v^{\mathrm{T}}(x,y)+\mu\boldsymbol{\Phi}_w^{(0,1)\mathrm{T}}(x,y) \\
-\mu k\boldsymbol{\Phi}_u^{\mathrm{T}}(x,y)+\mu\boldsymbol{\Phi}_w^{(1,0)\mathrm{T}}(x,y)
\end{bmatrix}.
\qquad (12.137)
$$

12.3 SOLVING FOR SOLUTIONS

Based on the discussions in the previous section, it is evident that the multiscale Fourier series solutions given by (12.135) have already satisfied all the differential equations which govern the propagations of elastic waves in a rectangular waveguide. If the series solutions are respectively truncated to $m = M$ and $n = N$ in the x- and y-direction, there is a total of $12M+12N+15$ undetermined expansion coefficients in $\mathbf{q}$. If the collocation method is used to derive the final system of equations, we need to select an adequate number, such as $12M+12N+15$, of collocation points on the boundaries. By equating the displacements (12.135) and/or the stresses (12.136) to the prescribed values at each collocation point, a system of $12M + 12N + 15$ homogeneous linear algebraic equations will be obtained about the expansion coefficients. Letting the determinant of the coefficient matrix be equal to zero, we thus yield a transcendental equation (often referred to as the frequency equation) which regulates the relations between the frequency ω and the wavenumber k. For instance, given a frequency, the frequency equation dictates that only the waves associated with the permissible wavenumbers can propagate in the waveguide. As the frequency changes, the wavenumbers will vary accordingly to form the so-called dispersion curves. In the $\omega-k$ plane, the frequency equation can lead to uncountable branches of the dispersion curves, and each of them corresponds to a cross-sectional deformation pattern or wave mode in the waveguide. The intersecting points of the branches and the ω-axis represent the cut-off frequencies.

As in studying the transverse vibrations of rectangular thin plates (Huang 1992), symmetric decomposition of the wave modes will be performed herein against the symmetry of the applied boundary conditions.

For instance, as shown in Figure 12.2(a), the boundary condition (CCFC) for the rectangular waveguide is symmetric about the x-axis. Thus, the wave modes can be decomposed into two types: the longitudinal wave (L) and the bending wave (B_x) about the x-axis. Specifically, for the family of longitudinal waves, the displacements u, v, and w are respectively even, odd and even with respect to y. For the family of bending waves, the displacements u, v, and w are respectively odd, even and odd with respect to y.

In the meantime, the other two boundary conditions, FFFF and FCFC, in Figure 12.2 are simultaneously symmetric about the x- and y-axis. Therefore, the wave modes can be decomposed into four types: the longitudinal wave (L), the torsional wave (T), and two bending waves (B_x and B_y) about the x- and y-axis, respectively. In particular, for a square cross-sectioned waveguide, the first two types of waves with FFFF boundary condition can also be decomposed into the symmetric waves (T_s, L_s) and asymmetric waves (T_a, L_a) with respect to the line $x = y$.

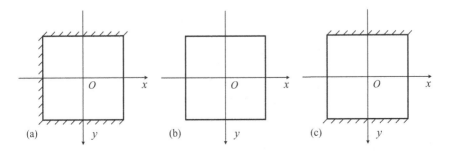

(a) (b) (c)

FIGURE 12.2 Three representative boundary conditions for a waveguide: (a) CCFC; (b) FFFF; and (c) FCC.

TABLE 12.2

Symmetric Decompositions of Wave Modes in a Rectangular Waveguide.

Cases	BC's	Wave Types	w.r.t. the x-axis			w.r.t. the y-axis			Numbers of Unknown Constants
			u	v	w	u	v	w	
1	CCFC	L	S	A	S				$6M + 6N + 7$
		B_x	A	S	A				$6M + 6N + 5$
2	FFFF	L	S	A	S	A	S	S	$3M + 3N + 4$
		T	A	S	A	S	A	A	$3M + 3N + 3$
		B_x	A	S	A	A	S	S	$3M + 3N + 4$
		B_y	S	A	S	S	A	A	$3M + 3N + 4$
3	FCFC	L	S	A	S	A	S	S	$3M + 3N + 4$
		T	A	S	A	S	A	A	$3M + 3N + 2$
		B_x	A	S	A	A	S	S	$3M + 3N + 3$
		B_y	S	A	S	S	A	A	$3M + 3N + 3$

BC: Boundary Condition, S: symmetric, A: asymmetric.

The numbers of unknown constants involved in the frequency equation are summarized in Table 12.2 for various types of wave modes. It should be noted that since $q_{3u,00}$, $q_{3v,00}$, and $q_{3w,00}$ vanish identically for the CCFC and FCFC boundary conditions, the total numbers of the unknown constants are correspondingly less than the full count $12M + 12N + 15$ by 3.

12.4 NUMERICAL EXAMPLES

As shown in Table 12.3, we have designated two groups of numerical tests to investigate the influences of some key factors, such as the side ratio b/a and boundary conditions, on the convergence characteristics and approximation accuracies of the multiscale Fourier series solutions for the current wave propagation problems. In the first group, by fixing the boundary condition to FFFF, the side ratio varies from 0.5 to 2.0 with three intermediate steps 0.67, 0.50, and 1.25. The second group of tests is used to examine the effects of the boundary conditions whilst the side ratio is fixed to $b/a = 1.0$.

In all these tests, the Poisson's ratio is set to $\upsilon = 0.3$. Additionally, the results are presented in terms of the non-dimensional wave frequency Ω and wavenumber K: $\Omega = \omega a/\pi c_T$ ($c_T = \sqrt{\mu/\rho}$) and $K = ka/\pi$.

First of all, the baseline case (the first test in Table 12.3) will be used to examine the convergence characteristics of the multiscale Fourier series solutions. By sequentially truncating the series solutions to $M = N = 4$, 8, 12, 16, and 20, the dispersion curves $\Omega_{M,N}^{l}(K)$-K for the first three types of waves are obtained from

TABLE 12.3
Testing Cases for Wave Propagations in a Rectangular Elastic Waveguide.

Groups	Cases	Boundary Conditions	Side Ratios a/b
1	a	FFFF	1.0
	b		0.67
	c		0.50
	d		1.25
	e		2.0
2	a	FFFF	1.0
	b	FCFC	
	c	CCFC	

the frequency equations, where the superscript l indicates the order of the specific wave traveling down the waveguide. If the dispersion curve determined by setting $M = N = 20$ is considered "exact," the error index for a dispersion curve over the wavenumber interval $[0,1]$ is here specifically defined as

$$e\left(\Omega_{M,N}^{l}\right) = \max_{K \in \Upsilon^l}\left|\Omega_{M,N}^{l}(K) - \Omega_{20,20}^{l}(K)\right|, \quad (12.138)$$

where the set of sampling points is

$$\Upsilon^l = \left\{(n-1)/100, n = 1, 2, \cdots, 101\right\}. \quad (12.139)$$

The corresponding results for the first order dispersion curves and the prediction errors for the various branches of dispersion curves over $K \in [0,1]$ are presented in Tables 12.4 and 12.5, respectively.

The first branches of dispersion curves for various types of mode waves converge consistently with the increase of the truncation numbers M and N. For the dispersion curves L_s^1, T_a^1, and B_x^1, the error indexes rapidly descend to about 4.0E−3 when only 4 terms are retained in the series expansions. For the other two dispersion curves L_a^1 and T_s^1, the errors are relatively larger at about 4.0E−2 for $M = N = 4$. In regards to the convergence characteristics, however, all these dispersion curves have displayed an impressive trend: as the truncation number is quadrupled to 16, the corresponding errors are all reduced approximately by an order of magnitude.

Noticeably, the current dispersion curves for L_s^1, T_a^1, B_x^1 agree well with the reference results (Fraser 1969), but significant differences are observed between the dispersion curves for L_a^1 and T_s^1. These deviations may be attributed to the disparity in selecting the forms of the basis functions because the basis functions used by Fraser are actually the analytical solutions for the wave propagations

TABLE 12.4
Convergence of the First Order Dispersion Curves in the Baseline Case.

Wave	K	Ω					Reference (Fraser 1969)
		$M = N = 4$	$M = N = 8$	$M = N = 12$	$M = N = 16$	$M = N = 20$	
L_s	0.3183	0.4937	0.4937	0.4937	0.4937	0.4937	0.4938
	0.5730	0.7668	0.7668	0.7668	0.7668	0.7668	0.7669
	0.8276	0.9339	0.9339	0.9339	0.9339	0.9339	0.9340
L_a	0.3183	0.5056	0.5261	0.5163	0.5114	0.5084	0.6890
	0.5730	0.5958	0.5815	0.5765	0.5740	0.5724	0.7273
	0.8276	0.6908	0.6841	0.6818	0.6806	0.6798	0.8490
T_s	0.3183	0.6818	0.6721	0.6679	0.6654	0.6637	0.6264
	0.5730	0.8058	0.7915	0.7857	0.7822	0.7799	0.7392
	0.8276	0.9818	0.9664	0.9604	0.9570	0.9546	0.9177
T_a	0.3183	0.2926	0.2923	0.2922	0.2922	0.2922	0.2923
	0.5730	0.5263	0.5257	0.5256	0.5255	0.5255	0.5255
	0.8276	0.7594	0.7585	0.7582	0.7581	0.7581	0.7581
B_x	0.3183	0.2047	0.2038	0.2035	0.2033	0.2032	0.2010
	0.5730	0.4575	0.4561	0.4555	0.4552	0.4550	0.4511
	0.8276	0.7124	0.7105	0.7097	0.7093	0.7090	0.7032

TABLE 12.5
Errors $e(\Omega_{M,N}^l)$ for the Dispersion Curves in the Baseline Case.

Wave	Branch Order l	$M = N = 4$	$M = N = 8$	$M = N = 12$	$M = N = 16$
L_s	1	0.0001	0.0001	0.0001	0.0000
	2	0.0014	0.0002	0.0001	0.0001
	3	0.0008	0.0003	0.0002	0.0001
	4	0.0013	0.0004	0.0001	0.0001
	5	0.0019	0.0003	0.0001	0.0001
	6	0.0010	0.0002	0.0001	0.0001
	7	0.0141	0.0004	0.0001	0.0001
	8	0.0004	0.0004	0.0002	0.0001
	9	0.0064	0.0013	0.0004	0.0001
L_a	1	0.0397	0.0190	0.0086	0.0033
	2	0.0138	0.0047	0.0021	0.0008
	3	0.0161	0.0397	0.0273	0.0117
	4	0.0392	0.0395	0.0211	0.0127
	5	0.0087	0.0022	0.0008	0.0003
	6	0.0022	0.0009	0.0004	0.0002
	7	0.0396	0.0104	0.0038	0.0014
	8	0.0099	0.0027	0.0012	0.0005
	9	0.0000	0.0000	0.0000	0.0000

(Continued)

TABLE 12.5 *(Continued)*
Errors $e(\Omega^l_{M,N})$ **for the Dispersion Curves in the Baseline Case.**

Wave	Branch Order l	$M = N = 4$	$M = N = 8$	$M = N = 12$	$M = N = 16$
T_s	1	0.0272	0.0119	0.0059	0.0024
	2	0.0388	0.0325	0.0161	0.0065
	3	0.0000	0.0000	0.0000	0.0000
	4	0.0105	0.0046	0.0022	0.0009
	5	0.0392	0.0393	0.0193	0.0078
	6	0.0395	0.0204	0.0096	0.0038
	7	0.0395	0.0188	0.0090	0.0036
	8	0.0114	0.0071	0.0034	0.0014
	9	0.0384	0.0247	0.0119	0.0048
	10	0.0396	0.0267	0.0135	0.0054
T_a	1	0.0016	0.0005	0.0002	0.0001
	2	0.0007	0.0007	0.0003	0.0001
	3	0.0034	0.0002	0.0001	0.0001
	4	0.0001	0.0001	0.0001	0.0001
	5	0.0021	0.0001	0.0001	0.0001
	6	0.0006	0.0002	0.0001	0.0001
	7	0.0143	0.0003	0.0002	0.0001
	8	0.0171	0.0007	0.0001	0.0001
	9	0.0090	0.0004	0.0001	0.0001
	10	0.0022	0.0002	0.0001	0.0001
	11	0.0001	0.0000	0.0000	0.0000
B_x	1	0.0041	0.0019	0.0010	0.0004
	2	0.0398	0.0218	0.0112	0.0046
	3	0.0396	0.0214	0.0110	0.0046
	4	0.0030	0.0007	0.0003	0.0001
	5	0.0147	0.0072	0.0038	0.0016
	6	0.0250	0.0194	0.0121	0.0059
	7	0.0390	0.0336	0.0166	0.0067
	8	0.0160	0.0059	0.0028	0.0011
	9	0.0168	0.0082	0.0041	0.0017
	10	0.0083	0.0044	0.0022	0.0009
	11	0.0060	0.0023	0.0011	0.0005
	12	0.0047	0.0027	0.0014	0.0006
	13	0.0251	0.0125	0.0064	0.0027
	14	0.0391	0.0293	0.0145	0.0059
	15	0.0364	0.0169	0.0084	0.0034
	16	0.0138	0.0046	0.0021	0.0008
	17	0.0393	0.0223	0.0131	0.0053
	18	0.0218	0.0114	0.0058	0.0023

TABLE 12.6

Uniform Convergence of the First Order Dispersion Curves B_x^1 with Different Side Ratios.

				Ω		
Side Ratio a/b	K	$M = N = 4$	$M = N = 8$	$M = N = 12$	$M = N = 16$	$M = N = 20$
1.0	0.1000	0.0282	0.0280	0.0280	0.0279	0.0279
	0.3000	0.1873	0.1865	0.1862	0.1860	0.1859
	0.5000	0.3842	0.3829	0.3824	0.3821	0.3819
	0.7000	0.5851	0.5834	0.5828	0.5824	0.5821
	0.9000	0.7842	0.7822	0.7813	0.7808	0.7805
0.67	0.1000	0.0394	0.0392	0.0392	0.0391	0.0391
	0.3000	0.2220	0.2213	0.2211	0.2209	0.2208
	0.5000	0.4224	0.4214	0.4210	0.4208	0.4206
	0.7000	0.6201	0.6188	0.6182	0.6178	0.6176
	0.9000	0.8144	0.8126	0.8118	0.8113	0.8109
0.50	0.1000	0.0486	0.0484	0.0483	0.0483	0.0483
	0.3000	0.2413	0.2408	0.2406	0.2405	0.2404
	0.5000	0.4402	0.4393	0.4389	0.4387	0.4386
	0.7000	0.6345	0.6332	0.6326	0.6322	0.6320
	0.9000	0.8253	0.8236	0.8227	0.8221	0.8217
1.25	0.1000	0.0231	0.0230	0.0229	0.0229	0.0229
	0.3000	0.1658	0.1649	0.1646	0.1644	0.1643
	0.5000	0.3563	0.3549	0.3543	0.3540	0.3538
	0.7000	0.5567	0.5549	0.5541	0.5537	0.5534
	0.9000	0.7576	0.7554	0.7545	0.7540	0.7536
2.0	0.1000	0.0149	0.0148	0.0148	0.0147	0.0147
	0.3000	0.1204	0.1196	0.1193	0.1191	0.1189
	0.5000	0.2850	0.2835	0.2829	0.2825	0.2823
	0.7000	0.4743	0.4723	0.4714	0.4709	0.4705
	0.9000	0.6728	0.6703	0.6692	0.6686	0.6682

in waveguides with a circular cross-section, rather than the rectangular cross-section herein.

Next, we will investigate the effects of the side ratio by varying it from 0.5 to 2.0, as specified in the first test group in Table 12.3. For conciseness, only the dispersion curves for B_x^1 are presented in Table 12.6. It is seen that as the truncation number increases from 4 to 16, the calculated frequencies for the wave mode B_x^1 have demonstrated consistent convergence characteristics in all these five cases.

The second test group is designated to investigate the impacts of the boundary conditions by considering two more cases, as illustrated in Figure 12.2: free at two opposite edges and clamped at the other two (FCFC), and free at one edge and clamped at other three (CCFC). The corresponding results are presented in Tables 12.7 and 12.8.

TABLE 12.7

Errors $e(\Omega^l_{M,N})$ for the Dispersion Curves in the FCFC Case.

Wave	Branch Order l	$M = N = 4$	$M = N = 8$	$M = N = 12$	$M = N = 16$
L	1	0.0062	0.0049	0.0045	0.0044
	2	0.0028	0.0008	0.0003	0.0001
	3	0.0036	0.0006	0.0002	0.0001
	4	0.0109	0.0025	0.0010	0.0004
	5	0.0088	0.0021	0.0008	0.0003
	6	0.0110	0.0034	0.0014	0.0005
	7	0.0074	0.0023	0.0009	0.0004
	8	0.0017	0.0003	0.0001	0.0001
	9	0.0063	0.0022	0.0009	0.0003
	10	0.0161	0.0020	0.0006	0.0002
	11	0.0275	0.0060	0.0023	0.0008
	12	0.0323	0.0024	0.0009	0.0003
	13	0.0323	0.0006	0.0003	0.0001
	14	0.0017	0.0002	0.0001	0.0001
T	1	0.0011	0.0002	0.0001	0.0001
	2	0.0384	0.0126	0.0051	0.0018
	3	0.0167	0.0057	0.0024	0.0009
	4	0.0162	0.0037	0.0013	0.0005
	5	0.0222	0.0055	0.0022	0.0008
	6	0.0073	0.0025	0.0017	0.0008
	7	0.0097	0.0030	0.0012	0.0004
	8	0.0014	0.0012	0.0007	0.0003
	9	0.0075	0.0051	0.0024	0.0009
	10	0.0124	0.0035	0.0014	0.0005
	11	0.0391	0.0157	0.0060	0.0021
	12	0.0282	0.0091	0.0036	0.0013
	13	0.0116	0.0003	0.0012	0.0006
B_x	1	0.0193	0.0071	0.0030	0.0011
	2	0.0090	0.0038	0.0016	0.0007
	3	0.0204	0.0047	0.0018	0.0007
	4	0.0234	0.0061	0.0025	0.0009
	5	0.0190	0.0060	0.0026	0.0009
	6	0.0203	0.0105	0.0049	0.0018
	7	0.0366	0.0047	0.0028	0.0027
	8	0.0339	0.0058	0.0021	0.0008
	9	0.0018	0.0002	0.0002	0.0001
	10	0.0332	0.0067	0.0026	0.0009
	11	0.0135	0.0025	0.0007	0.0003
	12	0.0397	0.0214	0.0079	0.0028
	13	0.0113	0.0127	0.0073	0.0030

(Continued)

TABLE 12.7 *(Continued)*
Errors $e(\Omega_{M,N}^l)$ **for the Dispersion Curves in the FCFC Case.**

Wave	Branch Order *l*	*M = N = 4*	*M = N = 8*	*M = N = 12*	*M = N = 16*
B_y	1	0.0112	0.0028	0.0010	0.0004
	2	0.0046	0.0013	0.0006	0.0002
	3	0.0394	0.0062	0.0019	0.0006
	4	0.0285	0.0079	0.0022	0.0007
	5	0.0234	0.0037	0.0012	0.0004
	6	0.0160	0.0027	0.0009	0.0003
	7	0.0179	0.0055	0.0023	0.0009
	8	0.0398	0.0064	0.0022	0.0007
	9	0.0394	0.0100	0.0022	0.0006
	10	0.0396	0.0254	0.0074	0.0023
	11	0.0077	0.0018	0.0006	0.0002
	12	0.0175	0.0024	0.0012	0.0004
	13	0.0080	0.0029	0.0012	0.0005
	14	0.0204	0.0016	0.0004	0.0001
	15	0.0398	0.0050	0.0020	0.0007

TABLE 12.8
Errors $e(\Omega_{M,N}^l)$ **for the Dispersion Curves in the CCFC Case.**

Wave	Branch Order *l*	*M = N = 4*	*M = N = 8*	*M = N = 12*	*M = N = 16*
L	1	0.0040	0.0010	0.0020	0.0017
	2	0.0038	0.0011	0.0006	0.0005
	3	0.0060	0.0036	0.0031	0.0027
	4	0.0019	0.0006	0.0003	0.0002
	5	0.0161	0.0050	0.0032	0.0030
	6	0.0062	0.0015	0.0005	0.0005
	7	0.0061	0.0014	0.0005	0.0004
	8	0.0118	0.0040	0.0015	0.0008
	9	0.0076	0.0022	0.0011	0.0005
	10	0.0043	0.0012	0.0006	0.0003
	11	0.0044	0.0014	0.0006	0.0004
	12	0.0047	0.0020	0.0013	0.0009
	13	0.0075	0.0028	0.0013	0.0005
	14	0.0032	0.0017	0.0006	0.0003
	15	0.0193	0.0045	0.0015	0.0007
	16	0.0043	0.0015	0.0004	0.0003
	17	0.0277	0.0069	0.0025	0.0009
	18	0.0326	0.0075	0.0027	0.0009
	19	0.0379	0.0062	0.0021	0.0008
	20	0.0329	0.0024	0.0008	0.0003

(Continued)

TABLE 12.8 *(Continued)*
Errors $e(\Omega^l_{M,N})$ for the Dispersion Curves in the CCFC Case.

Wave	Branch Order l	$M = N = 4$	$M = N = 8$	$M = N = 12$	$M = N = 16$
	21	0.0067	0.0015	0.0006	0.0003
	22	0.0109	0.0022	0.0008	0.0004
	23	0.0086	0.0020	0.0009	0.0006
	24	0.0085	0.0020	0.0008	0.0005
	25	0.0104	0.0028	0.0024	0.0003
B_x	1	0.0134	0.0052	0.0022	0.0008
	2	0.0019	0.0010	0.0006	0.0004
	3	0.0135	0.0046	0.0019	0.0007
	4	0.0134	0.0040	0.0017	0.0006
	5	0.0100	0.0031	0.0012	0.0006
	6	0.0107	0.0067	0.0040	0.0031
	7	0.0050	0.0014	0.0007	0.0005
	8	0.0060	0.0014	0.0004	0.0003
	9	0.0040	0.0026	0.0014	0.0006
	10	0.0036	0.0036	0.0022	0.0010
	11	0.0027	0.0159	0.0158	0.0006
	12	0.0014	0.0003	0.0001	0.0002
	13	0.0031	0.0007	0.0007	0.0004
	14	0.0046	0.0010	0.0005	0.0003
	15	0.0106	0.0030	0.0013	0.0005
	16	0.0051	0.0018	0.0009	0.0004
	17	0.0009	0.0002	0.0002	0.0001
	18	0.0065	0.0032	0.0014	0.0007
	19	0.0119	0.0038	0.0016	0.0007
	20	0.0048	0.0004	0.0002	0.0002
	21	0.0030	0.0017	0.0007	0.0003
	22	0.0336	0.0064	0.0019	0.0010
	23	0.0140	0.0087	0.0036	0.0017

Again, an excellent convergence characteristic is observed in both cases for all wave types: the solution errors are reduced, at least, by an order of magnitude as the truncation numbers increase from 4 to 16. It should be emphasized that this remark is insensitive to the boundary conditions, which may easily become an important merit of the current multiscale Fourier series method in contrast to other mostly BC-specific methods.

12.5 WAVE PROPAGATIONS IN A SQUARE WAVEGUIDE

In this section, some important aspects, such as frequency spectra and wave modes, of wave propagation in a square waveguide are investigated by using the multiscale Fourier series method. In the following calculations, the multiscale Fourier series solutions are invariably truncated to $M = N = 20$.

12.5.1 Frequency Spectra

Wave propagation in a square waveguide is considered for three different boundary conditions: FFFF, FCFC, and CCFC. The frequency spectra for several selected mode waves are plotted in Figures 12.3–12.5. It is seen that the whole suite of the dispersion curves can be well predicted for all these wave types and boundary conditions considered, except that the small segments of dispersion curves for L_a^3 and L_a^4 are missing in the region $K \in [0.43, 0.60]$ in the FFFF case. As shown in Figure 12.3(b), this blind region is roughly bounded by the two points where the dispersion curves for L_a^3 and L_a^4 intersect with each other.

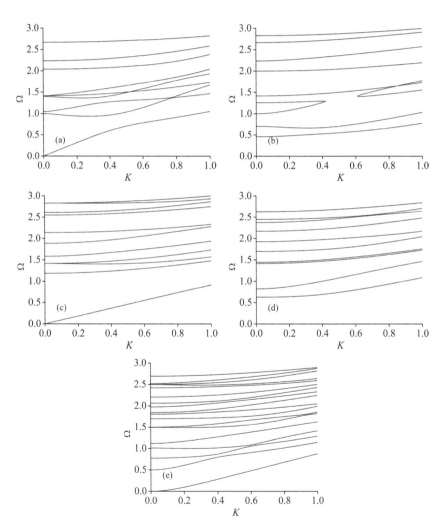

FIGURE 12.3 Frequency spectra corresponding to FFFF boundaries: (a) L_s waves; (b) L_a waves; (c) T_a waves; (d) T_s waves; and (e) B_x waves.

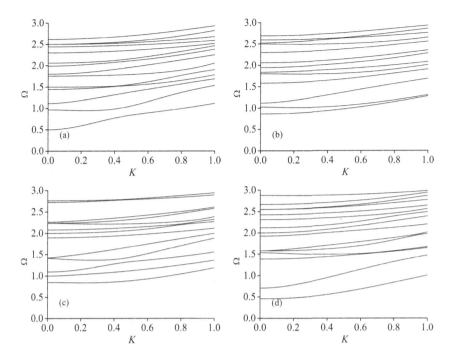

FIGURE 12.4   Frequency spectra corresponding to FCFC boundaries: (a) L waves; (b) T waves; (c) B_x waves; and (d) B_y waves.

12.5.2   WAVE MODES

The dispersion curves regulate which wave modes are permissible to travel down the waveguide. Additionally, for any given point on a dispersion curve, the corresponding wave mode is fully specified with respect to its deformation characteristic. For instance, presented in Tables 12.9–12.17 are the deformation patterns for several wave modes of various types and orders.

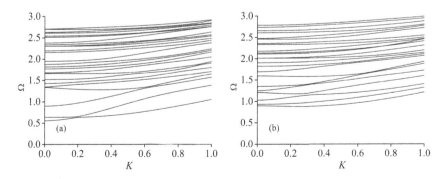

FIGURE 12.5   Frequency spectra corresponding to CCFC boundaries: (a) L waves and (b) B_x waves.

TABLE 12.9
Distortions and Out-of-Plane Displacements of the L_s and L_a Waves for the FFFF Boundary Condition.

Points	Distortion	Out-of-Plane Displacement	Points	Distortion	Out-of-Plane Displacement
L_s^1 $K = 0.1$			L_s^1 $K = 0.9$		
L_s^2 $K = 0.1$			L_s^2 $K = 0.9$		
L_s^3 $K = 0.1$			L_s^3 $K = 0.9$		
L_s^9 $K = 0.1$			L_s^9 $K = 0.9$		
L_a^1 $K = 0.1$			L_a^1 $K = 0.9$		
L_a^2 $K = 0.1$			L_a^2 $K = 0.9$		
L_a^3 $K = 0.1$			L_a^3 $K = 0.9$		

TABLE 12.10
Distortions and Out-of-Plane Displacements of the T_a and T_s Waves for the FFFF Boundary Condition.

Points	Distortion	Out-of-Plane Displacement	Points	Distortion	Out-of-Plane Displacement
T_a^1 $K = 0.1$			T_a^1 $K = 0.9$		
T_a^2 $K = 0.1$			T_a^2 $K = 0.9$		
T_a^3 $K = 0.1$			T_a^3 $K = 0.9$		
T_a^5 $K = 0.1$			T_a^5 $K = 0.9$		
T_s^1 $K = 0.1$			T_s^1 $K = 0.9$		
T_s^2 $K = 0.1$			T_s^2 $K = 0.9$		
T_s^5 $K = 0.1$			T_s^5 $K = 0.9$		
T_s^{10} $K = 0.1$			T_s^{10} $K = 0.9$		

TABLE 12.11
Distortions and Out-of-Plane Displacements of the B_x Waves for the FFFF Boundary Condition.

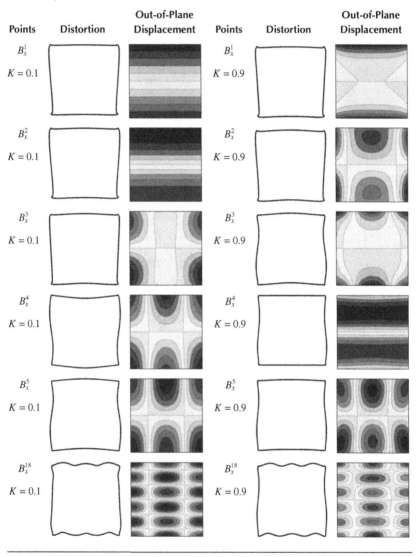

Points	Distortion	Out-of-Plane Displacement	Points	Distortion	Out-of-Plane Displacement
B_x^1 $K = 0.1$			B_x^1 $K = 0.9$		
B_x^2 $K = 0.1$			B_x^2 $K = 0.9$		
B_x^3 $K = 0.1$			B_x^3 $K = 0.9$		
B_x^4 $K = 0.1$			B_x^4 $K = 0.9$		
B_x^5 $K = 0.1$			B_x^5 $K = 0.9$		
B_x^{18} $K = 0.1$			B_x^{18} $K = 0.9$		

TABLE 12.12
Distortions and Out-of-Plane Displacements of the *L* Waves for the FCFC Boundary Condition.

Points	Distortion	Out-of-Plane Displacement	Points	Distortion	Out-of-Plane Displacement
L^1 $K = 0.1$			L^1 $K = 0.9$		
L^2 $K = 0.1$			L^2 $K = 0.9$		
L^3 $K = 0.1$			L^3 $K = 0.9$		
L^4 $K = 0.1$			L^4 $K = 0.9$		
L^5 $K = 0.1$			L^5 $K = 0.9$		
L^{14} $K = 0.1$			L^{14} $K = 0.9$		

TABLE 12.13

Distortions and Out-of-Plane Displacements of the T Waves for the FCFC Boundary Condition.

Points	Distortion	Out-of-Plane Displacement	Points	Distortion	Out-of-Plane Displacement
T^1 $K = 0.1$			T^1 $K = 0.9$		
T^2 $K = 0.1$			T^2 $K = 0.9$		
T^3 $K = 0.1$			T^3 $K = 0.9$		
T^4 $K = 0.1$			T^4 $K = 0.9$		
T^5 $K = 0.1$			T^5 $K = 0.9$		
T^{13} $K = 0.1$			T^{13} $K = 0.9$		

TABLE 12.14

Distortions and Out-of-Plane Displacements of the B_x Waves for the FCFC Boundary Condition.

Points	Distortion	Out-of-Plane Displacement	Points	Distortion	Out-of-Plane Displacement
B_x^1 $K = 0.1$			B_x^1 $K = 0.9$		
B_x^2 $K = 0.1$			B_x^2 $K = 0.9$		
B_x^3 $K = 0.1$			B_x^3 $K = 0.9$		
B_x^4 $K = 0.1$			B_x^4 $K = 0.9$		
B_x^5 $K = 0.1$			B_x^5 $K = 0.9$		
B_x^{13} $K = 0.1$			B_x^{13} $K = 0.9$		

TABLE 12.15

Distortions and Out-of-Plane Displacements of the B_y Waves for the FCFC Boundary Condition.

Points	Distortion	Out-of-Plane Displacement	Points	Distortion	Out-of-Plane Displacement
B_y^1 $K = 0.1$			B_y^1 $K = 0.9$		
B_y^2 $K = 0.1$			B_y^2 $K = 0.9$		
B_y^3 $K = 0.1$			B_y^3 $K = 0.1$		
B_y^4 $K = 0.1$			B_y^4 $K = 0.9$		
B_y^5 $K = 0.1$			B_y^5 $K = 0.9$		
B_y^{13} $K = 0.1$			B_y^{13} $K = 0.9$		

TABLE 12.16
Distortions and Out-of-Plane Displacements of the L Waves for the CCFC Boundary Condition.

Points	Distortion	Out-of-Plane Displacement	Points	Distortion	Out-of-Plane Displacement
L^1 $K = 0.1$			L^1 $K = 0.9$		
L^2 $K = 0.1$			L^2 $K = 0.9$		
L^3 $K = 0.1$			L^3 $K = 0.9$		
L^4 $K = 0.1$			L^4 $K = 0.9$		
L^5 $K = 0.1$			L^5 $K = 0.9$		
L^{25} $K = 0.1$			L^{25} $K = 0.9$		

TABLE 12.17

Distortions and Out-of-Plane Displacements of the B_x Waves for the CCFC Boundary Condition.

Points	Distortion	Out-of-Plane Displacement	Points	Distortion	Out-of-Plane Displacement
B_x^1 $K = 0.1$			B_x^1 $K = 0.9$		
B_x^2 $K = 0.1$			B_x^2 $K = 0.9$		
B_x^3 $K = 0.1$			B_x^3 $K = 0.9$		
B_x^4 $K = 0.1$			B_x^4 $K = 0.9$		
B_x^5 $K = 0.1$			B_x^5 $K = 0.9$		
B_x^{23} $K = 0.1$			B_x^{23} $K = 0.9$		

These plots carry some important information about the characteristics of the wave modes. However, it is not our intention here to try to fully comprehend the deformation patterns and their practical implications. Instead, the reason that they are exhaustively presented here is to demonstrate the potential complexities of the wave modes with respect to deformation patterns, dispersion behaviors, and the strong dependences on the boundary conditions, and to validate the usefulness and robustness of the multiscale Fourier series method as well.

The one-dimensional structural models, such as the longitudinal and torsional vibrations of rods, the transverse vibrations of beams, *etc.*, are mostly based on the hypothesis that the cross-section of a rod or beam remains to be planar in the process of deformation. It is seen from Tables 12.9–12.11 that this planer hypothesis is only roughly valid for the wave modes corresponding to the first branches of the dispersion curves (such as, the L_s^1, T_a^1, and B_x^1 waves), when propagating at small wavenumbers in a waveguide with the FFFF boundary condition. More specifically, as shown in Table 12.9, for the small wavenumber $K = 0.1$, the dominant component of the displacements in the L_s^1 wave is in the longitudinal (z-) direction, namely the out-of-plane displacement w. In the same time, within the cross-sectional area the values of the out-of-plane displacement are almost the same and axially symmetric; that is, the cross-section remains planer. Similarly, it is observed from Table 12.10 that for $K = 0.1$, the T_a^1 wave basically still keeps the cross-section to be planar (with some minor out-of-plane warping) and the in-plane motion is mainly a cross-section rotation about the z-axis. Moreover, as shown in Table 12.11, for $K = 0.1$ the out-of-plane displacement w of the B_x^1 wave varies linearly with the distance from the neutral axis (namely, the x-axis). During this deformation, the cross-section rotates about the x-axis and is still kept as a plane.

As the wavenumber increases, the wave modes, even those correspond to the lower order branches of dispersion curves, tend to become more complicated in terms of their deformation patterns. The wave modes corresponding to the higher order branches of dispersion curves are typically complicated in deformation patterns, regardless of whether the wavenumbers involved are actually small or not. Particularly, sharp edges, steep gradients or other near-singular behaviors start to show up as a typical multiscale behavior.

REFERENCES

Achenbach, J. D. 1973. *Wave Propagation in Elastic Solids*. New York: North-Holland Publishing Company.

Bullen, K. E. 1963. *An Introduction to the Theory of Seismology*. Cambridge: Cambridge University Press.

Fraser, W. B. 1969. Stress wave propagation in rectangular bars. *International Journal of Solids and Structures* 5: 379–397.

Huang, Y. 1992. *Theory of Elastic Thin Plates*. Changsha: Press of National University of Defense Technology (in Chinese).

Miklowitz, J. 1966. *Elastic Wave Propagation*, Applied Mechanics Surveys. Washington, DC: Spartan Books.

Index

A

Absolutely integrable 19, 21–29, 33
Approximation 31, 153, 154, 175, 178
  boundary approximation 154, 221, 222
  internal approximation 154, 222
Approximation errors 157, 184, 223, 225, 227

B

Beam equation 61, 63, 66, 70
Boundary conditions 61–65, 186–189, 207, 255,
  261, 266
  clamped-clamped 81
  simply supported 6, 63, 65
Bubble function method 227–229

C

Characteristic equation 70, 198, 213, 235, 258,
  284
Characteristic frequencies 65
Characteristic functions 198, 208, 257
Characteristic roots 198, 199, 203, 213, 235
Constitutive equations 280, 300
Continuous 17, 18
  piecewise continuous 18
Convection-diffusion-reaction equation 211,
  212, 219, 220, 233, 234
Convergence 17, 19, 24, 25, 31–34
Convergence rate 34, 39, 60, 130

D

Decomposition 207, 208, 211, 282
  structural decompositions 144,160, 163
Differentiable 31
  continuously differentiable 18, 145, 166,
    168, 170,172, 174
  termwise differentiable, 27, 103, 104, 131,
    133, 134, 143, 175, 186, 190, 259
Differentiations 23
  of the Fourier series 27
  term-by-term differentiations 24, 31, 67, 68,
    70, 103, 105, 124, 130, 133, 135, 143,
    144, 150, 152, 159
Discontinuity of the first kind 18

E

Eigenfunction method 66
Eigenfunctions 65, 66, 67, 75, 81
Equivalent transformation 204, 260
Error index 129, 153, 303
  boundary error indexes 154
  internal error indexes 154
Error norms 178
Euler-Bernoulli beam 8, 63, 64

F

Fourier coefficients 12, 13
  boundary Fourier coefficients 125, 166, 168,
    173, 174, 199, 202, 203, 236, 261,
    287, 293, 294
  convergence of 17, 31, 32
  for the derivatives 24, 91
  of higher order derivatives 89, 95
Fourier series 11–15
  full-range Fourier series 15, 17, 91, 92, 95,
    98, 108, 114, 118, 121, 130, 133, 163,
    185, 189
  half-range Fourier cosine series 15, 17, 144,
    148, 149, 187
  half-range Fourier series 108, 124, 126, 127,
    133, 162, 181, 188, 195
  half-range Fourier sine series 17, 144, 150,
    188
  half-range sine-sine series 174–177, 181,
    183, 192, 255
  multiscale Fourier series 196, 208,
    227, 299
Frequency spectra 309–311
Function 11
  basis functions 197, 213, 236, 296
  boundary functions 143–145, 148, 149, 161,
    162, 168
  Green's function 229–231
  harmonic function 11
  internal functions 143–147, 159–163
  period of 11
  periodic function 11
  supplementary function 70, 71,78
  trigonometric functions 12, 41, 95

Printed in the United States
by Baker & Taylor Publisher Services